CO-RE OF ELECTRICAL ENGINEERING

CO-RE OF ELECTRICAL ENGINEERING

By

S.P. JAIN

Professor

National Institute of Technology

Kurukshetra

Haryana

UNIVERSITY SCIENCE PRESS

(An Imprint of Laxmi Publications (P) Ltd.)

BANGALORE • CHENNAI • COCHIN • GUWAHATI • HYDERABAD

JALANDHAR • KOLKATA • LUCKNOW • MUMBAI • RANCHI

NEW DELHI • BOSTON, USA

Published by :
UNIVERSITY SCIENCE PRESS
(An Imprint of Laxmi Publications (P) Ltd.)
113, Golden House, Daryaganj,
New Delhi-110002
Phone : 011-43 53 25 00
Fax : 011-43 53 25 28
www.laxmipublications.com
info@laxmipublications.com

Price : **Rs. 250.00** *Only.* *First Edition : 2009*

OFFICES

India

© **Bangalore**	080-26 61 15 61
© **Chennai**	044-24 34 47 26
© **Cochin**	0484-239 70 04
© **Guwahati**	0361-254 36 69, 251 38 81
© **Hyderabad**	040-24 65 23 33
© **Jalandhar**	0181-222 12 72
© **Kolkata**	033-22 27 43 84
© **Lucknow**	0522-220 95 78
© **Mumbai**	022-24 91 54 15, 24 92 78 69
© **Ranchi**	0651-221 47 64

USA

Boston
11, Leavitt Street, Hingham,
MA 02043, USA

UCE-9263-250-CORE OF ELECTRICAL ENGG **C—16084/08/07**

Typeset at : Goswami Associates, Delhi. *Printed at* : Ajit Printers, Delhi.

CONTENTS

Chapters *Pages*

SECTION 1

CONSTANTS AND PROPERTIES

8. TRANSIENTS, THREE-PHASE SYSTEMS AND WAVES 57—62

9. NETWORKS AND FILTERS 63—75

10. ELEMENTRY NETWORK SYNTHESIS 76—77

SECTION 3

FIELDS AND WAVES

11. VECTOR ANALYSIS 81—85

15. ELECTROMAGNETIC WAVES 104—109

16. GUIDED WAVES 110—114

17. TRANSMISSION LINES 115—121

SECTION 4

ELECTRICAL ENGINEERING MATERIALS

18. ELECTRICAL ENGINEERING MATERIALS 125—133

SECTION 5

INSTRUMENTS AND MEASUREMENTS

SECTION 6

CONTROL SYSTEMS

21. CONTROL SYSTEMS BASICS 167—174

22. STATE VARIABLE CHARACTERIZATION OF LINEAR SYSTEMS 175—178

23. TRANSFER FUNCTIONS AND TIME DOMAIN ANALYSIS 179—187

24. STABILITY ANALYSIS OF CONTROL SYSTEMS 188—193

25. FREQUENCY RESPONSE CHARACTERISTICS 194—197

SECTION 7

ELECTRICAL MACHINES

29. ELECTROMECHANICAL ENERGY CONVERSION-GENERAL 219—228

30. DC MACHINES 229—235

31. INDUCTION MACHINES 236—243

32. SYNCHRONOUS MACHINES 244—268

SECTION 8

POWER SYSTEMS

36. SAG CALCULATIONS, INSULATORS AND CORONA 293—300

37. PER UNIT SYSTEM 301—304

38. UNDERGROUND CABLES 305—312

PREFACE

The author had since long been feeling the necessity of a document on constants and relations in various basic and important courses of Electrical Engineering so that one can refer to them for quickly solving problems on the subject.

The book 'CO-RE OF ELECTRICAL ENGINEERING' has been written with this objective. It is hoped that it will serve as a handy document for quickly finding the important constants and the relations needed to solve an Electrical Engineering problem in hand, or to develop other inter-relationships. The subjects covered are circuits and networks, field theory, electrical engineering materials, instruments and measurements, control systems, electrical machines, and power systems. The coverage of constants and the formulas is, of course, not exhaustive, but those which are basic in nature, or are quite often needed have been included. No detailed derivations of the relations have been given. Only very brief explanations or the key points of some of the relations have been presented. It has been assumed that the reader knows and understands the subject. He only needs to refresh.

Scanning this book just before appearing in competitive examinations or interviews can be of great help. The book will be useful to AMIE, BE/B. Tech. students of Electrical Engineering. The teachers and the field engineers too can be benefited.

I express my thanks to Prof. MN Bandyopadhyay, Director NIT Kurukshetra, who kindly granted me sabbatical leave for four months to take up this job; and the faculty of the Department of Electrical Engineering who shared my teaching load of that semester.

Errors and suggestions, if any, to improve the book are welcome and will be gratefully acknowledged.

—Author

Section 1 Constants and Properties

Chapter 1 UNITS AND DIMENSIONS

As Greek letters are frequently used as symbols for various quantities, the full greek alphabet is given below.

1.1 GREEK ALPHABET

Α α alpha	Β β beta	Γ γ gamma	Δ δ delta	Ε ε epsilon	Ζ ζ zeta
Η η eta	Θ θ theta	Ι ι iota	Κ κ kappa	Λ λ lambda	Μ μ mu
Ν ν nu	Ξ ξ xi	Ο ο omicron	Π π pi	Ρ ρ rho	Σ σ sigma
Τ τ tau	Υ υ upsilon	Φ φ phi	Χ χ chi	Ψ ψ psi	Ω ω omega

1.2 ROMAN NUMERALS

I	II	III	IV	V	VI	VII	VIII	IX	X	XX	XXX
1	2	3	4	5	6	7	8	9	10	20	30
XL	**L**	**LX**	**LXX**	**LXXX**	**XC**	**C**	**D**	**M**			
40	50	60	70	80	90	100	500	1000			

A letter placed before a letter of greater value subtracts, while a letter placed after a letter of greater value adds.

MDCCCXLIII	MCMLXXVII
1843	1977

1.3 BASE UNITS

S.No.	*Quantity*	*Usual Symbol*	*Dimension*	*Name of Unit*	*Unit Symbol*
1	Length	l	L	metre	m
2	Mass	m	M	kilogram	kg
3	Time	t	T	second	s
4	Electric Current	I	I	ampere	A
5	Temperature	T	θ	kelvin	K
6	Luminous Intensity	I	—	candela	cd
7	Amount of Substance	n	—	mole	mol

1.4 SUPPLEMENTARY UNITS

S.No.	*Quantity*	*Usual Symbol*	*Dimension*	*Name of Unit*	*Unit Symbol*
1	Plane Angle	θ	$[L]^o$	radian	rad
2	Solid Angle	ω	$[L^2]^o$	steradian	sr

1.5 DEFINITIONS/DEFINING EQUATIONS OF THE UNITS

1 m = 1650763.73 wavelengths in vacuum of radiation ($2p_{10} - 5d_5$) of Kr 86.

1 kg – International prototype kilogram.

1 s = 9192631770 periods of the radiation corresponding to the transition between the two hyperfine levels of the ground state of the caesium-133 atom.

1 A – The current which when flowing in each of the two infinitely long parallel conductors of negligible cross-section 1 metre apart in vacuum will produce on each a force of 2×10^{-7} N/m.

1 K – The kelvin is 1/273.16 of the thermodynamic temperature of the triple point of water.

1 cd – The luminous intensity of a black body radiator at the temperature of freezing platinum at a pressure of 1 std atm viewed normal to the surface is 6×10^5 Cd/m^2.

1 mol – The amount of substance of a system which contains as many elementary units as there are carbon atoms in 0.012 kg of ^{12}C; the elementary units must be specified (atom, molecule, ion, etc).

1 rad = A radian is equal to the angle subtended at the centre of a circle by an arc of the circle of length equal to the radius.

1 sr = A steradian is equal to the angle in three dimensions subtended at the centre of a sphere by a surface area of the sphere equal to the radius squared.

1.6 DERIVED UNITS

Note: The bold faced letters in the column 'Defining equation' are vector quantities.

S.No.	*Quantity*	*Usual symbol*	*Dimension*	*Defining equation*	*Name of the unit*	*Unit symbol*
1	Area	A	L^2	$A = l^2$	square metre	m^2
2	Volume	V	L^3	$V = l^3$	cubic metre	m^3
3	Velocity	v	LT^{-1}	$v = dl/dt$	metre/second	$m\ s^{-1}$
4	Acceleration	a	LT^{-2}	$a = d^2l/dt^2$	metre/second squared	$m\ s^{-2}$
5	Density	ρ	$L^{-3}M$	$\rho = m/V$	kilogram/cubic metre	$kg\ m^{-3}$
6	Momentum	p	LMT^{-1}	$p = mv$	kilogram metre/ second	$kg\ m\ s^{-1}$
7	Force	F	LMT^{-2}	$F = ma$	newton	N
8	Work, Energy	W	L^2MT^{-2}	$W = + \mathbf{f} . \mathbf{dl}$	joule	J
9	Power	P	L^2MT^{-3}	$P = dW/dt$	watt	W
10	Pressure, Stress	p	$L^{-1}MT^{-2}$	$p = F/A$	pascal	Pa
11	Moment of inertia	I	L^2M	$I = Mk^2$	kilogram metre squared	$kg\ m^2$
12	Surface tension	γ	MT^{-2}	$\gamma = F/l$	newton/metre	$N\ m^{-1}$
13	Viscosity	η	$L^{-1}MT^{-1}$	$F/A = \eta dv/dl$	pascal second	Pa s
14	Heat	H	L^2MT^{-2}		joule	J
15	Heat capacity, Entropy	C	$L^2MT^{-2}\theta^{-1}$	–	joule/kelvin	$J\ K^{-1}$
16	Angular velocity	ω	T^{-1}	$\omega = d\theta/dt$	radian/second	$rad\ s^{-1}$
17	Angular acceleration	α	T^{-2}	$\alpha = d^2\theta/dt^2$	radian/second squared	$rad\ s^{-2}$
18	Angular momentum	M	L^2MT^{-1}	$M = I\omega$	kilogram metre squared/second	$kg\ m^2\ s^{-1}$
19	Torque, moment of force	T	L^2MT^{-2}	$T = Fl$	newton metre	N m
20	Electric charge	Q	TI	$I = dQ/dt$	coulomb	C

21	Dipole moment	p	LTI	p = Ql	coulomb metre	C m
22	Electric potential	V	$L^2MT^{-3}I^{-1}$	V= – **E** . dl	volt	V
23	Current density	J	$L^{-2}I$	J = I/A	ampere/metre squared	A m^{-2}
24	Electric field strength	E	$LMT^{-3}I^{-1}$	E = – dV/dt	volt/metre	V m^{-1}
25	Electric resistance	R	$L^2MT^{-3}I^{-2}$	R = V/I	ohm	Ω
26	Conductance	G	$L^{-2}M^{-1}T^3I^2$	G = 1/R	siemens	S
27	Resistivity	ρ	$L^3MT^{-3}I^{-2}$	R = ρl/A	ohm metre	Ω m
28	Conductivity	σ	$L^{-3}M^{-1}T^3I^2$	σ = 1/ρ	mho/metre	S m^{-1}
29	Electric flux	ψ	TI	ψ = Q	coulomb	C
30	Electric flux density, electric displacement density	D	$L^{-2}TI$	D = dψ/dA	coulomb/metre squared	C m^{-2}
31	Electric capacitance	C	$L^{-2}M^{-1}T^4I^2$	C = Q/V	farad	F
32	Permittivity	ε	$L^{-3}M^{-1}T^4I^2$	ε = D/E	farad/metre	F m^{-1}
33	Polarization	P	$L^{-2}TI$	**P** = **D** – ε**E**	coulomb/metre squared	C m^{-2}
34	Magnetic flux	φ	$L^2MT^{-2}I^{-1}$	Φ = – e dt	weber	Wb
35	Magnetic flux density	B	$MT^{-2}I^{-1}$	B = dφ/dA	tesla	T
36	Magnetic field strength	H	$L^{-1}I$	dH= idl sin θ /4πr²	ampere/metre	A m^{-1}
37	Magnetomotive force	F	I	F = **H** . dl	ampere	A
38	Magnetic dipole moment	m	L^2I	–	ampere metre squared	A m^2
39	Magnetization	M	$L^{-1}I$	**M** = **B**/μ_0 – **H**	ampere/metre	A m^{-1}
40	Magnetic vector potential	**A**	$LMT^{-2}I^{-1}$	d**A** = μ_0 id**l**/4πr²	weber/metre	Wb m^{-1}
41	Inductance	L	$L^2MT^{-2}I^{-2}$	e = Ldi/dt	henry	H
42	Permeability	μ	$LMT^{-2}I^{-2}$	μ = B/H	henry/metre	H m^{-1}
43	Reluctance	R	$L^{-2}M^{-1}T^2I^2$	R = F/φ	ampere/weber	A Wb^{-1}
44	Permeance	P	$L^2MT^{-2}I^{-2}$	P =1/R	weber/ampere	Wb A^{-1}
45	Reluctivity	r	$L^{-1}M^{-1}T^2I^2$	R = 1/μ	metre/henry	m H^{-1}

46	Radian frequency	ω	T^{-1}	$\omega = d\theta/dt$	radian/second	rad s^{-1}
47	Reactance	X	$L^2MT^{-3}I^{-2}$	$X = L\omega$ or $-1/C\omega$	ohm	Ω
48	Susceptance	B	$L^{-2}M^{-1}T^3I^2$	$B = -1/X$	siemens	S
49	Frequency	f	T^{-1}	–	hertz	Hz
50	Impedance	Z	$L^2MT^{-3}I^{-2}$	$Z = (R^2 + X^2)$	ohm	Ω
51	Admittance	Y	$L^{-2}M^{-1}T^3I^2$	$Y = 1/Z$	siemens	S
52	Volt-ampere	S	L^2MT^{-3}	$\lvert S \rvert = \lvert VI^* \rvert$	volt- ampere	VA
53	Reactive volt-ampere	Q	L^2MT^{-3}	$Q^2 = \lvert S \rvert^2 - P^2$	volt-ampere reactive	VAr
54	Attenuation constant	α	L^{-1}	–	neper/metre	Np m^{-1}
55	Phase constant	B	L^{-1}	–	radian/metre	rad m^{-1}
56	Poynting vector	P	MT^{-3}	$\mathbf{P} = \mathbf{E} \times \mathbf{H}$	watt/metre squared	W m^{-2}
57	Candle power	I	cd	–	candela	cd
58	Luminous flux	Φ	cd	$\Phi = I\,\omega$	lumen	lm
59	Illumination	E	L^{-2}cd	$E = \Phi/A$	lux	lx
60	Luminance, brightness	B	L^{-2}cd	$B = I/A$	candle/metre squared	cd m^{-2}

1.7 CONVENTIONS TO BE FOLLOWED WHILE USING UNITS AND SYMBOLS

1. All unit symbols of proper names are written in capitals; *e.g.*, *W*, *J*, *N* etc. When these are written as words, the lowercase is used throughout; *e.g.*, watt, joule, newton etc.
2. A unit symbol is the same for the singular and the plural; for example, 5 N, 1 J, and 15 *H*. However, when unit name is used, plural is used where required; for example, 5 newtons, 1 joule, 15 henrys. No full stops to be used with units or their symbols.
3. No degree symbol for kelvin; write as *K*, and not as *K*.
4. One space between two basic abbreviations for clarity; *e.g.*, N m, and not Nm, W s, and not Ws.

Chapter 2

MULTIPLES, CONVERSION FACTORS AND CONSTANTS

2.1 MULTIPLE AND SUBMULTIPLES PREFIXES

S. No.	*Factor*	*Name*	*Symbol*
1	10^{18}	exa	E
2	10^{15}	peta	P
3	10^{12}	tera	T
4	10^{9}	giga	G
5	10^{6}	mega	M
6	10^{3}	kilo	k
7	10^{2}	hecto	h
8	10^{1}	deca	da
9	10^{-1}	deci	d
10	10^{-2}	centi	c
11	10^{-3}	milli	m
12	10^{-6}	micro	μ
13	10^{-9}	nano	n
14	10^{-12}	pico	p
15	10^{-15}	femto	f
16	10^{-18}	atto	a

2.2 CONVENTIONS TO BE FOLLOWED WHILE USING PREFIXES

1. No space between prefix symbol and the unit symbol, millisecond is expressed as ms and not as m s.
2. Prefix symbols for multiples and submultiples are preferred in steps of 10^3.
3. Only one prefix should be applied; *e.g.*, kkN is not in order, it should be expressed as MN.
4. In a number, for clarity, three digits are grouped together on either side of the decimal point. However, a four-digit number need not be so grouped.

2.3 CONVERSION FACTORS

- The exact values of numbers in 'Multiply by' column are printed in **bold** face.
- The numbers which are not exact have been rounded to six decimal digits.
- A number is expressed as $a \times 10^p$. Further, a is written as S.M, where *S* is a single digit, and *M* is a multi-digit string, depending upon the accuracy of the number. We may also express it as S.M *E p*. However, a number between 10^{-3} and 10^3 may be written as it is.

S. No.	*Quantity*	*Given in*	*To find in*	*Multiply by* $a \times 10^p$		*Inverse* $b \times 10^q$	
				a	*p*	*b*	*q*
1	Length	Å	m	**1**	– 10	1	10
2		in	mm	**25.4**	00	0.039 370	00
3		ft	m	**0.3048**	00	3.280 840	00
4		yd	m	**0.9144**	00	1.093 613	00
5		mile	km	**1.609 344**	00	0.621 371	00
6		nautical mile	km	**1.852**	00	0.539 957	00
7	Area	in^2	cm^2	**6.4516**	00	0.155 000	00
8		cir mil	in^2	7.853 981	– 07	0.127 324	07
9		cir mil	mm^2	5.067 075	– 04	0.197 353	04
10		ft^2	m^2	0.092 903	00	10.763 91	00
11		yd^2	m^2	0.836 127	00	1.195 99	00
12		$mile^2$	km^2	2.589 988	00	0.386 102	00
13	volume	litre	m^3	**0.001**	00	1000	00
14		gallon	litre	**4.546 09**	00	0.219 969	00
15		in^3	cm^3	16.387 06	00	0.061 024	00
16		ft^3	m^3	0.028 317	00	35.314 667	00

17	mass	lb	kg	**0.453 592 37**	00	2.204 623	00
18		slug	kg	14.5939	00	0.068 522	00
19		metric ton	kg	1000	00	0.001	00
20		uk ton	kg	1016.05	00	0.000 984 204	00
21		short ton	kg	907.185	00	0.001 102 311	00
22	density	lb/ft^3	kg/m^3	16.0185	00	0.062 428	00
23		gm/cm^3	kg/m^3	1000	00	0.001	00
24	speed	ft/s	mile/h	0.681 818	00	1.466 667	00
25		m/s	km/h	**3.6**	00	0.277 778	00
26		knot	km/h	**1.852**	00	0.539 957	00
27	acceleration	ft/s^2	m/s^2	**0.3048**	00	3.280 840	00
28	force	dyne	N	1	– 05	1	05
29		kgf	N	**9.806 65**	00	0.101 972	00
30		lbf	N	4.448 22	00	0.224 809	00
31	pressure	lbf/in^2	N/m^2 (Pa)	6894. 76	00	0.000 145 0 38	00
32		lbf/in^2	kgf/cm^2	0.070 307	00	14.223 34	00
33		kgf/cm^2	N/m^2	**9.806 65**	04	0.101 972	– 04
34		mm Hg (Torr)	N/ m^2	133.322	00	0.007 500 63	00
35		bar	N/ m^2	1	05	1	– 05
36		std atm	mm Hg	760	00	0.001 315 78	00
37		std atm	N/ m^2	**1.013 25**	05	0.986 923	– 05
38		std atm	bar	1.01325	00	0.986 923	00
39		std atm	kgf/cm^2	1.03323	00	0.967 839	00
40		std atm	lbf/in^2	14.695 874	00	0.068 046	00
41		tonf/in^2	N/ m^2	15.46	06	0.064 683	– 06
42	work,	ft lbf	m N(J)	1.355 82	00	0.737 561	00
43	energy	erg	J	1	– 07	1	07
44		kWh	J	**3.6**	06	0.277 778	– 06
45		cal	J	**4.1868**	00	0.238 846	00
46		Btu	cal	251.997	00	0.003 968 3	00
47		Btu	J	1055.06	00	0.000 947 813	00
48		therm	Btu	1	05	1	– 05
49		eV	J	1.6021	– 19	0.624 181	19
50		amu	eV	0.931 506	09	1.073 530	– 09
51		amu	J	0.149 232	– 09	6.700 976	09

52	power	hp	ft lbf/s	550	00	0.001 818 182	00
53		hp	kW	0.745 700	00	1.341 022	00
54		metric hp	m kgf/s	75	00	0.013 333	00
55		metric hp	kW	0.735 499	00	1.359 621	00
56		kcal/h	W	1.163	00	0.859 845	00
57		Btu/h	W	0.293 071	00	3.412 142	00
58	angle	rad	degree	57.2958	00	0.017 453	00
59	angular speed	rpm	rad/s	0.104 720	00	9.549 274	00
60	viscosity	poise	Pa s	0.1	00	10	00
61	mmf	AT	gilbert	0.4π = 1.256 637	00	0.795 775	00
62	magnetizing force	A/m	oersted	4 π = 12.56371	– 03	0.079 577	03
63	flux	Wb	M	1	08	1	– 08
64	flux density	T	G	1	04	1	– 04
65		Lines/in^2	T	1.55	– 05	0.645 161	05
66	logarithm	Lg	Ln	2.302 585 (= Ln 10)	00	0.434 294	00
67	power ratio	neper	db	8.686	00	0.1151	00
68	refrigeration	ton	BthU/h	1.2	04	0.8333	– 04
69		ton	Cal/s	840	00	0.0012	00

Temperature in $K = (T + 273.15)$, T is in °C

2.4 CONVERSION RELATION

If a quantity has the dimension $L^a M^b T^c I^d$, and the ratios of the magnitudes of the units for L, M, T, and I in the given system of units to those in the new system of units are k_L, k_M, k_T, k_I respectively; then the magnitude of the quantity in the new system is related to the magnitude of that quantity in the given system as

$$Q_N = Q_G \, k_L{}^a \, k_M{}^b \, k_T{}^c \, k_I{}^d$$

2.5 COMMONLY USED SPECIAL NUMBERS

Quantity	*Value*	*Reciprocal*	*Quantity*	*Value*	*Reciprocal*
$\sqrt{2}$	1.414 214	0.707 107	π^2	9.869 604	0.101 321
$\sqrt{3}$	1.732 051	0.577 350	$\pi^2\sqrt{2}$	6.978 864	0.143 290
$\sqrt{5}$	2.236 068	0.447 214	e	2.718 282	0.367 879
$\sqrt{6}$	2.449 490	2.408 248	e^2	7.389 056	0.135 335
$\sqrt{10}$	3.162 278	0.316 228	ln 2	0.693 147	1.442 695
π	3.141 593	0.318 310	ln 10	2.302 585	0.434 294
$\sqrt{\pi}$	1.772 454	0.564 190	lg 2	0.30103	3.321 928
$\sqrt{(2\pi)}$	2.506 628	0.398 942	g	9.806 65	0.101 353
$\sqrt{2}\pi$	4.442 883	0.225 079	g^2	96.170 384	0.010 398
$\pi/\sqrt{2}$	2.221 441	0.450 158	$\sqrt{g}$	3.131 557	0.319 330

2.6 PHYSICAL CONSTANTS

S. No.	Quantity	Symbol	Value- $a \times 10^p$ *a*	*p*	Unit
1	Acceleration due to gravity	g	9.806 65	00	m s^{-2}
2	Atomic mass unit	amu	1.660 43	– 27	kg
3	Avogadro's number	N_A	6.022 52	23	mol^{-1}
4	Binding energy of electron in a hydrogen atom	E	2.1724	– 18	J
5	Bohr radius	a_0	5.291 67	– 11	m
6	Boltzmann constant	k	1.380 54	– 23	J K^{-1}
7	Charge to mass ratio for an electron	e/m_e	1.758 796	11	C kg^{-1}
8	Electron charge	e	1.602 189	– 19	C
9	Electron rest mass	m_e	9.109 534	– 31	kg
10	Electron radius	r_e	2.817 77	– 15	m
11	Faraday constant	F	9.648 70	14	C mol^{-1}
12	Gas constant	R_0	8.314 34	00	J K^{-1} mol^{-1}
13	Gravitational constant	G	6.670	– 11	N m^2 kg^{-2}
	Mass of earth	M	5.98	24	Kg
	Mean radius of earth	R	6.38	06	m

14	Neutron rest mass	m_n	1.674 82	– 27	kg
15	Permeability of free space	μ_0	4 pi = 12.566	– 07	H m^{-1}
16	Permittivity of free space	ε_0	1/36 pi = 8.854 189	– 09 – 12	F m^{-1}
17	Planck's constant	h	6.625 59	– 34	J s
18	Proton rest mass	m_p	1.672 52	– 27	kg
19	Solar year	Y	365 d 5 h 48 min 45.5 s		
20	Standard atmosphere pressure	p_0	1.013 25	05	Pa
21	Stefan-Boltzmann constant	σ	5.669 7	– 08	W m^{-2} K^{-4}
22	Triple point of water		0.0100 273.15	00 00	°C K
23	Velocity of light in vacuum	c	2.997 925	08	m s^{-1}
24	Velocity of sound at sea level at 0 °C	v	331.7	00	m s^{-1}

2.7 ELECTROMAGNETIC SPECTRUM

S. No.	*Frequency range*	*Wavelength range*	*Band class*
1	30 Hz to 300Hz	1 Mm to 10 Mm	Power, servomechanisms, sonar and audio
2	0.3 kHz to 3 kHz	0.1 Mm to 1 Mm	Audio
3	3 kHz to 30 kHz	10 km to 100 km	VLF, Audio
4	30 kHz to 300 kHz	1 km to 10 km	LF
5	0.3 MHz to 3 MHz	0.1 km to 1 km	MF, Medium wave
6	3 MHz to 30 MHz	10 m to 100 m	HF, Short waves
7	30 MHz to 300 MHz	1 m to 10 m	VHF, Short waves
8	0.3 GHz to 3 GHz	0.1 m to 1 m	UHF, Short waves
9	3 GHz to 30 GHz	10 mm to 100 mm	SHF, Micro waves
10	30 GHz to 300 GHz	1 mm to 10 mm	EHF, Micro waves
11	0.3 THz to 0.3 PHz	1 μm to 1 mm	Infrared, Heat waves
12	0.375 PHz to 0.75 PHz	0.4 μm to 0.8 μm	Light, visible
13	0.75 PHz to 30 PHz	10 nm to 0.4 μm	Ultraviolet
14	30 PHz to 30 EHz	10 pm to 10 nm	X-rays
15	30 EHz to 3000 EHz	0.1 pm to 10 pm	Gamma rays
16	3000 EHz and above	0.1 pm and less	Cosmic rays

Please note that E is 10^{18} and P is 10^{15}.

The range of frequencies from that at S. No. 4 (LF) to that at S. No. 10 (EHF) is the **radio waves** range.

The demarcation between the adjacent classes is not rigid, there can be a narrow range of frequencies overlapping the neighboring classes.

2.8 WAVELENGTH AND COLOUR

Wavelength, μm :	0.4		0.5		0.6		0.7	0.8
Colour : ultra-violet		violet	blue	green	yellow	orange	red	infrared

2.9 STRUCTURE OF ATMOSPHERE

The sequence of various layers from the earth's surface, and the approximate range of their thicknesses in km are given below.

Troposphere	0	to	15
D layer	35	to	65
E layer	0	to	20
F_1 layer	0	to	110
F_2 layer	30	to	180
G region (outer atmosphere)	All the region above F_2 layer		

The region from 50 km to 400 km is known as the ionosphere.

The region above 400 km is known as the outer atmosphere.

2.10 LOUDNESS OF SOUNDS

Sound pressure level in db = 20 lg (p/p_o), where p is the sound pressure in μ bar, and p_o equals 2×10^{-4} μ bar. (1 μ bar = 1 dyne / cm^2). The sound pressure p_o is the threshold of hearing. A pressure of about 1000 dynes / cm^2 causes serious discomfort or damage to the ear. Some typical sound conditions and the corresponding db levels are given below.

db	Sound condition
0	Threshold of hearing
10	Winter night with snow and no wind
20	Pocket watch (1m away)
30	Quiet interior

40	Insect noises in summer meadow
50	Buzz of conversation
60	Vacuum cleaner
70	Rattling in light signaling system (1 m away)
80	Heavy town traffic
90	Freight train climbing hill (3 m away)
100	Compressed air hammer (3 m away)
110	High volume bell (3 m away)
120	Vehicle siren (3 m away)
130	Sensation of pain

2.11 RELATIVE AND ABSOLUTE AIR HUMIDITY AT ATMOSPHERE TEMPERATURES

	Absolute humidity in gm / m³								
Temp. °C \ *Relative humidity, %*	*20*	*30*	*40*	*50*	*60*	*70*	*80*	*90*	*100*
0	1.0	1.5	1.9	2.4	2.9	3.4	3.9	4.3	4.8
2	1.1	1.7	2.2	2.8	3.3	3.9	4.4	5.0	5.6
4	1.3	1.9	2.5	3.2	3.8	4.5	5.1	5.7	6.4
6	1.5	2.2	2.9	3.6	4.4	5.1	5.8	6.5	7.3
8	1.7	2.5	3.3	4.1	5.0	5.8	6.6	7.4	8.3
10	1.9	2.9	3.8	4.7	5.6	6.5	7.5	8.5	9.4
12	2.1	3.2	4.3	5.3	6.4	7.4	8.5	9.6	10.7
14	2.4	3.6	4.8	6.0	7.3	8.5	9.7	10.9	12.1
16	2.7	4.1	5.5	6.8	8.2	9.5	10.9	12.3	13.6
18	3.1	4.6	6.2	7.7	9.2	10.7	12.3	13.8	15.4
20	3.5	5.2	6.9	8.6	10.4	12.1	13.8	15.6	17.3
22	3.9	5.9	7.8	9.7	11.7	13.6	15.5	17.5	19.4
24	4.4	6.5	8.7	10.9	13.1	15.3	17.4	19.6	21.8
26	4.9	7.4	9.8	12.2	14.6	17.1	19.6	22.0	24.4

28	5.5	8.2	10.9	13.6	16.3	19.1	21.8	24.5	27.3
30	6.1	9.2	12.1	15.2	18.2	21.3	24.3	27.4	30.4
32	6.7	10.1	13.5	16.9	20.3	23.7	27.1	30.5	33.8
34	7.5	11.3	15.0	18.8	22.6	26.3	30.1	33.9	37.7
36	8.3	12.5	16.7	20.9	25.0	29.2	33.4	37.6	41.7
38	9.2	13.9	18.6	23.2	27.8	32.4	37.0	41.6	46.2
40	10.2	15.3	20.4	25.6	30.7	35.8	40.9	46.0	51.1

2.12 PAPER SIZES

The next lower sheet sizes is derived from the current size by halving it. For each sheet, the ratio of the sides is $1 : \sqrt{2}$. The A_0 size equals 1 m^2.

Series A		*Series B*		*Series C*	
Size	*mm × mm*	*Size*	*mm × mm*	*Size*	*mm × mm*
A_0	1189 × 841	B_0	1414 × 1000	C_0	1297 × 917
A_1	841 × 594	B_1	1000 × 707	C_1	917 × 648
A_2	594 × 420	B_2	707 × 500	C_2	648 × 458
A_3	420 × 297	B_3	500 × 353	C_3	458 × 324
A_4	297 × 210	B_4	353 × 250	C_4	324 × 229
A_5	210 × 148	B_5	250 × 176	C_5	229 × 162
A_6	148 × 105	B_6	176 × 125	C_6	162 × 114
A_7	105 × 74	B_7	125 × 88	C_7	114 × 81
A_8	74 × 52	B_8	88 × 62	C_8	81 × 57

Chapter 3 PROPERTIES OF MATERIALS

3.1 PERIODIC TABLE

In chemistry, the generalization that there is a recurring pattern in the properties of the elements when they are arranged in order of increasing atomic number—*i.e.*, the total number of protons in the atomic nucleus, is depicted in the form of periodic table. The periods (horizontal rows) of the periodic table illustrate these relationships.

The periodic system of the elements contains all of the chemical elements that have been discovered or made ; they are arranged in the order of their atomic numbers, in seven horizontal periods, with the lanthanons (lanthanum, 57, to lutetium, 71) and the actions (actinium, 89, to lawrencium, 103) indicated separately below. The periods are of varying lengths. First there is the hydrogen period, consisting of the two elements hydrogen, 1, and helium, 2. Then there are two periods of eight elements each: the first short period, from lithium, 3, to neon, 10; and the second short period, from sodium, 11, to argon, 18. There follow two periods of 18 elements each: the first long period, from potassium 19, to krypton, 36; and the second long period, from rubidium, 37, to xenon, 54. The very long period of 32 elements, from cesium, 55, to radon, 86, is condensed into 18 columns by the omission of the lanthanons (which are indicated separately below), permitting the remaining 18 elements, which are closely similar in their properties to corresponding elements of the first and second long periods, to be placed directly below these elements. The last period is incomplete. The properties of these elements indicate that this period is a second very long period of 32 elements, which would be completed at element 118.

Periodic Table

Within the framework of the table it is customary to recognize certain groupings of elements; viz. *transition elements* (Sc—Zn, Y—Cd, La—Hg); *lanthanides* (or *rare earth group;* Ce—Lu) occupying a distinctive place; and the *actinides* (Th, etc.). More recently discovered elements (artificial) are Rutherfordium (Rf; at. no. 104) and Hahnium (Ha; at. no. 105).

		Transition Elements															
H 1 1.008 1																	He 2 4.003 2
Li 3 6.941 2, 1	Be 4 9.012 2, 2											B 5 10.81 2, 3	C 6 12.011 2, 4	N 7 14.007 2, 5	O 8 16.0 2, 6	F 9 18.998 2, 7	Ne 10 20.179 2, 8
Na 11 22,990 2, 8, 1	Mg 12 24.305 2, 8, 2											Al 13 26.982 2, 8, 3	Si 14 28.086 2, 8, 4	P 15 30.974 2, 8, 5	S 16 32.06 2, 8, 6	Cl 17 35.453 2, 8, 7	Ar 18 39.948 2, 8, 8
K 19 39.102 2, 8, 8, 1	Ca 20 40.08 2, 8, 8, 2	Sc 21 44.956 2, 8, 9, 2	Ti 22 47.90 2, 8, 10, 2	V 23 50.941 2, 8, 11, 2	Cr 24 51.996 2, 8, 13, 1	Mn 25 54.938 2, 8, 13, 2	Fe 26 55.847 2, 8, 14, 2	Co 27 58.933 2, 8, 15, 2	Ni 28 58.71 2, 8, 16, 2	Cu 29 63.546 2, 8, 18, 1	Zn 30 65.37 2, 8, 18, 2	Ga 31 69.72 2, 8, 18, 3	Ge 32 72.59 2, 8, 18, 4	As 33 74.922 2, 8, 18, 5	Se 34 78.96 2, 8, 18, 6	Br 35 79.904 2, 8, 18, 7	Kr 36 83.80 2, 8, 18, 8
Rb 37 85.468 2, 8, 18, 8, 1	Sr 38 87.62 2, 8, 18, 8, 2	Y 39 88.906 2, 8, 18, 9, 2	Zr 40 91.22 2, 8, 18, 10, 2	Nb 41 92.906 2, 8, 18, 12, 1	Mo 42 95.94 2, 8, 18, 13, 1	Tc 43 98.906 2, 8, 18, 14, 1	Ru 44 101.07 2, 8, 18, 15, 1	Rh 45 102.906 2, 8, 18, 16, 1	Pd 46 106.4 2, 8, 18, 18, 0	Ag 47 107.868 2, 8, 18, 18, 1	Cd 48 112.40 2, 8, 18, 18, 2	In 49 114.82 2, 8, 18, 18, 3	Sn 50 118.69 2, 8, 18, 18, 4	Sb 51 121.75 2, 8, 18, 18, 5	Te 52 127.60 2, 8, 18, 18, 6	I 53 126.905 2, 8, 18, 18, 7	Xe 54 131.30 2, 8, 18, 18, 8
Cs 55 132.906 2, 8, 18, 18, 8, 1	Ba 56 137.34 2, 8, 18, 18, 8, 2	La 57 138.905 2, 8, 18, 18, 9, 2	Hf 72 178.49 2, 8, 18, 32, 10, 2	Ta 73 180.948 2, 8, 18, 32, 11, 2	W 74 183.85 2, 8, 18, 32, 12, 2	Re 75 186.2 2, 8, 18, 32, 13, 2	Os 76 190.2 2, 8, 18, 32, 14, 2	Ir 77 192.22 2, 8, 18, 32, 17, 0	Pt 78 195.09 2, 8, 18, 32, 17, 1	Au 79 196.967 2, 8, 18, 32, 18, 1	Hg 80 200.59 2, 8, 18, 32, 18, 2	Tl 81 204.37 2, 8, 18, 32, 18, 3	Pb 82 207.2 2, 8, 18, 32, 18, 4	Bi 83 208.981 2, 8, 18, 32, 18, 5	Po 84 — 2, 8, 18, 32, 18, 6	At 85 — 2, 8, 18, 32, 18, 7	Rn 86 — 2, 8, 18, 32, 18, 8
Fr 87 — 2, 8, 18, 32,18, 8, 1	Ra 88 226.025 2, 8, 18, 32, 18, 8, 2	Ac 89 — 2, 8, 18, 32, 18, 9, 2															

Lanthanides:	Ce 58 140.12 2, 8, 18, 20, 8, 2	Pr 59 140.908 2, 8, 18, 21, 8, 2	Nd 60 144.24 2, 8, 18, 22, 8, 2	Pm 61 — 2, 8, 18, 23, 8, 2	Sm 62 150.4 2, 8, 18, 24, 8, 2	Eu 63 151.96 2, 8, 18, 25, 8, 2	Gd 64 157.25 2, 8, 18, 25, 9, 2	Tb 65 158.925 2, 8, 18, 27, 8, 2	Dy 66 162.50 2, 8, 18, 28, 8, 2	Ho 67 164.930 2, 8, 18, 29, 8, 2	Er 68 167.26 2, 8, 18, 30, 8, 2	Tm 69 168.934 2, 8, 18, 31, 8, 2	Yb 70 173.04 2, 8, 18, 32, 8, 2	Lu 71 174.97 2, 8, 18, 32, 9, 2
Actinides:	Th 90 232.038 2, 8, 18, 32, 18, 10, 2	Pa 91 231.036 2, 8, 18, 32, 20, 9, 2	U 92 238.029 2, 8, 18, 32, 21, 9, 2	Np 93 237.048 2, 8, 18, 32, 22, 9, 2	Pu 94 — 2, 8, 18, 32, 24, 8, 2	Am 95 — 2, 8, 18, 32, 25, 8, 2	Cm 96 — 2, 8, 18, 32, 25, 9, 2	Bk 97 — 2, 8, 18, 32, 26, 9, 2	Cf 98 — 2, 8, 18, 32, 28, 8, 2	Es 99 — 2, 8, 18, 32, 29, 8, 2	Fm 100 — 2, 8, 18, 32, 30, 8, 2	Md 101 — 2, 8, 18, 32, 31, 8, 2	No 102 — 2, 8, 18, 32, 32, 8, 2	Lr 103 — 2, 8, 18, 32, 32 9. 2

3.2 CONDUCTIVITIES, RESISTIVITIES, RELATIVE RESISTIVITIES (WITH RESPECT TO COPPER), AND RESISTANCE TEMPERATURE COEFFICIENTS OF ELEMENTS AND MATERIALS

A material of resistivity less than 10^{-4} Ω m is considered a conductor, the one having resistivity between 10^{-4} Ω m and 10^{4} Ω m is a semiconductor, and materials of resistivity greater than 10^{4} Ω m are classed as insulators.

S.No.	Element / Material	Conductivity, σ, $S\ m^{-1}$ $a \times 10^p$		Resistivity, ρ, Ω m $a \times 10^p$		Relative resistivity, $\rho_r = \rho/\rho_{Cu}$ $a \times 10^p$		Temperature coefficient, α_{20}, K^{-1} $a \times 10^p$	
		a	*p*	*a*	*p*	*a*	*p*	*a*	*p*
1	Silver	6.17	07	1.62	– 08	0.94	00	0.40	– 02
2	Copper	5.80	07	1.72	– 08	1.00	00	0.43	– 02
3	Gold	4.10	07	2.43	– 08	1.41	00	0.35	– 02
4	Aluminium	3.54	07	2.83	– 08	1.65	00	0.42	– 02
5	Molybdenum	2.04	07	4.9	– 08	2.85	00	0.50	– 02
6	Tungsten	1.82	07	5.49	– 08	3.19	00	0.45	– 02
7	Zinc	1.67	07	5.99	– 08	3.48	00	0.42	– 02
8	Brass	1.50	07	6.67	– 08	3.88	00	0.20	– 02
9	Nickel	1.30	07	7.69	– 08	4.47	00	0.60	– 02
10	Iron	1.03	07	9.71	– 08	5.64	00	0.45	– 02
11	Phosphor bronze	1.00	07	1.00	– 07	5.80	00		
12	Platinum	9.17	06	1.09	– 07	6.34	00	0.37	– 02
13	Solder	7.00	06	1.43	– 07	8.31	00		
14	Tantalum	6.45	06	1.55	– 07	9.01	00	0.32	– 02
15	Carbon steel	6.00	06	1.66	– 07	9.69	00		
16	Lead	4.80	06	2.08	– 07	12.11	00	0.42	– 02
17	German silver	3.00	06	3.33	– 07	19.37	00		
18	Manganin	2.27	06	4.41	– 07	25.61	00	0.1	– 04
19	Constantan	2.26	06	4.43	– 07	25.72	00	0.3	– 04
20	Silicon steel	2	06	5	– 07	29.07	00		
21	Stainless steel	1.10	06	9.09	– 07	52.85	00		
22	Mercury	1	06	1	– 06	58.13	00	0.99	– 02
23	Nichrome	8.9	05	1.12	– 06	65.32	00	0.1	– 03
24	Kanthal	7.4	05	1.35	– 06	78.48	00	0.3	– 04
25	Tin	5	05	2	– 06	112	00		
26	Graphite	7	04	1.42	– 05	826	00		

27	Carbon	3	04	3.33	– 05	1.936	03	– 0.5	– 03
28	Ferrite Fe_3O_4	100	00	1	– 02	5.8	05		
29	Water (Sea)	3 – 5	00	0.25	00	1.45	07		
30	Germanium (Intrinsic)	2.2	00	0.46	00	2.67	07		
31	Limestone	1	– 02	1	02	5.8	09		
32	Clay	5	– 03	2	02	1.2	10		
33	Water (Fresh)	1	– 03	1	03	5.8	10		
34	Silicon (Intrinsic)	3.33	– 04	3	03	1.7	11		
35	Water (Distilled)	2	– 04	5	03	3	11		
36	InSb	5	– 05	2	04	1.2	12		
37	Soil (Sandy or Dry)	1	– 05	1	05	6	12		
38	Glass (Window)	1	– 05	1	05	6	12		
39	Granite	1	– 06	1	06	6	13		
40	Marble	1	– 08	1	08	6	15		
41	Wood	1	– 08 to – 11	1	08 to 11	6	15 to 18		
42	Bakelite	1	– 09	1	09	6	16		
43	Glass	1	– 10 to – 14	1	10 to 14	6	17 to 18		
44	Mica	1	– 11 to – 15	1	11 to 15	1	18 to 22		
45	Porcelain	2	– 13	5	12	3	19		
46	Diamond	2	– 13	5	12	3	19		
47	Paraffin	1	– 15	1	15	6	22		
48	Rubber, PVC	1	– 15	1	15	6	22		
49	Polysterene	1	– 16	1	16	6	23		
50	Fused quartz (SiO_2)	4	– 18	2.5	17	1.5	25		

3.3 SUPERCONDUCTING MATERIALS

Substance	*Transition Temperature K*	*Critical magnetic Flux density*
Hg	3.95	
Ta	4.48	
V	5.13	
Pb	7.2	0.055
Nb	9.2	0.29
NbTi 50	9.7	11
V_3Si	17.0	
Nb $(C_{0.35}N_{0.65})$	17.8	15
V_3Ga	14.5	20
Nb_3Sn	18.3	22
Nb_3Ge	22.8	35

Note : Values of critical magnetic flux density are nominal only, as these vary with impurities

3.4 COLOUR CODE FOR CARBON RESISTORS

There are generally 4 bands on a carbon resistor; the first three are for the value of the resistor, and the 4th one specifies the tolerance in the value. The first 2 bands colour yields 2-digit number. This number is to be multiplied by 10^p, p being given by the 3rd band colour.

Colour	*Value*	*Colour*	*Value*	*Colour*	*Value*	*Remarks*
Black	0	Green	5	Gold	– 1	Only for the 3rd band
Brown	1	Blue	6	Silver	– 2	
Red	2	Violet	7	Gold	± 5%	Only for the 4th band
Orange	3	Grey	8	Silver	± 10%	
Yellow	4	White	9	None	± 20%	

3.5 RELATIVE PERMITTIVITY AND APPROXIMATE DIELECTRIC STRENGTH OF MATERIALS

The dielectric strength of some materials vary greatly with the composition of its constituents. Values given are approximate only.

Note : The dielectric strength, in general, is not linearly proportional to material thickness; for instance, to cause breakdown of an air gap of 1 mm, the voltage required is 4.36 kV and not 3 kV.

S. No.	*Material*	*Relative permittivity*	*Dielectric strength kV/cm*
1	Air	1.0006	30
2	Alcohol, Ethyl	25	
3	Aluminium oxide	8.8	
4	Amber	2.7	900
5	Bakelite	4.5-5.5	250
6	Barium titanate	1200	
7	Carbon dioxide	1.001	
8	Earth, dry	5	
9	Earth, wet	10	
10	Germanium	16	
11	Glass	5-10	300
12	Ice	4.2	
13	Mica	3-7	2000
14	Neoprene	6.6	
15	Nylon	3.5	
16	Oil, mineral	2.2-2.3	150
17	Paper	2-3	150
18	Paraffin	2.1	300
19	Plexiglass	2.6-3.5	
20	Polyethelene	2.26	400
21	Polypropylene	2.25	
22	Polystrene	2.55	200
23	Porcelain	6-7	250
24	PVC	3.8-4.3	
25	Pyranol	4.4	
26	Pyrex glass	4	
27	Quartz, fused	3.8	300
28	Rubber	2-3.5	210

29	Silica, fused	3.8	
30	Silicon	11.8	1000-1200
31	Snow	3.3	
32	Sodium chloride	5.9	
33	Steatite	5.8	
34	Styrofoam	1.03	
35	Teflon	2.1	600
36	Titanium dioxide	100	60
37	Water, distilled or fresh or sea	80	
38	Water dehydrated	1	
39	Wood, dry	1.5-4	

3.6 MAGNETIC SUSCEPTIBILITY OF MATERIALS

S. No.	*Material*	*Susceptibility*
1	Air	0.4×10^{-6}
2	Aluminium	21×10^{-6}
3	Bismuth	-166×10^{-6}
4	Carbon (diamond)	-26×10^{-6}
5	Carbon (graphite)	-99×10^{-6}
6	Carbon dioxide	-0.23×10^{-3}
7	Cerium	1.3×10^{-3}
8	Copper	-9.6×10^{-6}
9	Gold	-36×10^{-6}
10	Hydrogen	-0.99×10^{-3}
11	Lead	-16×10^{-6}
12	Magnesium	12×10^{-6}
13	Mercury	-28×10^{-6}
14	Nitrogen	-0.5×10^{-3}
15	Oxygen	1.8×10^{-6}
16	Silver	-26×10^{-6}
17	Sodium	-2.4×10^{-6}
18	Titanium	70.6×10^{-6}
19	Tungsten	68×10^{-6}

3.7 PHYSICAL PROPERTIES OF ELEMENTS AND MATERIALS

S. No.	Element / material	Symbol	Relative density	Melting point °C	Thermal expansion coefficient 10^{-6} K^{-1}	Specific heat cal g^{-1}	Tensile strength GN m^{-2}
1	Aluminium	Al	2.70	660	23.1	0.216	0.060
2	Aluminium oxide	Al_2O_3	3.8	–	9.0	--	0.06
3	Antimony	Sb	6.62	630	10.9	0.050	–
4	Argon	Ar	–	– 189	–	–	–
5	Arsenic	As	5.72	Volatile, 450	4.28	–	–
6	Barium	Ba	3.75	850	18.8	0.068	–
7	Beryllium	Be	1.93	1281	11.5	0.277	–
8	Bismuth	Bi	9.80	271	13.41	0.029	–
9	Boron	B	2.5	2300	8.3	–	–
10	Brass	–	8.4-8.9	900	19.8	0.0917	0.40
11	Bronze	–	8.8	–	18.0	0.086	0.60
12	Cadmium	Cd	8.65	321	30.6	0.056	–
13	Calcium	Ca	1.58	851	22.4	0.157	–
14	Carbon	C	3.52	(Infusible) 3550	3.8	–	–
15	Cesium	CS	1.9	28	97	–	–
16	Chlorine	Cl	3.21E-03	– 101	–	–	–
17	Chromium	Cr	6.5	1510	8.4	0.107	–
18	Cobalt	Co	8.6	1490	12.4	–	–
19	Copper	Cu	8.96	1083	16.7	0.0921	0.210
20	Diamond	–	3.5	(infusible) 3600	–	0.121	
21	Fluorine	F	–	– 223	–	–	–
22	Gallium	Ga	5.95	30	18.1	0.089	–
23	Germanium	Ge	5.47	958	5.75	0.077	–
24	Glass (Borosilicate)	–	2.4	–	2.7	0.18	0.05
25	Gold	Au	19.32	1063	14.1	0.031	0.138
26	Graphite	–	1.9	–	5.4	0.17	–
27	Helium	He		– 270	–	–	–
28	Hydrogen	H	0.00009	– 259	–	–	–
29	Ice	–	0.916	0	–	2.09	–

30	Indium	In	7.4	155	31.4	0.0556	–
31	Iridium	Ir	22.4	2375	6.63	0.0311	–
32	Iron	Fe	7.86	1525	11.7	0.107	0.60
33	Lead	Pb	11.36	327	29.0	0.031	0.015
34	Lithium	Li	0.585	186	45	–	–
35	Manganese	Mn	7.39	1220	22.6	0.114	–
36	Manganin	–	8.2	910	–	0.097	0.42
37	Mercury	Hg	13.55	– 38.9	61	0.0333	–
38	Molybdenum	Mo	10.22	2500	4.98	0.059	0.7
39	Neon	Ne	–	– 249	–	–	–
40	Nichrome	–	8.41	–	–	–	1.0
41	Nickel	Ni	8.90	1453	12.7	0.106	0.322
42	Nitrogen	N	–	– 210	–	–	–
43	Nylon	–	1.15	–	100	–	0.08
44	Oxygen	O	0.00143	– 227	–	–	–
45	Phosphorus	P	1.8-2.3	44	124	–	–
46	Platinum	Pt	21.45	1755	8.95	0.0315	0.35
47	Plutonium	Pu	19.5	640	55	0.032	–
48	Porcelain	–	2.3	–	–	–	–
49	Potassium	K	0.87	63	83	0.18	–
50	Polysterene	–	1.05	–	60	–	0.055
51	Selenium	Se	4.5	170-220	36.9	–	–
52	Silicon	Si	2.0-2.4	1420	3.07	0.168	–
53	Silver	Ag	10.49	961	19.2	0.0565	0.125
54	Sodium	Na	0.978	97	70.6	0.292	–
55	Strontium	Sr	2.54	800	20	–	–
56	Sulphur	S	2.07	115-119	64	0.168	–
57	Tantalum	Ta	16.6	2900	6.55	0.0336	0.49
58	Thorium	Th	11.00	1750	11.2	0.0281	–
59	Tin	Sn	7.30	232	5.3	0.0531	–
60	Titanium	Ti	4.54	1850	8.35	–	–
61	Tungsten	W	19.27	3500	4.59	0.0322	2.80
62	Uranium	U	18.7	1132	12.6	0.0278	–
63	Vulcanized rubber	–	1.2	–	80	–	–
64	Xenon	Xe	–	– 140	–	–	–
65	Zinc	Zn	7.13	420	29.7	0.0931	–
66	Zirconium	Zr	4.15	2130	5.78	0.0659	–

Chapter 4 USEFUL FUNCTIONS, INTEGRALS AND LAPLACE TRANSFORMS

4.1 EXPONENTIAL, TRIGONOMETRIC, HYPERBOLIC FUNCTIONS AND INTER-RELATIONS

$$e^x = 1 + x + x^2/2! + x^3/3! + \ldots$$

$$e^{-x} = 1 - x + x^2/2! - x^3/3! + \ldots$$

$$\cosh x = (e^x + e^{-x})/2$$

$$= 1 + x^2/2! + x^4/4! + \ldots$$

$$\sinh x = (e^x - e^{-x})/2$$

$$= x + x^3/3! + x^5/5! + \ldots$$

$$\cosh x + \sinh x = e^x$$

$$\cosh x - \sinh x = e^{-x}$$

$$e^{jx} = 1 + jx - x^2/2! - j\,x^3/3! + x^4/4! + j\,x^5/5! - \ldots$$

$$e^{-jx} = 1 - jx - x^2/2! + j\,x^3/3! + x^4/4! - j\,x^5/5! - \ldots$$

$$\cosh jx = (e^{jx} + e^{-jx})/2$$

$$= 1 - x^2/2! + x^4/4! - x^6/6! + \ldots = \cos x$$

$$\sinh jx = (e^{jx} - e^{-jx})/2$$

$$= j\,(x - x^3/3! + x^5/5! - \ldots)$$

$$= j \sin x$$

$$\cos x + j \sin x = e^{jx}$$

$$\cos x - j \sin x = e^{-jx}$$

$$\cosh (A + B) = \cosh A \cosh B + \sinh A \sinh B$$

$$\cosh (A - B) = \cosh A \cosh B - \sinh A \sinh B$$

$$\sinh (A + B) = \sinh A \cosh B + \cosh A \sinh B$$

$$\sinh(A-B) = \sinh A \cosh B - \cosh A \sinh B$$
$$\cosh 2A = \cosh^2 A + \sinh^2 A$$
$$\cosh(A-A) = \cosh^2 A - \sinh^2 A$$

As $\cosh 0 = 1$,

$$\cosh^2 A - \sinh^2 A = 1,$$
$$\cosh 2A = 2\cosh^2 A - 1, \text{ and also}$$
$$= 2\sinh^2 A + 1$$

Thus,

$$\cosh^2 A = (\cosh 2A + 1)/2, \text{ and}$$
$$\sinh^2 A = (\cosh 2A - 1)/2$$
$$\cosh(A + jB) = \cosh A \cosh jB + \sinh A \sinh jB$$
$$= \cosh A \cos B + j \sinh A \sin B$$
$$\cosh(A - jB) = \cosh A \cosh jB - \sinh A \sinh jB$$
$$= \cosh A \cos B - j \sinh A \sin B$$
$$\sinh(A + jB) = \sinh A \cosh jB + \cosh A \sinh jB$$
$$= \sinh A \cos B + j \cosh A \sin B$$
$$\sinh(A - jB) = \sinh A \cosh jB - \cosh A \sinh jB$$
$$= \sinh A \cos B - j \cosh A \sin B$$
$$\cos(A+B) = \cos A \cos B - \sin A \sin B$$
$$\cos(A-B) = \cos A \cos B + \sin A \sin B$$
$$\sin(A+B) = \sin A \cos B + \cos A \sin B$$
$$\sin(A-B) = \sin A \cos B - \cos A \sin B$$
$$\cos(A+A) = \cos^2 A - \sin^2 A$$
$$\cos(A-A) = \cos^2 A + \sin^2 A$$

As $\cos 0 = 1$,

$$\cos^2 A + \sin^2 A = 1,$$
$$\cos 2A = 2\cos^2 A - 1, \text{ and also}$$
$$= -2\sin^2 A + 1$$

Thus,

$$\cos^2 A = (\cos 2A + 1)/2, \text{ and}$$
$$\sin^2 A = -(\cos 2A - 1)/2$$
$$= (1 - \cos 2A)/2$$
$$\sin A + \sin B = 2\sin(A+B)/2 . \cos(A-B)/2$$
$$\sin A - \sin B = 2\cos(A+B)/2 . \sin(A-B)/2$$
$$\cos A + \cos B = 2\cos(A+B)/2 . \cos(A-B)/2$$
$$\cos A - \cos B = -2\sin(A+B)/2 . \sin(A-B)/2$$
$$\sin 3A = 3\sin A - 4\sin^3 A;$$
$$\cos 3A = 4\cos^3 A - 3\cos A;$$
$$\tan 3A = (3\tan A - \tan^3 A)/(1 - 3\tan^2 A)$$

Taylor's series expansion of a function $f(x)$ around x_o –

$$f(x_o + h) = f(x_o) + f'(x_o)\,h + f''(x_o)\,h^2/2! + f'''(x_o)\,h^3/3! + f''''(x_o)\,h^4/4! + \ldots$$

$$(x + a)^n = \sum_{r=0}^{r=n} {}^nC_r\, x^{(n-r)}\, a^r \text{ ; where } {}^nC_r = n\,!/((n-r)\,!\,r\,!)$$

$$Ln\,(1 + x) = x - x^2/2 + x^3/3 - x^4/4 + \ldots$$

$$Ln\,(1 - x) = -(x + x^2/2 + x^3/3 + x^4/4 + \ldots)$$

4.2 SOME USEFUL INTEGRALS

$$\int x^n\,dx = \frac{x^{n+1}}{n+1}$$

$$\int (ax+b)^n\,dx = \frac{(ax+b)^{n+1}}{a\,(n+1)}$$

$$\int \frac{1}{x}\,dx = Ln\,|x|$$

$$\int \frac{1}{ax+b}\,dx = \frac{Ln\,|ax+b|}{a}$$

$$\int a^x\,dx = \frac{a^x}{Lna}$$

$$\int a^{mx+b}\,dx = \frac{a^{mx+b}}{mLna}$$

$$\int \cos(ax+b)dx = \frac{\sin\,(ax+b)}{a}$$

$$\int \sin(ax+b)dx = \frac{-\cos\,(ax+b)}{a}$$

$$\int \tan(ax+b)dx = \frac{Ln\,\sec\,(ax+b)}{a}$$

$$\int \cot(ax+b)dx = \frac{Ln\,\mathrm{cosec}\,(ax+b)}{a}$$

$$\int \sec(ax+b)dx = \frac{Ln\,[\sec\,(ax+b) + \tan\,(ax+b)]}{a}$$

$$\int \mathrm{cosec}(ax+b)dx = \frac{Ln\,[\mathrm{cosec}\,(ax+b) - \cot\,(ax+b)]}{a}$$

$$\int \sec^2(ax+b)dx = \frac{\tan\,(ax+b)}{a}$$

$$\int \mathrm{cosec}^2(ax+b)dx = -\frac{\cot\,(ax+b)}{a}$$

$$\int \sec(ax+b)\tan(ax+b)dx = \frac{\sec\,(ax+b)}{a}$$

$$\int \frac{1}{\sqrt{1-x^2}}\,dx = \sin^{-1} x$$

$$\int \frac{1}{\sqrt{1-(ax+b)^2}}\,dx = \frac{\sin^{-1}(ax+b)}{a}$$

$$\int \frac{1}{\sqrt{a^2-x^2}}\,dx = \sin^{-1}\frac{x}{a}$$

$$\int -\frac{1}{\sqrt{a^2-x^2}}\,dx = \cos^{-1}\frac{x}{a}$$

$$\int \frac{1}{1+(ax+b)}\,dx = \frac{\tan^{-1}(ax+b)}{a}$$

$$\int \frac{1}{x^2+a^2}\,dx = \frac{1}{a}\tan^{-1}\frac{x}{a}$$

$$\int \frac{1}{x\sqrt{x^2-1}}\,dx = \sec^{-1} x$$

$$\int \tan x dx = Ln\,|\sec x|$$

$$\int \sec x dx = Ln\;|\sec x + \tan x|$$

$$\int \frac{1}{x^2-a^2}\,dx = \frac{1}{2a}\,Ln\,\frac{x-a}{x+a}$$

$$\int \frac{1}{\sqrt{x^2+a^2}}\,dx = Ln\left|x+\sqrt{x^2+a^2}\right| = \sinh^{-1}\frac{x}{a}$$

$$\int \frac{1}{\sqrt{x^2-a^2}}\,dx = Ln\left|x+\sqrt{x^2-a^2}\right| = \cosh^{-1}\frac{x}{a}$$

$$\int \ln x\,dx = x\,Ln\,x - x$$

$$\int e^{ax}\sin bx\,dx = \frac{e^{ax}}{a^2+b^2}\,(a\sin bx - b\cos bx)$$

$$\int e^{ax}\cos bx\,dx = \frac{e^{ax}}{a^2+b^2}\,(a\cos bx + b\sin bx)$$

$$\int uvdx = u\int vdx - \int \frac{du}{dx}\left(\int vdx\right)dx$$

$$\int_0^{2\pi} \cos m\theta\,\sin n\theta\,d\theta = 0$$

$$\int_0^{2\pi} \cos m\theta \cos n\theta \, d\theta = 0 \quad \text{for} \quad m \neq n$$

$$\int_0^{2\pi} \sin m\theta \sin n\theta \, d\theta = 0 \quad \text{for} \quad m \neq n$$

4.3 TRIANGLE RELATIONS

$$\frac{a}{\sin A} = \frac{b}{\sin B} = \frac{c}{\sin C}$$

$$c = b \cos A + a \cos B$$

$$c^2 = a^2 + b^2 - 2ab \cos C$$

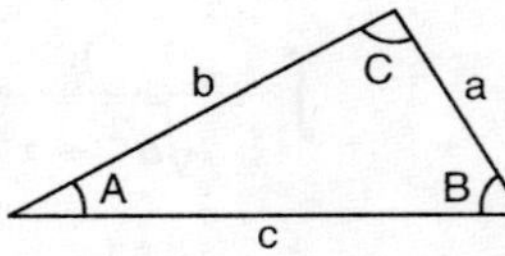

FIGURE 4.1. Side lengths and angles of a triangle.

Area of a triangle $= \frac{1}{2} ab \sin C$

$$= \sqrt{s(s-a)(s-b)(s-c)},$$

where s is the sum of the lengths of the three sides of the triangle.

4.4 PLANE ANGLE AND SOLID ANGLE RELATION

The plane angle θ (in radians) subtended at point O by a small length Δl is $\frac{\Delta l \, \boldsymbol{i}_n . \boldsymbol{i}_r}{r}$; where r is the length of the straight line joining O and P, P being a point at the middle of Δl, $\boldsymbol{i}_n$ is a unit vector normal to Δl at point P, and $\boldsymbol{i}_r$ is a unit vector along OP. $\Delta l \, \boldsymbol{i}_n$ and $\boldsymbol{i}_r$ are to be in the same plane. The solid angle ω (in steradians) subtended at point O by a small surface Δs is $\frac{\Delta s \, \boldsymbol{i}_n . \boldsymbol{i}_r}{r^2}$; where r is the length of the straight line joining O and P, P being a point at the centre of Δs, $\boldsymbol{i}_n$ is a unit vector drawn normal to Δs at point P, and $\boldsymbol{i}_r$ is a unit vector along OP.

The plane angle θ, subtended by the radius of a circular disc at any point O on its axis, and the solid angle ω subtended by the disc at the same point O are related as

$$\boldsymbol{\omega = 2\pi (1 - \cos \theta)}$$

4.5 LAPLACE TRANSFORMS

S. No.	*Time function f(t)*	*Laplace transform F(s)*
1	$f(t)$	$F(s) = \int_{0_-}^{\infty} f(t) e^{-st} dt$
2	$\delta(t)$	1
3	$u(t)$	$\frac{1}{s}$

4	$t\,u(t)$	$\frac{1}{s^2}$
5	$\frac{t^n}{n!}$	$\frac{1}{s^{n+1}}$
6	$e^{-\alpha t}$	$\frac{1}{s+\alpha}$
7	$\sin at$	$\frac{a}{s^2+a^2}$
8	$\cos at$	$\frac{s}{s^2+a^2}$
9	$\sinh at$	$\frac{a}{s^2-a^2}$
10	$\cosh at$	$\frac{s}{s^2-a^2}$
11	$e^{\alpha t} f(t)$	$F(s-\alpha)$
12	$e^{-\alpha t} \sin at$	$\frac{a}{(s+\alpha)^2+a^2}$
13	$e^{-\alpha t} \cos at$	$\frac{(s+\alpha)}{(s+\alpha)^2+a^2}$
14	$\frac{e^{-\alpha t}\, t^n}{n!}$	$\frac{1}{(s+\alpha)^{n+1}}$
15	$\frac{1}{2a} \sin at$	$\frac{s}{(s^2+a^2)^2}$
16	$\frac{d}{dt} f(t)$	$sF(s) - f(0_-)$
17	$\frac{d^n}{dt^n} f(t)$	$s^n F(s) - \sum_{j=1}^{n} s^{n-j} f^{j-1}(0_-)$
18	$\int_{0_-}^{t} f(\tau)\, d\tau$	$\frac{F(s)}{s}$
19	$(-t)^n f(t)$	$\frac{d^n}{ds^n} F(s)$
20	$f(t-a)u(t-a)$	$e^{-as} F(s)$

21	a periodic function $f(t)$ of time period T	$\frac{1}{1-e^{-sT}}\int_{0_-}^{T} f(t)\, e^{-st}\, dt$
22	$f_1(t)*f_2(t)$ $= \int_0^t f_1(t-\tau)\, f_2(\tau) d\tau$ $= \int_0^t f_1(\tau)\, f_2(t-\tau) d\tau$	$F_1(s)F_2(s)$

Initial value theorem is that $\lim_{t \to 0_+} f(t) = \lim_{s \to \infty} s\, F(s)$

$F(s)$ must be a proper fraction.

Final value theorem is that $\lim_{t \to \infty} f(t) = \lim_{s \to 0_+} s\, F(s)$

The poles of $F(s)$ must not be in the right half of the complex frequency s-plane.

4.6 INVERSE LAPLACE TRANSFORMS OF SOME STANDARD FUNCTIONS

S. No.	*Laplace transform F(s)*	*Time function f(t)*
1	$\frac{1}{(s+a)(s+b)}$	$\frac{e^{-at} - e^{-bt}}{b-a}$
2	$\frac{\omega_n^{\,2}}{s^2 + 2\delta\omega_n s + \omega_n^{\,2}}$	$\frac{\omega_n}{\sqrt{1-\delta^2}} e^{-\delta\omega_n t} \sin(\omega_n \sqrt{1-\delta^2}\, t)$
3	$\frac{1}{(1+sT)^n}$	$\frac{1}{T^n\,(n-1)!} t^{n-1} e^{-t/T}$
4	$\frac{1}{s(1+sT)}$	$1 - e^{-t/T}$
5	$\frac{1}{s(1+sT)^2}$	$1 - \frac{t+T}{T} e^{-t/T}$
6	$\frac{1}{s^2(1+sT)^2}$	$t - 2T + (t + 2T)e^{-t/T}$
7	$\frac{1+as}{s^2(1+sT)}$	$t + (a - T)(1 - e^{-t/T})$

8	$\dfrac{s\omega_n^{\,2}}{s^2 + 2\delta\omega_n s + \omega_n^{\,2}}$	$\dfrac{\omega_n^{\,2}}{\sqrt{1-\delta^2}}\, e^{-\delta\omega_n t} \sin(\omega_n \sqrt{1-\delta^2}\, t + \phi)$, $\phi = \tan^{-1} \dfrac{\sqrt{1-\delta^2}}{-\delta}$

The solution of *ordinary second order differential equation* $\dfrac{d^2 y}{dx^2} - k^2 y = 0$ is

$$y = A\, e^{kx} + B\, e^{-kx}$$

The other form of this solution is

$$y = C \sinh kx + D \cosh kx$$

The solution of $\dfrac{d^2 y}{dx^2} + k^2 y = 0$, is $y = E\, e^{jkx} + F e^{-jkx}$

The other form of this solution

$$y = G \sin kx + H \cos kx$$

The *Bessel's differential equation* is

$$x^2 \frac{d^2 y}{dx^2} + x \frac{dy}{dx} + (x^2 - n^2)\, y = 0.$$

Chapter 5

CONDUCTOR SIZES, INSULATION CLASSES AND CAPACITOR RATINGS

5.1 STANDARD WIRE GAUGES, DIAMETERS AND CROSS-SECTIONAL AREAS

SWG No.	*Diameter in*		*Cross-sectional area, mm²*	*SWG No.*	*Diameter in*		*Cross-sectional area, mm²*
	milli-inches	*mm*			*milli-inches*	*mm*	
0	324.0	8.230	53.1973	26	18.0	0.457	0.1640
1	300.0	7.620	45.6037	27	16.4	0.4166	0.1363
2	276.0	7.010	38.5945	28	14.8	0.3759	0.1110
3	252.0	6.401	32.1800	29	13.6	0.3454	0.0937
4	232.0	5.893	27.2745	30	12.4	0.3150	0.0779
5	212.0	5.385	22.7752	31	11.6	0.2946	0.0682
6	192.0	4.877	18.6808	32	10.8	0.2743	0.0591
7	176.0	4.470	15.6930	33	10.0	0.2540	0.0507
8	160.0	4.064	12.9717	34	9.2	0.2337	0.0429
9	144.0	3.658	10.5094	35	8.4	0.2134	0.0358
10	128.0	3.251	8.3009	36	7.6	0.1930	0.02926
11	116.0	2.946	6.8164	37	6.8	0.1727	0.02342
12	104.0	2.642	5.4822	38	6.0	0.1524	0.01824
13	92.0	2.337	4.2895	39	5.2	0.1321	0.01371
14	80.0	2.032	3.2429	40	4.8	0.1219	0.01167

15	72.0	1.829	2.6273	41	4.4	0.1118	0.00982
16	64.0	1.626	2.0765	42	4.0	0.1016	0.00811
17	56.0	1.422	1.5881	43	3.6	0.0914	0.00656
18	48.0	1.219	1.1671	44	3.2	0.0813	0.00519
19	40.0	1.016	0.811107	45	2.8	0.0711	0.00397
20	36.0	0.914	0.6561	46	2.4	0.0610	0.00292
21	32.0	0.813	0.5191	47	2.0	0.0508	0.00203
22	28.0	0.711	0.3970	48	1.6	0.0406	0.00129
23	24.0	0.610	0.2922	49	1.2	0.0305	0.00073
24	22.0	0.559	0.2454	50	1.0	0.0254	0.00051
25	20.0	0.508	0.2027	-	-	-	-

5.2 ACSR CONDUCTORS PROPERTIES

Al wire dia mm	steel wire dia mm	overall dia mm	Al cross-sectional area mm^2	ac resistance at 20° C ohms per km	current carryiny capacity A	ultimate strength kN	name

2 layers. The central layer of one steel wire, and the 2nd layer of 6 Aluminium conductors; all wires of the same diameter. Overall dia is 3 × wire dia.

Al wire dia mm	steel wire dia mm	overall dia mm	Al cross-sectional area mm^2	ac resistance at 20° C ohms per km	current carryiny capacity A	ultimate strength kN	name
2.11	2.11	6.33	20.98	1.403	125	7.61	Squirrel
2.36	2.36	7.08	25.91	1.098	148	9.37	Gopher
2.59	2.59	7.77	31.61	0.935	160	11.12	Weasel
3.0	3.0	9.00	41.87	0.679	190	14.95	Ferret
3.35	3.35	10.05	52.88	0.556	218	18.25	Rabbit
3.66	3.66	10.98	62.32	0.456	236	21.70	Mink
3.99	3.99	11.97	74.07	0.384	260	25.80	Beaver
4.09	4.09	12.27	78.83	0.374	270	26.91	Racoon
4.22	4.22	12.66	83.92	0.343	280	28.67	Otter
4.50	4.50	13.55	94.21	0.302	300	32.60	Cat

3 layers. The central layer and the 2nd layer of 1 and 6 steel wires respectively, and the 3rd layer of 6 Aluminium conductors. Overall dia is 3 × steel wire dia + 2 × Aluminium conductor dia.

Al wire dia mm	steel wire dia mm	overall dia mm	Al cross-sectional area mm^2	ac resistance at 20° C ohms per km	current carryiny capacity A	ultimate strength kN	name
4.72	1.57	14.15	105	0.281	320	32.41	Dog
5.28	1.76	15.84	131.37	0.219	410	40.58	Leopard

4 layers. The central layer and the 2nd layer of 1 and 6 steel wires respectively, 3rd and the 4th layers of 12 and 18 Aluminium conductors respectively; all wires of the same diameter. Overall dia is 7 × wire dia.

2.36	2.36	16.52	128.1	0.222	400	56.50	Tiger
2.59	2.59	18.13	158.1	0.188	455	67.34	Wolf
2.79	2.79	19.53	183.4	0.159	500	77.98	Lynx
3.0	3.0	21.0	212.1	0.140	540	89.67	Panther
3.18	3.18	22.26	238.3	0.122	605	100.1	Lion
3.35	3.35	23.45	264.4	0.11	640	110.9	Bear
3.71	3.71	25.97	316.5	0.09	720	135.5	Goat
3.99	3.99	27.93	366.1	0.078	800	157	Sheep
4.27	4.27	29.89	429.6	0.068	920	178.8	Deer

5 layers. The central layer and the 2nd layer of 1 and 6 steel wires respectively, 3rd, 4th and the 5th layers of 12, 18 and 24 Aluminium conductors respectively; all wires of the same diameter. Overall dia is 9 × wire dia.

3.18	3.18	28.62	428.9	0.069	905	130.32	Zebra
3.35	3.35	30.17	475.96	0.0623	930	144.7	Camel
3.53	3.53	31.77	528.5	0.0563	970	159.6	Moose

5.3 INSULATION CLASSES ON THE BASIS OF MAXIMUM TEMPERATURE

Class of insulation	Y	A	E	B	F	H	C
Max temperature, °C	90	105	120	130	155	180	Above 180

5.4 CAPACITOR RATINGS FOR POWER FACTOR IMPROVEMENT

The capacitor kVAr/kW of the load is given by the relation $Q_C = (\sin(\varphi_1 - \varphi_2))/\cos\varphi_1 . \cos\varphi_2$

	Capacitor kVAr rating per kW of the load for				
New power factor, cos φ_2 / *Existing power factor, cos φ_1*	0.80	0.85	0.90	0.95	1.0
0.40	1.54	1.67	1.81	1.96	2.29
0.42	1.41	1.54	1.68	1.83	2.26
0.44	1.29	1.42	1.56	1.71	2.04
0.46	1.18	1.31	1.45	1.60	1.93

0.48	1.08	1.21	1.34	1.50	1.83
0.50	0.98	1.11	1.25	1.40	1.73
0.52	0.89	1.03	1.16	1.31	1.64
0.54	0.81	0.94	1.08	1.23	1.56
0.56	0.73	0.86	1.00	1.15	1.48
0.58	0.66	0.78	0.92	1.08	1.41
0.60	0.58	0.71	0.85	1.01	1.33
0.62	0.52	0.65	0.78	0.94	1.27
0.64	0.45	0.58	0.72	0.87	1.20
0.66	0.39	0.52	0.66	0.81	1.14
0.68	0.33	0.46	0.59	0.75	1.08
0.70	0.27	0.40	0.54	0.69	1.02
0.72	0.21	0.34	0.48	0.64	0.96
0.74	0.16	0.29	0.43	0.58	0.91
0.76	0.11	0.23	0.37	0.53	0.86
0.78	0.05	0.18	0.32	0.47	0.80
0.80	–	0.13	0.27	0.42	0.75
0.82	–	0.08	0.21	0.37	0.70
0.84	–	0.03	0.16	0.32	0.65
0.86	–	–	0.11	0.26	0.59
0.88	–	–	0.06	0.21	0.54
0.90	–	–	–	0.16	0.48
0.92	–	–	–	0.10	0.43

Section 2 CIRCUITS AND NETWORKS

Chapter 6 BASICS OF CIRCUITS

6.1 BASIC TWO-TERMINAL CIRCUIT ELEMENTS

The passive elements are—Resistor, Inductor, and Capacitor.

The ***resistance*** of an element is a measure of the opposition offered to the flow of electric current through it.

In all the relations listed below, the symbols have their usual meanings in context with the topic being considered.

dc resistance of an element, $R = \dfrac{\rho l}{a}$

The resistance of a conductor changes with temperature, the change can be found by using the ***resistance temperature coefficient*** of the material, α. α also is dependent on temperature. Assuming linearity,

$$\Delta R = R\,\alpha\,\Delta t;$$

where R and α are respectively the resistance and the resistance temperature coefficient of the conductor, both at the initial temperature and Δt is the increase in temperature.

We have, $R_0\alpha_0 = R_1\alpha_1 = R_2\alpha_2 = \ldots = R_n\alpha_n$

$$R_t = R_0\,(1 + \alpha_0 t),\ \text{or}\ \ R_2 = R_1(1 + \alpha_1(t_2 - t_1) \quad \alpha_2 = \frac{\alpha_1}{1 + \alpha_1\,(t_2 - t_1)}$$

t's are the temperatures in °C.

The self ***inductance*** of an element is a measure of the opposition offered to the change of flow of electric current through it.

It equals the magnetic flux linkages λ of the element per unit current through it, *i.e.*,

$$L = \frac{\lambda}{I},$$

Self inductance of a coil of N turns, also designated as simply inductance,

$L = N^2 \times$ permeance of the closed magnetic flux path, the path over which the flux φ produced by the coil current exists.

$L = N^2\,\mu\,a/l$, μ assumed constant, and a, uniform. It is also assumed that each turn of the inductor is threaded by the same magnetic flux φ.

Inductance is also obtained through the relations

$$L = \frac{N\phi}{I}, e = -L\, di/dt.$$

The ***Capacitance*** of an element is a measure of the opposition offered to the change of voltage across it.

It is also the electric charge storing capacity of the element per unit voltage difference across it, *i.e.*, $C = Q/V$.

Capacitance of a parallel plate capacitor— $\frac{a\varepsilon}{d}$

Capacitance of a sphere— $4\pi\varepsilon R$

Capacitance of a concentric spherical capacitor— $4\pi\varepsilon \frac{R_1R_2}{R_2 - R_1}$

Capacitance per meter of a coaxial cylindrical capacitor— $\frac{2\pi\varepsilon}{Ln\left(\frac{R_2}{R_1}\right)}$

Capacitance per meter of a two-wire line—

$$\frac{\pi\varepsilon}{Ln\left(\frac{D}{d} + \sqrt{\left(\frac{D}{d}\right)^2 - 1}\right)} = \cosh^{-1}\frac{D}{d} \approx \frac{\pi\varepsilon}{Ln\left(\frac{D}{d}\right)}$$

A practical ***voltage source*** is represented by an ideal voltage source of steady emf **E** in series with an impedance element $\mathbf{Z}_s$, representing the internal impedance (will soon meet the complex quantities) of the practical source.

A practical ***current source*** is represented by an ideal current source of steady current **I**, having across it an admittance element $\mathbf{Y}_s$, representing the internal admittance of the practical source.

The two sources are interconvertible.

$$\mathbf{E} = \mathbf{I}\,\mathbf{Z}_s \quad \text{and} \quad \mathbf{Z}_s = 1/\,\mathbf{Y}_s$$

For *dc* systems, **E, I,** $\mathbf{Z}_s$ and $\mathbf{Y}_s$ are all *dc* quantities, *i.e.*, E, I, R_s, and G_s respectively.

6.2 TERMINAL RELATIONS FOR BASIC ELEMENTS

Resistor : $v(t) = R\, i(t)$ $\qquad i(t) = \frac{1}{R} v(t)$

Inductor : $v(t) = L\frac{di(t)}{dt}$ $\qquad i(t) = \frac{1}{L}\int_0^t v(t)dt + i(0)$

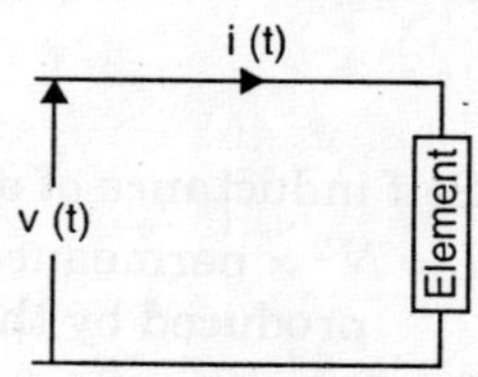

FIGURE 6.1. Voltage polarity and the current direction for a passive element.

Capacitor : $v(t) = \frac{1}{C}\int_0^t i(t)\,dt + v(0) \qquad i(t) = C\,\frac{dv(t)}{dt}$

An element is said to be ***linear*** if its response-excitation pair follows the superposition, and the proportionality principles.

An element is ***bilateral*** if its response is the same for either direction of the current flow.

An element is ***causal*** if before the application of the excitation there is no response.

An element is ***time-invariant*** if the time of application of an excitation has no effect on the response, initial conditions remaining the same.

A circuit, or a network is linear, bilateral, causal and time invariant if it is composed of linear, bilateral, causal and time invariant elements.

A circuit, or a network will always be assumed so unless stated otherwise.

6.3 KIRCHHOFF'S LAWS

Kirchhoff's current law states that at any node in a circuit, the algebraic sum of all the currents is zero. To take the sum of the currents at a node, a current towards the node may be reckoned positive, and that away from the node would then be reckoned negative. We can equally well reckon the currents the other way round.

Kirchhoff's voltage law states that in any loop of a circuit, the algebraic sum of all the element voltages is zero. To take the sum of the voltages in a loop, we may traverse the loop in either sense. While traversing the loop, a voltage rise may be reckoned positive, and the voltage fall would then be reckoned negative. We can equally well reckon the voltages the other way round.

The currents and voltages may be instantaneous currents and voltages, or complexor currents and voltages.

6.4 FLUXES, INDUCTANCES, COEFFICIENT OF COUPLING, AND ENERGY STORED

A coil j of N_j turns, when carrying current I_j produces flux φ_{jj} ; of which the part φ_{jk} links the coil k of N_k turns, and the remaining φ_{jk}' does not link the coil k. Then

the self-inductance of the coil j is $\quad L_{jj} = N_j\,\varphi_{jj}/I_j$

the mutual inductance of the coil j with coil k is $\quad L_{jk} = N_k\,\varphi_{jk}/I_j$

the leakage inductance of the coil j with respect to coil k is $\quad L_{jk}' = N_j\,\varphi_{jk}'/I_j$

the common flux factor is $\quad K_{jk} = \varphi_{jk}/\varphi_{jj}$

φ_{jk}/N_jI_j is the permeance P_{jk} of the flux path over which the common magnetic flux φ_{jk} exists. If μ is assumed to be constant, it is a fixed quantity. Similarly, φ_{jk}'/N_jI_j is the permeance P_{jk}' of the flux path over which the leakage magnetic flux φ_{jk}' exists.

Therefore,

$$L_{jk} = N_j N_k P_{jk}, \qquad L_{jj} = N_j^2 P_{jk} + N_j^2 P_{jk}' = (N_j/N_k) L_{jk} + L_{jk}' \text{ and}$$

$$L_{jk}' = L_{jj} - (N_j/N_k)\, L_{jk} \approx L_{jj} - \sqrt{(L_{jj}/L_{kk})}\, L_{jk}$$

Similarly, the coil k, when carrying current I_k produces flux φ_{kk}; of which the part φ_{kj} links the coil j, and the remaining φ_{kj}' does not link coil j. Then

the self-inductance of the coil k is $L_{kk} = N_k\, \varphi_{kk}/I_k$

the mutual inductance of the coil k with coil j is $L_{kj} = N_j\, \varphi_{kj}/I_k$

the leakage inductance of the coil k with respect to coil j is $L_{kj}' = N_k\, \varphi_{kj}'/I_k$

the common flux factor is $K_{kj} = \varphi_{kj}/\varphi_{kk}$

We have the following relations similar to those for coil j

$$L_{kj} = N_k N_j P_{kj}, \qquad L_{kk} = N_k^2 P_{kj} + N_k^2 P_{kj}' = (N_k/N_j) L_{kj} + L_{kj}' \text{ and}$$

$$L_{kj}' = L_{kk} - (N_k/N_j)\, L_{kj} \approx L_{kk} - \sqrt{(L_{kk}/L_{jj})}\, L_{kj}$$

The flux path for the common or the mutual flux, whether it is φ_{jk} or φ_{kj} is the same.

Hence, P_{jk} and P_{kj} are identical and therefore, $L_{kj} = L_{jk} = N_j N_k P_{jk}$.

However, L_{jk}' and L_{kj}' are, in general, not equal. Also, the paths of φ_{jk}' and φ_{kj}' are not the same.

$$K_{jk} K_{kj} = (\varphi_{jk}/\varphi_{jj})(\varphi_{kj}/\varphi_{kk}) = (L_{jk}\, I_j/N_k\, \varphi_{jj})\,(L_{jk}\, I_k/N_j\, \varphi_{kk}) = L_{jk}^2/(L_{jj}\, L_{kk})$$

The ***coefficient of coupling*** $k_{jk} = \sqrt{(K_{jk} K_{kj})} = L_{jk}/\sqrt{(L_{jj} L_{kk})}$

If we have only two coils, we may use the notation L_1, L_2, M, and k. Then the coefficient of coupling between the two mutually coupled inductors, $k = \dfrac{M}{\sqrt{L_1 L_2}}$

M is taken positive if φ_{12} and φ_{21} are in the same direction, else it is negative.

$$L_1 = L_1 - (N_1/N_2)\, M \approx L_1 - (L_1/L_2)\, M = L_1(1 - k)$$

$$L_2 = L_2 - (N_2/N_1)\, M \approx L_2 - (L_2/L_1)\, M = L_2(1 - k)$$

Leakage coefficient, $\sigma = (1 - k^2)$

Leakage factor $= \dfrac{\textit{Total flux}}{\textit{Useful flux}}$

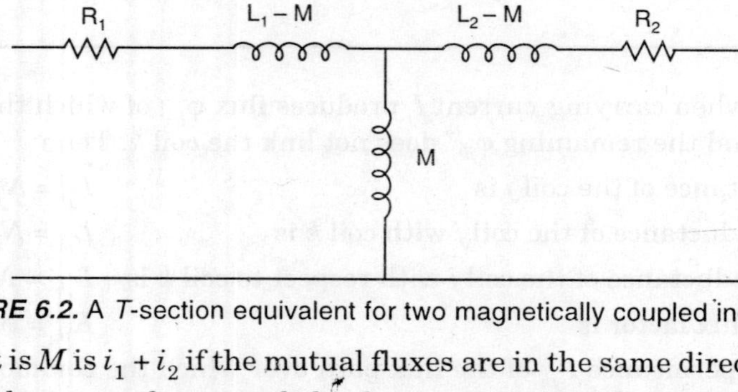

FIGURE 6.2. A *T*-section equivalent for two magnetically coupled inductors.

(The current is M is $i_1 + i_2$ if the mutual fluxes are in the same direction, else it is $i_1 - i_2$)

Resultant inductance of two coupled inductors in series $= L_1 + L_2 \pm 2M$; + sign for aiding fluxes, and – sign for the opposing fluxes.

Resultant inductance of two coupled inductors in parallel = $\dfrac{L_1 L_2 - M^2}{L_1 + L_2 \mp 2M}$; – sign for aiding fluxes, and + sign for the opposing fluxes.

Energy stored in an inductor = $\frac{1}{2} L\, i^2$, i is the current through the inductor L.

Energy stored in two mutually coupled inductors = $\frac{1}{2} L_1 i_1^{\,2} + \frac{1}{2} L_2 i_2^{\,2} + L_{12}\, i_1 i_2$

6.5 CAPACITORS IN SERIES/PARALLEL AND ENERGY STORED

The resultant capacitance, C_s, of n capacitors in series is given by

$$\frac{1}{C_s} = \frac{1}{C_1} + \frac{1}{C_2} + \ldots + \frac{1}{C_n}$$

The resultant capacitance, C_p, of n capacitors in parallel is given by

$$C_p = C_1 + C_2 + \ldots + C_n$$

Energy stored in a capacitor = $\frac{1}{2} C\, v^2$, v is the voltage across the capacitor C.

Energy radiated on sudden change in the voltage across the capacitor = $\frac{1}{2} C\, (\Delta v)^2$

6.6 SOME STANDARD SIGNALS

Step function $Au(t - T)$: It is a function whose value is zero when the argument is negative, and A when it is zero or positive.

$$Au(t - T) = 0 \text{ for } t < T \quad \text{and}$$
$$= A \text{ for } t \geq T$$

The above is in fact, a shifted step function of magnitude A

If $A = 1$, and $T = 0$, then it is simply a unit step function.

$u(t - T)\, f(t - T)$ is the function $f(t)$ shifted by T. The value of $u(t - T)\, f(t - T)$ is zero for $t < T$.

Ramp function $Ar(t - T)u(t - T)$: It is a function whose value is zero when the argument is negative, and increases linearly with t thereafter, with a rate of increase A per second.

$$Ar(t - T)u(t - T) = 0 \text{ for } t < T \quad \text{and}$$
$$= At \text{ for } t \geq T$$

The above is in fact, a shifted ramp function having slope A.

If $A = 1$, and $T = 0$, then it is simply a unit ramp function.

Impulse, or delta function $A\delta(t - T)$: It is a function whose value is infinite (undefined) at $t = T$, zero for all other times, and the area under the function is A.

$$A\delta(t - T) = \infty \quad \text{for } t = T,$$
$$= 0 \quad \text{for } t \neq T, \quad \text{and}$$

$$\int_{-\infty}^{+\infty} A\delta(t - T)\, dt = A$$

The above is in fact, a shifted delta function having area under it A.

If $A = 1$, and $T = 0$, then it is simply a unit delta function.

Shifting property of the delta function is

$$\int_{-\infty}^{+\infty} A\delta(t - T)\, f(t) dt = Af(T)$$

These functions are useful in constructing the signals which are made up of straight line segments.

The delta function is the derivative of the step function, and the step function is the derivative of the ramp function.

The derivative of the delta function is called the ***'doublet'***.

$$\int_{-\infty}^{+\infty} A\delta'(t - T) dt = 0$$

A property of the doublet is

$$\int_{-\infty}^{+\infty} A\delta'(t - T)\, f(t) dt = -Af'(T)$$

In general,

$$\int_{-\infty}^{+\infty} \delta^n (t - T)\, f(t) dt = (-1)^n f^n(T)$$

6.7 WAVES AND WAVE VALUES

A signal $f(t)$ is periodic, with time period T, if for all t, $f(t \pm nT) = f(t)$, n being any integer number.

The ***average***, or *dc* value of a periodic wave is given by

$$\frac{1}{T}\int_{t_0}^{t_0+T} f(t) dt$$

The ***mean*** value of a periodic wave is given by

$$\frac{1}{T}\int_{t_0}^{t_0+T} |f(t)|\, dt$$

The ***root mean square*** (rms) value of a periodic wave is given by

$$\sqrt{\frac{1}{T}\int_{t_0}^{t_0+T} f^2(t) dt}$$

For a sinusoidal wave, or a full rectified sinusoidal wave,

the mean value is $\frac{2}{\pi}$ times the peak value, and

the rms value is $\frac{1}{\sqrt{2}}$ times the peak value.

$$E_{\text{mean}} = \frac{2}{\pi} E_m, \quad \text{and} \quad E_{\text{rms}} = \frac{1}{\sqrt{2}} E_m$$

For a half rectified sinusoidal wave,

$$E_{\text{mean}} = \frac{1}{\pi} E_m, \quad \text{and} \quad E_{\text{rms}} = \frac{1}{2} E_m$$

The ***form factor*** of a periodic wave is its $\frac{\textit{rms value}}{\textit{mean value}}$; and the crest, or the ***peak factor*** is its $\frac{\textit{peak value}}{\textit{rms value}}$, the rms value determined with *dc* component removed.

Chapter* 7

CIRCUIT ANALYSIS AND THEOREMS

7.1 PHASORS AND RELATED QUANTITIES

For sinusoidal time variations of voltages and currents, we analyze the circuits, for steady-state conditions, through the use of phasors of the corresponding quantities. Using phasors we are able to suppress the time variation, and hence the convenience in analyzing. Yet we obtain the results as if we analyzed the circuits in time domain.

The phasor representing $V_m \cos(\omega t + \alpha)$ is denoted by **V**. **V** is $V e^{j\alpha}$, where $V = V_m/\sqrt{2}$. Alternatively, **V** is expressed as $V \angle \alpha$. We may also express it in $(a + jb)$ form; where a is $V \cos \alpha$ and b is $V \sin \alpha$. The phasor representing d^n/dt^n of a sinusoidal variable is simply obtained as $(j\omega)^n$ times the phasor of the original sinusoidal variable. Similarly, the phasor of nth integration is obtained by dividing the original phasor by $(j\omega)^n$.

In the light of above, the phasor relations for the three types of passive elements are as under :

$$\mathbf{V} = R\mathbf{I} \qquad \mathbf{I} = \frac{1}{R}\mathbf{V}$$

$$\mathbf{V} = jL\omega\,\mathbf{I} \qquad \mathbf{I} = \frac{1}{jL\omega}\mathbf{V}$$

$$\mathbf{V} = \frac{1}{jC\omega}\mathbf{I} \qquad \mathbf{I} = jC\omega\,\mathbf{V}$$

The reactance of an inductor, X_L, is $L\omega$; while that of a capacitor, X_C, is $-\frac{1}{C\omega}$.

On the phasor diagram, the phasor $kj^n\mathbf{V}$ means a phasor whose magnitude is k times that of **V** and is ahead of (*i.e.*, leads) **V** by $n \times 90^0$. Similar interpretation can be given to $\mathbf{V}/(kj^n)$.

The KCL and KVL hold even when currents and voltages are expressed as phasors, provided the phasors are of the same angular frequency.

The complex impedance $\mathbf{Z} = \mathbf{V/I}$. $\mathbf{Z} = R + jX$. The impedance magnitude, Z, is $\sqrt{(R^2 + X^2)}$ and the impedance angle, θ_Z, is $\tan^{-1}(X/R)$. This is the angle by which the current $\mathbf{I}$ lags the voltage $\mathbf{V}$. The phase difference between the applied voltage $\mathbf{V}$ and the resulting current $\mathbf{I}$ into the circuit, is denoted by φ, and is taken positive when $\mathbf{I}$ lags $\mathbf{V}$. The power factor of the circuit is $\cos\varphi$, or $\cos\theta_Z$.

The susceptence of a capacitor, B_C, is $C\omega$; while that of an inductor, B_L, is $-\dfrac{1}{L\omega}$.

The complex admittance $\mathbf{Y} = \mathbf{I/V}$. $\mathbf{Y} = G + jB$. The admittance magnitude, Y, is $\sqrt{(G^2 + B^2)}$ and the admittance angle, θ_y, is $\tan^{-1}(B/G)$. This is the angle by which the voltage $\mathbf{V}$ lags the current $\mathbf{I}$.

The impedance, or the admittance of an element or any two-terminal 'black box', in general, depends on the frequency of the current which flows through it.

Whenever we are to see the phase difference between sinusoidally time-varying quantities, all of these should first be expressed as either cosine functions, or as sine functions.

In ac circuits, unless stated otherwise, it is understood that the time variation of voltage or current is sinusoidal of steady single angular frequency, ω. Also, whenever we speak of voltage or current magnitude, it is understood to be the rms value.

$$\mathbf{Y} = 1/\mathbf{Z},\ Y = 1/Z,\ \theta_y = -\theta_Z.\ G = R/(R^2 + X^2),\ B = -X/(R^2 + X^2).$$

Thus, a series $R_s - L_s$ circuit is equivalent to a parallel $R_p - L_p$ circuit in which the resistance R_p is $R_s \dfrac{Z_s^2}{R_s^2} = \dfrac{R_s}{\cos^2\theta_{z_s}}$ and the inductance L_p is $L_s \dfrac{Z_s^2}{X_s^2} = \dfrac{L_s}{\sin^2\theta_{z_s}}$; and a series $R_s - C_s$ circuit is equivalent to a parallel $R_p - C_p$ circuit in which the resistance is R_p is $R_s \dfrac{Z_s^2}{R_s^2} = \dfrac{R_s}{\cos^2\theta_{z_s}}$ and the capacitance C_p is $C_s \dfrac{Y_s^2}{B_s^2} = C_s \sin^2\theta_{zs}$, where $B_s = C_s\,\omega$.

Similarly, $\mathbf{Z} = 1/\mathbf{Y}$, $Z = 1/Y$, $\theta_z = -\theta_y$. $R = G/(G^2 + B^2)$, $X = -B/(G^2 + B^2)$.

Thus, a parallel $R_p - L_p$ circuit is equivalent to a series $R_s - L_s$ circuit in which the resistance R_s is $R_p \dfrac{Z_p^2}{R_p^2} = R_p \cos^2\theta_y$ and the inductance L_s is $L_p \dfrac{Z_p^2}{X_p^2} = L_p \sin^2\theta_y$, where $X_p = L_p\omega$; and a parallel $R_p - C_p$ circuit is equivalent to a series $R_s - C_s$ circuit in which the resistance R_s is $R_p \dfrac{Z_p^2}{R_p^2} = R_p \cos^2\theta_y$ and the capacitance C_s is $C_p \dfrac{Y_p^2}{B_p^2} = \dfrac{C_p}{\sin^2\theta_y}$.

If the power factor angle magnitude of a series $R - X$ circuit is φ, then the power factor angle magnitude of the same two elements connected in parallel is $90° - \varphi$.

The ***apparent power*** to a circuit, S, is VI; and the complex power, $\mathbf{S}$ (= $P + jQ$), is $\mathbf{VI}^*$.

The ***real***, or the ***active***, or the ***average*** power, or simply power, P, is VI cos φ; and the ***reactive*** power, Q, is VI sin φ.

A pure resistor draws active power only, and the power factor is unity, φ = 0°.

A pure inductor draws reactive power only, and the power factor is zero, φ = 90°.

A pure capacitor draws negative reactive power only, *i.e.*, it supplies reactive power, and the power factor is zero, φ = – 90°.

However, all the practical components are associated with some small residuals; the relative effects of which depend on how fast the current through it, or the voltage across it changes.

When the current flowing through the impedance Z is I, the real power loss is $I^2 R$, and the reactive power loss is $I^2 X$. Similarly, when the voltage across the admittance Y is V, the real power loss is $V^2 G$, and the reactive power loss is $-V^2 B$.

The instantaneous power into an element, or into a single-phase circuit, pulsates at double the frequency of the voltage (or the current). The expression for the instantaneous power into an element of impedance $Z \angle\theta_z = R + jX$, carrying current I (rms value) of angular frequency ω is

$$p(t) = I^2 Z\,[\cos\theta_z + \cos(2\omega t - \theta_z)]$$

$p(t)$ can be expressed in various other forms.

$$\begin{aligned} p(t) &= I^2 R[1 + \cos 2(\omega t - \theta_z)] - I^2 X \sin 2(\omega t - \theta_z) \\ &= VI\cos\hat{\theta}_z\,[1 + \cos 2(\omega t - \theta_z)] - VI\sin\theta_z \sin 2(\omega t - \theta_z); \\ &= VI\cos\varphi\,[1+\cos 2(\omega t - \varphi)] - VI\sin\varphi \sin 2(\omega t - \varphi), \text{ } V \text{ being the applied voltage} \end{aligned}$$

The first term is the instantaneous power into the resistive part of Z, and the second term is instantaneous power into its reactive part.

The average value of the first part of the expression, or the whole expression over a cycle is VI cos φ. The average value of the second part of the expression is zero. Thus, the average power into the resistance, or into the whole of the circuit is VI cos φ; and the average power into the reactance is zero. We also see that VI sin φ is the amplitude of the pulsating power into the reactance. Thus, the reactive power of a circuit is actually the amplitude of the instantaneous power to the reactive part of the circuit.

7.2 CIRCUIT ANALYSIS

In the given circuit if there are series, or parallel elements, and if these can be reduced to single equivalent elements; the reduction should be affected first. If e is the number of elements of a network, and n is the number of nodes, then there are $e - (n - 1)$ independent loop equations, and $(n - 1)$ independent node pair equations.

A network having specified passive elements and voltage/current sources may be solved for the desired quantities using either the loop method, or the nodal method of analysis. In the loop method of analysis all the sources should be voltage sources; and in the nodal method, all the sources should be current sources. The matrix form of equations of the two methods is

$$\mathbf{Z}_{\text{LOOP}}\,\mathbf{I}_{\text{LOOP}} = \mathbf{E}_{\text{LOOP}}, \text{ and}$$

$$\mathbf{Y}_{\text{NODE}}\mathbf{V}_{\text{NODE}} = \mathbf{I}_{\text{NODE}}$$

The equations can be applied to dc networks, as well as to ac networks.

For ac networks, all the quantities involved are, in general, complex quantities.

If there are l independent loops, then $\mathbf{Z}_{\text{LOOP}}$ is an $l \times l$ square matrix, and $\mathbf{I}_{\text{LOOP}}$ and $\mathbf{E}_{\text{LOOP}}$ are both column vectors of size l each. Assuming no couplings, a diagonal element Z_{jj} of $\mathbf{Z}_{\text{LOOP}}$, is the sum of the impedances of all the elements in the loop j, an off-diagonal element Z_{jk} is the impedance of the element common to the two loops and carries negative sign if the two loop currents I_j and I_k flow in opposite directions in the common element, else it carries positive sign. $Z_{jk} = Z_{kj}$. I_j is the assumed loop current in loop j, and E_j is the algebraic sum of all the sources' emfs in loop j tending to send current in the assumed direction of the loop current I_j.

If there are m nodes(excluding the reference node), then $\mathbf{Y}_{\text{NODE}}$ is an $m \times m$ square matrix, and $\mathbf{V}_{\text{NODE}}$ and $\mathbf{I}_{\text{NODE}}$ are both column vectors of size m each. Again assuming no couplings, diagonal element Y_{jj} of $\mathbf{Y}_{\text{NODE}}$, is the sum of the admittances of all the elements incident at node j, an off-diagonal element Y_{jk} is the admittance of the element connecting nodes j and k and carries negative sign. $Y_{jk} = Y_{kj}$. V_j is the assumed voltage of node j with respect to the reference node, and I_j is the algebraic sum of all the sources' currents into node j.

If couplings are present, even then we can write KVL and KCL equations and proceed further; but then some of the diagonal, and off-diagonal elements of $\mathbf{Z}_{\text{LOOP}}$ and $\mathbf{Y}_{\text{NODE}}$ are not to be found as explained above.

Equations such as above, or any other type, can be written in Laplace transform (complex frequency s) domain also, thereby enabling to obtain complete solution (*i.e.*, including the transient part).

7.3 CURRENT DIVISION BETWEEN TWO PARALLEL BRANCHES

Let branch 1 comprise an impedance $\mathbf{Z}_1$ in series with a voltage source of voltage $\mathbf{E}_1$, and similarly, let branch 2 comprise an impedance $\mathbf{Z}_2$ in series with a voltage source of voltage $\mathbf{E}_2$. Let the current entering the combination be $\mathbf{I}$, and the branch currents be $\mathbf{I}_1$, and $\mathbf{I}_2$, then

$$\mathbf{I}_1 = \frac{\mathbf{Z}_2}{(\mathbf{Z}_1 + \mathbf{Z}_2)}\mathbf{I} + \frac{(\mathbf{E}_2 - \mathbf{E}_1)}{(\mathbf{Z}_1 + \mathbf{Z}_2)} \quad \text{and} \quad \mathbf{I}_2 = \frac{\mathbf{Z}_1}{(\mathbf{Z}_1 + \mathbf{Z}_2)}\mathbf{I} + \frac{(\mathbf{E}_1 - \mathbf{E}_2)}{(\mathbf{Z}_1 + \mathbf{Z}_2)}$$

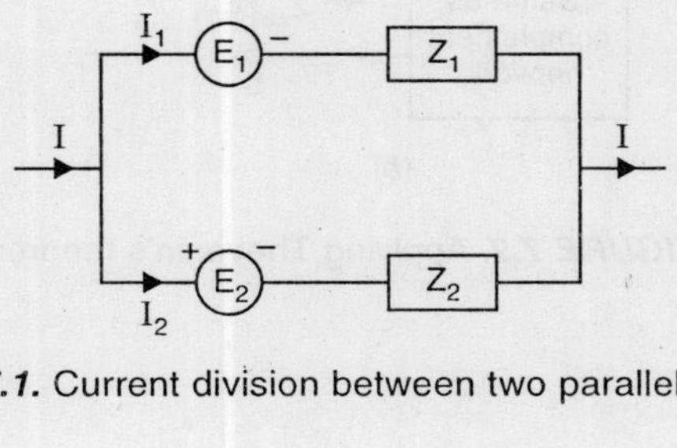

FIGURE 7.1. Current division between two parallel branches.

7.4 STAR-DELTA OR T-π CONVERSION

The general star-delta, and delta-star conversion relations are as under.

Star to delta :

$$Z_{ab} = \frac{\sum Z_a Z_b}{Z_c} = Z_a + Z_b + \frac{Z_a Z_b}{Z_c}$$

Delta to star

$$Z_a = \frac{Z_{ab} Z_{ca}}{\sum Z_{ab}}$$

In above relations, mathematical operations are to be performed keeping in mind that in general, the quantities involved are complex quantities.

These are also termed as $T-\pi$ and $\pi-T$ conversions. For equal impedances, the star impedance is 1/3 of the delta impedance value.

7.5 THEVENIN'S THEOREM

Thevenin's theorem is the most widely used theorem to simplify determining the current in some sub network of a given network. The sub network may be just a two-terminal element, or a group of connected elements with two terminals of interest available. As already stated, the network is assumed to be linear, bilateral, causal, and time invariant.

The sub network in which current is to be found is disconnected from the network at its terminals. Let the opened terminals in the leftover network be designated as a and b. Now we look into this leftover network at the terminals a and b, find the voltage difference $\mathbf{E}_{ab}$, and the impedance $\mathbf{Z}_{ab}$. Then the required current is the same as would flow when the sub network is connected to the source of emf $\mathbf{E}_{ab}$ in series with the impedance $\mathbf{Z}_{ab}$. The Thevenin's voltage-source equivalent can be transformed into a current-source equivalent; which then is called **Norton's** equivalent circuit.

The Thevenin's voltage-source equivalent or the current-source equivalent at terminals a-b remains the same irrespective of the sub network in which current is to be found.

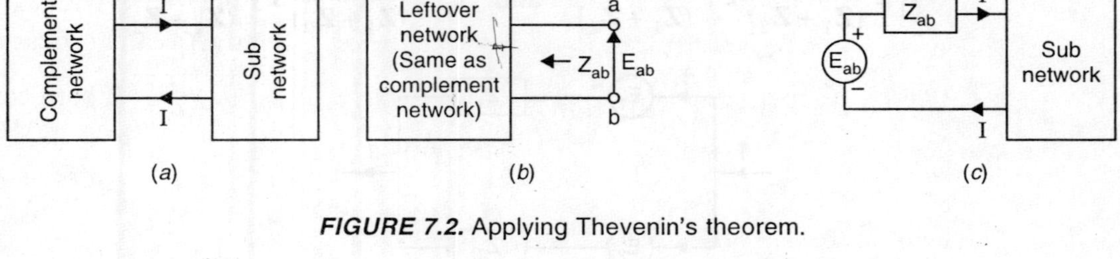

FIGURE 7.2. Applying Thevenin's theorem.

7.6 MILLMAN'S THEOREM

This theorem enables us to find an equivalent current source or voltage source of n sources in parallel.

Let the current sources of currents $\mathbf{I}_1, \mathbf{I}_2, \ldots, \mathbf{I}_n$ with accompanying shunt admittances $\mathbf{Y}_1, \mathbf{Y}_2, \ldots, \mathbf{Y}_n$ respectively be in parallel. Then the equivalent current source has the current value $\mathbf{I} = \sum_{j=1}^{n} \mathbf{I}_j$, and the shunt admittance value $\mathbf{Y} = \sum_{j=1}^{n} \mathbf{Y}_j$

For voltage sources of voltages $\mathbf{E}_1, \mathbf{E}_2, \ldots, \mathbf{E}_n$, and the corresponding series impedances $\mathbf{Z}_1, \mathbf{Z}_2, \ldots, \mathbf{Z}_n$; the equivalent voltage is $\mathbf{E} = \dfrac{\sum_{j=1}^{n} \dfrac{\mathbf{E}_j}{\mathbf{Z}_j}}{\sum_{j=1}^{n} \dfrac{1}{\mathbf{Z}_j}}$, and the series impedance $\mathbf{Z}$ is given by $\dfrac{1}{\mathbf{Z}} = \sum_{j=1}^{n} \dfrac{1}{\mathbf{Z}_j}$. Expressed in terms of admittances, $\mathbf{E} = \dfrac{\sum_{j=1}^{n} \mathbf{E}_j \mathbf{Y}_j}{\sum_{j=1}^{n} \mathbf{Y}_j}$, and $\mathbf{Y}$ (to be connected in series with $\mathbf{E}$) $= \sum_{j=1}^{n} \mathbf{Y}_j$.

7.7 MAXIMUM POWER TRANSFER THEOREM

This theorem has applications in electronics circuits and communication circuits, where amount of power transfer, rather than the efficiency of power transfer is more important.

The conditions of maximum power transfer for various situations are summarized below.

(1) The source has resistance only, R_S.

(*a*) The load also comprises only a resistance R_L. $R_L = R_S$.

(*b*) The load comprises only a reactance X_L. No question of any power transfer.

(*c*) The load comprises a resistance R_L and a reactance X_L. If only X_L can be varied, X_L should be zero. If only R_L can be varied, then $R_L = \sqrt{(R_S{}^2 + X_L{}^2)}$. If both can be varied then $X_L = 0$, and $R_L = R_S$.

(2) The source has only reactance X_S.

(*a*) The load comprises only a resistance R_L. $R_L = X_S$.

(*b*) The load also comprises only a reactance X_L. No question of any power transfer.

(*c*) The load comprises a resistance R_L and a reactance X_L. If only X_L can be varied, $X_L = -X_S$. If only R_L can be varied, then $R_L = (X_S + X_L)$. If both can be varied then $X_L = -X_S$, and R_L indeterminate.

(3) The source has resistance R_S and reactance X_S.

(*a*) The load comprises only a resistance R_L. $R_L = (R_S^2 + X_S^2)$

(*b*) The load comprises only a reactance X_L. No question of any power transfer.

(*c*) The load comprises a resistance R_L and a reactance X_L. If only X_L can be varied, then $X_L = -X_S$. If only R_L can be varied, then $R_L = \sqrt{\{(X_S + X_L)^2 + R_S^2\}}$. If the load impedance magnitude and the angle both can be varied, then $R_L = R_S$ and $X_L = -X_S$. If only the load impedance magnitude can be varied, the angle remaining fixed, then $(R_L^2 + X_L^2) = (R_S^2 + X_S^2)$.

Thus, if only the load resistance R_L can be varied, it should equal the magnitude of the impedance seen by it. If only the load reactance X_L can be varied, it should be such that the total reactance in the circuit is zero. If R_L and X_L can both be varied independently, then $R_L = R_S$ $(R_S \neq 0)$, and $X_L = -X_S$. Finally, if only Z_L can be varied, then $Z_L = Z_S$.

7.8 RESONANCE

In an *R-L-C* circuit, resonance is said to occur if the voltage across the circuit and the current into it come in phase. Resonance can be achieved either by varying the inductance, or the capacitance, or the frequency of the excitation.

Following statements hold for a ***series R-L-C resonant circuit***.

The inductive reactance and the capacitive reactance cancel each other, and the circuit impedance is simply R and is minimum. The whole of the applied voltage V is opposed by R alone.

The resonance frequency (angular), $\omega_0 = \dfrac{1}{\sqrt{LC}}, f_0 = \dfrac{\omega_0}{2\pi}$

For a given applied voltage, the current is maximum.

The voltages across the inductor and across the capacitor are equal, and in phase opposition, and can be very large; even several times the applied voltage depending upon the value of either L or C. It is actually Q_o V, Q_o being the quality factor of the circuit.

$$Q_o = L\,\omega_o/R = 1/(RC\,\omega_o) = \frac{1}{R}\sqrt{\frac{L}{C}} = \frac{R_o}{R}.$$

The half power points are given by

$$\omega_{1,2} = \mp\frac{R}{2L} + \sqrt{\left(\frac{R}{2L}\right)^2 + \frac{1}{LC}}$$

The radian frequency bandwidth is $\omega_{bw} = R/L$.

The selectivity of the circuit is $(\omega_2 - \omega_1)/\ \omega_o = (R/L)/\ \omega_o = 1/\ Q_o$.

The impedance at frequencies close to the resonance frequency is given by

$$\mathbf{Z} = R[1 + j\,Q_o\,\delta(2-\delta)],\ \text{where } \delta = (\omega - \omega_o)/\ \omega_o$$

The frequency at which the voltage across the capacitor is maximum,

$$f_C = \frac{1}{2\pi}\sqrt{\frac{1}{LC} - \frac{R^2}{2L^2}}$$

The frequency at which the voltage across the inductor is maximum,

$$f_L = \frac{1}{2\pi}\sqrt{\frac{1}{LC - \frac{C^2R^2}{2}}}$$

$$f_C < f_o < f_L$$

For a ***parallel*** *L-C* resonant circuit, each of *L* and *C* having a lossy component,

$$f_{ar} = \frac{1}{2\pi}\sqrt{\frac{1}{LC}\left(\frac{L - CR_L{}^2}{L - CR_C{}^2}\right)}$$

When R_C is zero,

$$f_{ar} = \frac{1}{2\pi}\sqrt{\frac{1}{LC} - \frac{R_L{}^2}{L^2}} = f_0\sqrt{1 - \frac{1}{Q_o{}^2}} < f_0$$

$$I_L = Q_{ar}\,I_s, \quad \text{and} \quad I_C = I_L\sqrt{1 - \frac{1}{Q_{ar}{}^2}}$$

$$X_L = -X_C\,(1 - 1/Q_o{}^2)$$

The dynamic resistance,

$$R_{ar} = \frac{L}{CR_L} = R_L\,(1 + Q^2{}_{ar}) = R_L\,Q^2{}_o$$

The impedance at frequencies close to the antiresonance frequency is given by

$$\frac{Z}{Z_{ar}} = \left|\frac{1}{1 + j2Q_{ar}\delta}\right|,\ \text{where } \delta = (\omega - \omega_{ar})/\ \omega_{ar}$$

The condition to achieve maximum impedance by varying frequency is

$$\omega_f^2 = \frac{1}{LC}\sqrt{1+\frac{2CR_L^{\ 2}}{L}} - \frac{R_L^{\ 2}}{L^2}$$

The condition to achieve maximum impedance by varying capacitance is

$$\omega_C^{\ 2} = \frac{1}{LC} - \frac{R_L^{\ 2}}{L^2}$$, which is the same as $\omega_{ar}^{\ 2}$.

The condition to achieve maximum impedance by varying inductance is

$$\omega_L^{\ 2} = \frac{1}{LC} + \frac{R_L^{\ 2}}{L^2}$$

The bandwidth is

$$f_{BW} = (f_{ar} / Q_{ar})(1 + L/(CR_L R_s))$$
$$= 2 f_{ar} / Q_{ar},$$

with source resistance R_s matched to R_{ar} $(= L/CR_L)$.

Chapter 8

TRANSIENTS, THREE-PHASE SYSTEMS AND WAVES

8.1 DC TRANSIENTS

Assuming zero initial conditions, for a ***series R-L*** circuit, the expression for the resulting current on sudden application of a dc voltage V is

$$i(t) = \frac{V}{R}\left(1 - e^{-\frac{t}{T}}\right), T \text{ is the time constant of the circuit and equals } L/R.$$

$$v_L(t) = Ve^{-\frac{t}{T}}$$

For a circuit which is explicitly not a series circuit, R is substituted by the resistance seen by the inductor.

Assuming zero initial conditions, for a ***series R-C*** circuit, the expression for the resulting current on sudden application of a dc voltage V is

$$i(t) = \frac{V}{R}e^{-\frac{t}{T}}, T \text{ is the time constant of the circuit and equals } RC.$$

$$v_C(t) = V\left(1 - e^{-\frac{t}{T}}\right).$$

For a circuit which is explicitly not a series circuit, R is substituted by the resistance seen by the capacitor.

Assuming an initial charge q_o on the capacitor and no initial current in the inductor, for a ***series R-L-C*** circuit, the expression for the resulting current on sudden application of a dc voltage V is

$$i(t) = \frac{V - \frac{q_0}{C}}{\omega_d L}e^{-\delta\omega_n t} \sin \omega_d t, \quad \text{where } \omega_n = \sqrt{\frac{1}{LC}}, \omega_d = \omega_n\sqrt{1-\delta^2}, \text{ and } \delta = \frac{R}{2\sqrt{\frac{L}{C}}}.$$

8.2 AC TRANSIENTS

When, at $t = t_0$, a sinusoidal voltage $v(t) = V_m \sin \omega t$ is switched on to a series $R - L$ circuit, the resulting current is given by

$$i(t) = \frac{V_m}{|Z|}\left[\sin(\omega t - \theta) - \sin(\omega t_0 - \theta)e^{-\frac{R}{L}(t-t_0)}\right],$$

where, $|Z| = \sqrt{R^2 + (L\omega)^2}$ and $\theta = \tan^{-1}\frac{L\omega}{R}$.

Alternatively, when at $t = 0$, a sinusoidal voltage $v(t) = V_m \sin(\omega t + \alpha)$ is switched on to a series $R - L$ circuit, the resulting current is given by

$$i(t) = \frac{V_m}{|Z|}\left[\sin(\omega t + \alpha - \theta) - \sin(\alpha - \theta)e^{-\frac{R}{L}t}\right]$$

Similarly, the corresponding expressions for a series $R - C$ circuit are

$$i(t) = \frac{V_m}{|Z|}\left[\sin(\omega t - \theta) - \left(\sin(\omega t_0 - \theta) - \frac{|Z|}{R}\right)e^{-\frac{t-t_0}{RC}}\right],$$

where $|Z| = \sqrt{R^2 + \left(\frac{1}{C\omega}\right)^2}$ and $\theta = \tan^{-1} - \frac{1}{C\omega R}$ and

$$i(t) = \frac{V_m}{|Z|}\left[\sin(\omega t + \alpha - \theta) - \left(\sin(\alpha - \theta) - \frac{|Z|}{R}\sin\alpha\right)e^{-\frac{R}{L}t}\right].$$

8.3 THREE-PHASE CIRCUITS

A three-phase circuit, or a system comprises a three-phase load and a three-phase supply. The three phases are generally denoted by the letters R, Y, and B. A three-phase system may be a balanced one, or an unbalanced one.

Further, a three-phase connection may be a star connection, or a delta connection.

In three-phase circuits, we speak of line voltages and the line currents; phase voltages and phase currents. Another voltage which we may speak of is line to neutral voltage.

In a balanced system, all the three line voltages $\mathbf{V}_{RY}$, $\mathbf{V}_{YB}$, and $\mathbf{V}_{BR}$ are equal in magnitude and differ in phase from each other by 120°, *i.e.*, $\mathbf{V}_{YB} = \mathbf{V}_{RY}\angle -120°$, $\mathbf{V}_{BR} = \mathbf{V}_{YB}\angle -120°$. Same is true for the line currents $\mathbf{I}_R$, $\mathbf{I}_Y$, and $\mathbf{I}_B$; and also for the phase voltages, and the phase currents. While dealing with three-phase phasor quantities, it is usual to take line R to neutral voltage, $\mathbf{V}_{RN}$, as reference.

A system which is not balanced is an unbalanced system.

In a three-phase system whenever we speak of a voltage or a current, it is understood to be a line voltage or a line current.

A balanced three-phase impedance load implies equal impedances (in magnitude as well as in angle) in all the three phases.

8.4 THREE-PHASE BALANCED SYSTEM

In a ***star-connected*** balanced system, following relations hold.

$$\mathbf{V}_{RY} = \sqrt{3}\ \mathbf{V}_{RN} \angle 30°, \mathbf{I}_R \text{ lags } \mathbf{V}_{RY} \text{ by } (30° + \varphi).$$

The line current and the corresponding phase current are one and the same thing.

Similarly, the line to neutral voltage and the corresponding phase voltage are one and the same thing.

In a ***delta-connected*** balanced system, following relations hold.

$$\mathbf{I}_R = \sqrt{3}\ \mathbf{I}_{RY} \angle -30°, \mathbf{V}_{RY} = \sqrt{3}\ \mathbf{V}_{RN} \angle 30°, \mathbf{I}_R \text{ lags } \mathbf{V}_{RY} \text{ by } (30° + \varphi).$$

The line voltage and the corresponding phase voltage are one and the same thing.

In ***both types*** of the connections, the line voltage and the line current magnitudes are denoted by V_L and I_L respectively; and the three-phase power is given by 3 $V_L\ I_L \cos\varphi$.

The three-phase instantaneous power is steady and is not a function of time, though the individual phase powers pulsate at double the frequency as in any single-phase circuit.

In the ***two-wattmeter method*** of three-phase power measurement, we usually connect the wattmeters so that one carries the R-phase current and is across lines R and Y, while the other is in B-phase and is across lines B and Y. Then the line R meter shows $V_L\ I_L \cos(30° + \varphi)$ and the line B meter, $V_L\ I_L \cos(30° - \varphi)$. The sum of the wattmeter readings is $\sqrt{3}\ V_L\ I_L \cos\varphi$.

It also gives the relation $\tan\varphi = \sqrt{3}\ (W_B - W_R)/(W_B + W_R)$.

The two-wattmeter method of measuring power is applicable to an unbalanced system also. The method measures power correctly even if the wave shape of the applied voltages and the resulting currents are not sinusoidal.

A balanced three-phase circuit can be solved using its ***per phase circuit*** (say, phase R). If it is a delta-connected circuit, it should first be converted into a star-connected circuit.

8.5 UNBALANCED SYSTEMS

An unbalanced system, whether a 3-wire system or a 4-wire system, may be solved using the loop method.

In a 3-wire, star-connected system, the voltage of the load neutral n' wrt the source neutral n, is given by

$$\mathbf{V}_{n'n} = (\mathbf{E}_R\,\mathbf{Y}_R + \mathbf{E}_Y\,\mathbf{Y}_Y + \mathbf{E}_B\,\mathbf{Y}_B)/(\mathbf{Y}_R + \mathbf{Y}_y + \mathbf{Y}_B)$$

The admittances should include the respective line impedances effect also, if these are not impedanceless.

8.6 RELATION BETWEEN DELTA LINE CURRENTS AND LIMB CURRENTS

$$\mathbf{I}_R = \mathbf{I}_{RY} - \mathbf{I}_{BR}$$
$$\mathbf{I}_y = \mathbf{I}_{YB} - \mathbf{I}_{RY}$$

$$\mathbf{I}_B = \mathbf{I}_{BR} - \mathbf{I}_{YB}$$

$$\mathbf{I}_{RY} = \frac{\mathbf{I}_R\,\mathbf{Z}_{BR} - \mathbf{I}_Y\mathbf{Z}_{YB}}{\sum \mathbf{Z}_{RY}}$$

$$\mathbf{I}_{YB} = \frac{\mathbf{I}_Y\,\mathbf{Z}_{RY} - \mathbf{I}_B\mathbf{Z}_{BR}}{\sum \mathbf{Z}_{RY}}$$

$$\mathbf{I}_{BR} = \frac{\mathbf{I}_B\,\mathbf{Z}_{YB} - \mathbf{I}_R\mathbf{Z}_{RY}}{\sum \mathbf{Z}_{RY}}$$

If all the delta impedances are equal, the relation can readily be simplified to

$$\mathbf{I}_{RY} = \frac{\mathbf{I}_R - \mathbf{I}_Y}{3},$$

the relations for the other two currents can be easily written.

8.7 COUPLED CIRCUITS

In coupled circuits analysis, the usual loop method can be applied; however, while writing the voltage terms, care should be taken in placing the sign of mutual coupling voltage term as per the dot convention. If, at the dot marked ends, both the currents labeled on the two mutually coupled inductors, either enter them or leave them, then the sign of the mutually-induced voltage term is the same as that of the self-induced voltage term. If one current enters and the other leaves, then the two voltage terms carry opposite signs.

In communication engineering, tuned coupled circuits are used to obtain largest output signal. The coupling is by means of an air-core transformer. When there is a capacitor in the secondary circuit alone, it's a singly tuned circuit; and if capacitors are included in both the circuits, it's a doubly tuned circuit.

For a singly tuned circuit, the resonance condition in the secondary circuit is given by $\omega L_2 = \dfrac{1}{\omega C_2}$. Further, the critical coupling condition (for the largest value of the output voltage) is given by $\omega M = \sqrt{(R_1 R_2)}$. R_1 and R_2 are the primary circuit and the secondary circuit resistances.

8.8 FOURIER SERIES

A non-sinusoidal periodic wave $f(t)$ can be expressed as a summation series of a constant term and sine and cosine waves of fundamental and multiple frequencies. This is called Fourier series. ***Fundamental frequency*** is the frequency of the given non-sinusoidal wave, and the other frequencies, termed ***harmonics***, are integer multiples of the fundamental frequency.

$$f(t) = \frac{a_0}{2} + \sum_{n=1}^{\infty} (a_n \cos n\omega t + b_n \sin n\omega t)$$

The ***constants*** a_o, a_n, b_n can be found from the following relations.

$$a_o = \frac{2}{T}\int_{t_0}^{t_0+T} f(t)\,dt$$

$$a_n = \frac{2}{T}\int_{t_0}^{t_0+T} f(t)\cos n\omega t\,dt$$

$$b_n = \frac{2}{T}\int_{t_0}^{t_0+T} f(t)\sin n\omega t\,dt$$

The ***series*** may also be expressed as a ***function of*** $\theta = \omega t$. Then

$$f(\theta) = \frac{a_0}{2} + \sum_{n=1}^{\infty} (a_n \cos n\theta + b_n \sin n\theta)$$

$$a_o = \frac{1}{\pi}\int_{\alpha}^{\alpha+2\pi} f(\theta)\,d\theta$$

$$a_n = \frac{1}{\pi}\int_{\alpha}^{\alpha+2\pi} f(\theta)\cos n\theta\,d\theta$$

$$b_n = \frac{1}{\pi}\int_{\alpha}^{\alpha+2\pi} f(\theta)\sin n\theta\,d\theta$$

The series can also be expressed so that it consists of the constant term, and ***modified cosine terms***; or the constant term, and ***modified sine terms***.

$$f(\theta) = \frac{a_0}{2} + \sum_{n=1}^{\infty} c_n \cos(n\theta - \alpha_n)$$

$$f(\theta) = \frac{a_0}{2} + \sum_{n=1}^{\infty} c_n \sin(n\theta + \delta_n)$$

$$C_n = \sqrt{a_n^2 + b_n^2}, \ \tan\alpha_n = \frac{b_n}{a_n}, \ \tan\delta_n = \frac{a_n}{b_n}$$

The ***complex form*** of the series is

$$f(t) = \sum_{n=-\infty}^{n=\infty} \beta_n e^{jn\omega t}$$

$$\beta_n = \frac{1}{T}\int_0^T f(t)\,e^{-jn\omega t}\,dt$$

β_n is the ***discrete Fourier transform*** of $f(t)$, and $f(t)$ is the inverse transform of β_n. The amplitude spectrum of the series is given by

$|\beta_n| = (\text{Re}^2\,\beta_n + \text{Im}^2\,\beta_n)^{1/2}$, and the phase spectrum by

$$\phi_n = \arctan\frac{\text{Im}\,\beta_n}{\text{Re}\,\beta_n}$$

For ***aperiodic functions***, we have ***Fourier integral*** or transform, given by

$$F(f) = \int_{-\infty}^{+\infty} f(t)\, e^{-j2\pi ft}\, dt,$$ and the inverse transform is given by

$$f(t) = \int_{-\infty}^{+\infty} F(f)\, e^{j2\pi ft}\, dt.$$

8.9 WAVE SYMMETRIES

If we can identify the symmetry in the wave, a particular type of coefficients may be found to be absent, and therefore those need not be calculated.

1. Even function, or mirror symmetry, characterized by $f(-\alpha) = f(\alpha)$

   The series then consists of cosine terms only.
2. Odd function, or rotation symmetry, characterized by $f(-\alpha) = -f(\alpha)$

   The series then consists of sine terms only.
3. Half wave, or alternation symmetry, characterized by $f(-\alpha) = -f(\alpha + \pi)$

   The series then consists of odd harmonics only (cosine, and sine both).

   The presence of a constant term, or an even harmonic term makes the wave unsymmetrical.
4. Quarter wave symmetry, characterized by either conditions 1 and 3 together, when it shall consist of odd cosine terms only; or conditions 2 and 3 together, when it shall consist of odd sine terms only.

8.10 RMS VALUE AND POWER FACTOR

The ***rms value of the wave*** is given by the expression

$$\sqrt{\frac{a_0^{\,2}}{4} + \frac{c_1^{\,2}}{2} + \frac{c_2^{\,2}}{2} + \ldots + \frac{c_n^{\,2}}{2}}$$

The ***power*** in an element is caused by the voltage and current terms having the same frequency. For example, a 3rd harmonic voltage term cannot interact with a 5th harmonic current term to produce power; however a 3rd harmonic voltage term and a 3rd harmonic current term do interact to produce power. Thus, the real power may be expressed as

$$P = E_o I_o + E_1 I_1 \cos \varphi_1 + E_2 I_2 \cos \varphi_2 + \ldots + E_n I_n \cos \varphi_n$$

E_j and I_j are the rms values of the jth harmonic voltage and current, and φ_j is the phase difference between the two.

The ***power factor*** of the circuit is given by

(Total power in the circuit)/((rms value of the voltage) × (rms value of the current)).

Chapter 9

NETWORKS AND FILTERS

9.1 ANALYSIS IN COMPLEX FREQUENCY *s* DOMAIN

The Laplace variable s is a complex quantity; as such all the functions, expressions, and equations involving s are also complex. However, for ease, no bold face typing is used for them.

The laplace transform of the excitation, which may either be a voltage signal, or a current signal applied to a pair of terminals; will be denoted by $E(s)$, and that of the response which again can be a voltage across some element, or a current through an element, by $R(s)$.

$R(s)/E(s)$ is denoted by $H(s)$, and is called the ***system or the network function***. Thus,

$$R(s) = H(s)\,E(s)$$

For a given network, the system function will depend on the excitation port and the response port.

For the three basic passive elements, the terminal relations become

$$V(s) = R\,I(s) \qquad I(s) = V(s)\,1/R$$

$$V(s) = Ls\,I(s) - Li\,(0\_) \qquad I(s) = (1/Ls)\,V(s) + i\,(0\_)/s$$

$$V(s) = (1/Cs)\,I(s) + v\,(0\_)/s \qquad I(s) = CsV(s) - C\,v\,(0\_)$$

These equations are obtained from their counterparts in time domain.

We can draw the corresponding ***'transformed' circuit***, in which all the elements and quantities are shown labeled with their s-domain symbols and values. We can write transform voltage term in terms of transform current and impedance, or current term in terms of voltage and admittance for any element and for any circuit. We are, therefore, able to write loop equations, or nodal equations in s-domain. These equations are algebraic in nature and can be solved for loop currents $I(s)$, or for node voltages $V(s)$. Taking the Laplace inverse of the expression for the transform of the required quantity gives the complete expression for that required quantity in time domain.

Also, all the network analyzing/simplification theorems can be applied to transformed networks.

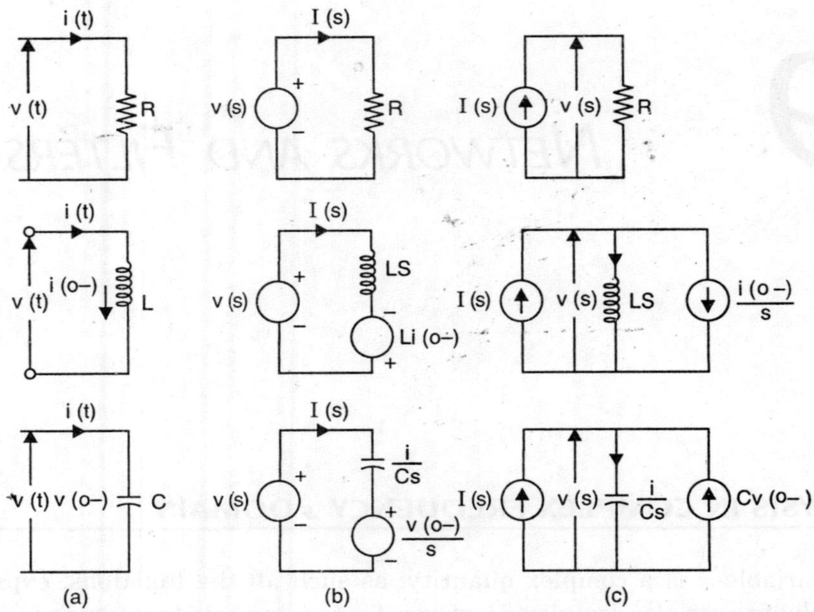

FIGURE 9.1. The time domain circuits and the corresponding *s*-domain voltage source, and current source circuits for the three passive elements.

In general, $H(s)$ is written in the form of the ratio of two polynomials in s.

$$H(s) = \frac{a_n s^n + a_{n-1}s^{n-1} + \ldots + a_1 s + a_0}{b_m s^m + b_{m-1}s^{m-1} + \ldots + b_1 s + b_0} = \frac{P(s)}{Q(s)} = A\,\frac{(s-z_1)(s-z_2)\ldots(s-z_{n-1})(s-z_n)}{(s-p_1)(s-p_2)\ldots(s-p_{m-1})(s-p_m)}$$

The n z's are the roots of $P(s) = 0$, and are called the ***zeros of H(s)***, and the m p's are the roots of $Q(s) = 0$, and are called the ***poles of H(s)***. The relation between m and n depends upon the type of $H(s)$, *i.e.*, whether it is a driving-point function, or a transfer function, or a current ratio, or a voltage ratio function.

9.2 NECESSARY CONDITIONS FOR DRIVING POINT FUNCTIONS

1. The coefficients a's and b's in $P(s)$ and $Q(s)$ must be real and positive.
2. There can not be any missing term between the highest degree term and the lowest degree term in $P(s)$ and $Q(s)$, unless all the even order terms, or all the odd order terms are missing.
3. The degree of $P(s)$ and $Q(s)$ may differ at most by unity.
4. The lowest degree terms in $P(s)$ and $Q(s)$ also may differ at most by unity.
5. The real parts of all poles and zeros must be either zero or negative.
6. Complex, or imaginary poles and zeros must occur in conjugate pairs.
7. The poles and zeros on the imaginary axis, if any, must be simple.

9.3 NECESSARY CONDITIONS FOR TRANSFER FUNCTIONS

1. The coefficients a's and b's in $P(s)$ and $Q(s)$ must be real and b's must be positive as well.
2. There cannot be any missing term between the highest degree term and the lowest degree term in $Q(s)$, unless all the even order terms, or all the odd order terms are missing.
3. There may be missing terms between the highest degree term and the lowest degree term in $P(s)$.
4. The degree of $P(s)$ may be as small as zero; independent of the degree of $Q(s)$.
5. For voltage ratio transfer functions and current ratio transfer functions, the degree of $P(s)$ can at most equal the degree of $Q(s)$.
6. For transfer impedance or transfer admittance functions, the degree of $P(s)$ can at most equal one plus the degree of $Q(s)$.
7. The real parts of all the poles must be either zero or negative.
8. Complex, or imaginary poles and zeros must occur in conjugate pairs.
9. The poles on the imaginary axis, if any, must be simple.

9.4 POLE-ZERO PLOTS

In the complex s-plane, zeros are shown by o's and the poles are shown as x's; the resulting plot is the pole-zero plot of the function $H(s)$.

The response of a network can be determined graphically from the pole-zero plot of the corresponding network function.

$R(s)$ can be expressed in the partial fraction expansion form as

$$\sum_i \frac{A_i}{(s - s_i)} + \sum_j \frac{B_j}{(s - s_j)}$$

where s_i are the poles of $H(s)$, and s_j are the poles of $E(s)$.

Taking the Laplace inverse of $R(s)$, we get the response $r(t)$.

$$r(t) = \sum_i A_i e^{s_i t} + \sum_j B_j e^{s_j t}$$

The frequencies s_i are the ***natural frequencies*** of the system, whereas the frequencies s_j are the ***forced frequencies***; the former being associated with the system function, and the latter with the excitation, and hence called the forced ones.

A transformed function $F(s)$ can be expressed in the ***partial fraction expansion*** form as $\sum_{i=1}^{i=m} \frac{K_i}{(s - p_i)}$; the p's are the m poles, and the K's are the corresponding m ***residues*** of the function $F(s)$. In general, the p's as well as the K's have complex values. The K's can be determined graphically as given below.

$K_i = A \times$ (product of the vectors drawn from all the zeros to the pole p_i)/(product of the vectors drawn from all the poles (except p_i) to the pole p_i). A is a constant.

If $E(s)$ is 1, then $R(s)$ is $H(s)$. 1 is the Laplace transform of the unit impulse function. Therefore, $h(t)$, the Laplace inverse of $H(s)$, is the ***impulse response*** of the network. Knowing $h(t)$, the ***step response*** can be obtained simply by integrating the impulse response.

The ***steady-state sinusoidal response*** of the network can be obtained from the corresponding network function by replacing s by $j\omega$ in that function. This response is in the complexor, or the phasor form.

The steady-state response $H(j\omega)$ is expressed as $M(\omega)\, e^{j\varphi(\omega)}$. $M(\omega)$ is the ***amplitude response function***, and $\varphi(\omega)$ is the ***phase response function***.

As an example, consider a simple series R-C circuit, with voltage across the capacitor as the response. The system function $\dfrac{V_2(s)}{V_1(s)}$ is $\dfrac{1}{RC\,s+1}$.

$$\frac{V_2(j\omega)}{V_1(j\omega)} = \frac{1}{j\omega RC+1} \approx \frac{1}{j\omega RC}, \text{ if } \omega RC \text{ is} >> 1. \quad \text{Thus}$$

$$\frac{V_2(j\omega)}{V_1(j\omega)} = \frac{1}{\omega RC} e^{-j\frac{\pi}{2}}. \quad M(\omega) = \frac{1}{\omega RC}, \text{ and } \varphi(\omega) = -\frac{\pi}{2}$$

By examining this we find that the circuit (*i*) causes a phase lag of 90°, (*ii*) acts as an integrator, and (*iii*) acts as a low pass filter.

Similarly, when the voltage across the resistor is considered as response, then the system function

$$\frac{V_2(s)}{V_1(s)} \text{ is } \frac{RC\,s}{RC\,s+1}. \quad \frac{V_2(j\omega)}{V_1(j\omega)} = \frac{j\omega RC}{j\omega RC+1} \approx j\omega RC, \text{ if } \omega RC \text{ is} << 1.$$

$$M(\omega) = \omega RC, \quad \text{and} \quad \varphi(\omega) = \frac{\pi}{2}$$

By examining this we find that the circuit (*i*) causes a phase lead of 90°, (*ii*) acts a differentiator, and (*iii*) acts as a high pass filter.

$M(\omega)$ and $\varphi(\omega)$ can also be found graphically. This is a ***point by point*** exercise on the $j\omega$-axis.

$M(\omega) = A \times$ (Product of magnitudes of all the vectors drawn from the zeros to the point ω on the $j\omega$-axis)/ (Product of magnitudes of all the vectors drawn from the poles to the point ω on the $j\omega$-axis).

$\varphi(\omega) = \Sigma$ the angles of all the vectors drawn from the zeros to the point ω on the $j\omega$-axis $- \Sigma$ the angles of all the vectors drawn from the poles to the point ω on the $j\omega$-axis.

9.5 FOUR-TERMINAL (TWO-PORT) NETWORKS

Various circuit elements interconnected to serve some useful purpose is a network. A network would have several branches and junction points of branches. Usually, one may not be interested in knowing how the various elements of the network are behaving, but is interested to know the network performance at some selected junction points. These junction points are, therefore, accessible and are called terminals. If n terminals of the network are brought out, then it is an n-terminal network.

A common class of networks is a four-terminal network. Two of the terminals serve as the excitation, or the input, or the driving port; and the remaining two as the response, or the output, or the transfer port. A four-terminal network is thus also called a two-port network. Further, one of the terminals of the input port, and of the output port is generally common; thus effectively it's a three-terminal network.

A two-port network is represented by a box showing input port on the left side, and the output port on the right side.

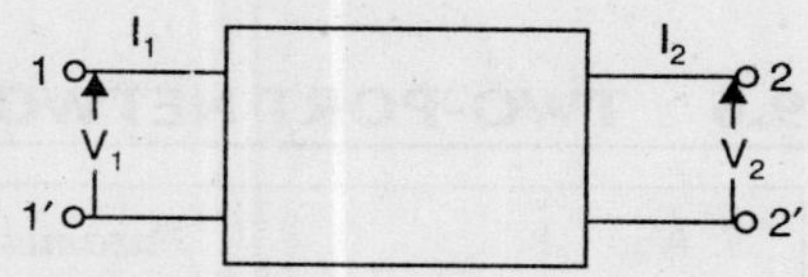

FIGURE 9.2. A two-port network.

We will deem the network as an ac network; all the quantities involved, therefore are, In general, complex quantities. However, the relations given can readily be simplified for dc networks.

Z_{1oc}, Z_{1sc} are the impedances seen at port 1; with port 2 open-circuited, and short-circuited respectively. Similarly, Z_{2oc}, Z_{2sc} are the impedances seen at port 2; with port 1 open-circuited, and short-circuited respectively.

A two-port network can be reduced to an equivalent T-network, or to an equivalent π-network.

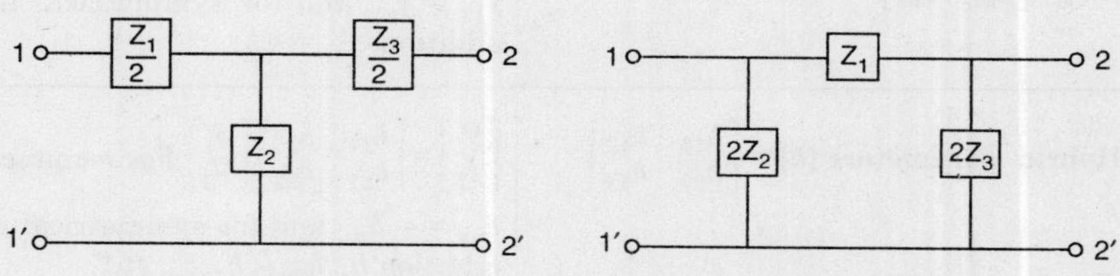

(*a*) (*b*)

FIGURE 9.3. A T- network and a π-network.

To find the ***equivalent T-network***

$$Z_2 = \pm\sqrt{Z_{2oc}\,(Z_{1oc} - Z_{1sc})}$$

$$\frac{Z_1}{2} = Z_{1oc} - Z_2$$

$$\frac{Z_3}{2} = Z_{2oc} - Z_2$$

To find the ***equivalent π-network***

$$2Z_2 = \frac{Z_{2oc}\,Z_{1sc}}{Z_{2oc} - \sqrt{Z_{2oc}\,(Z_{1oc} - Z_{1sc})}}$$

$$Z_1 = \frac{Z_{2oc}\,Z_{1sc}}{\sqrt{Z_{2oc}\,(Z_{1oc} - Z_{1sc})}}$$

$$2Z_3 = \frac{Z_{2oc}\, Z_{1sc}}{Z_{1oc} - \sqrt{Z_{2oc}\,(Z_{1oc} - Z_{1sc})}}$$

The values will hold for the frequency at which the open-circuit, and short-circuit impedances are found. If the frequency is changed, the equivalent network parameters will also change.

All the above impedances may be the transformed impedances, *i.e.*, in terms of s; or may be the usual impedances, *i.e.*, in terms of $j\omega$. Mathematical operations should be carried out keeping in mind that, in general, all the quantities involved are complex.

9.6 TWO-PORT NETWORK PARAMETERS

S. No.	*Parameters*	*Defining equations*
1.	Open-circuit impedance parameters $[z] = \begin{bmatrix} z_{11} & z_{12} \\ z_{21} & z_{22} \end{bmatrix}$	$\begin{bmatrix} V_1 \\ V_2 \end{bmatrix} = \begin{bmatrix} z_{11} & z_{12} \\ z_{21} & z_{22} \end{bmatrix} \begin{bmatrix} I_1 \\ I_2 \end{bmatrix}$ For reciprocal networks, $z_{12} = z_{21}$, and for symmetrical networks, in addition, $z_{11} = z_{22}$.
2.	Short-circuit admittance parameters $[y] = \begin{bmatrix} y_{11} & y_{12} \\ y_{21} & y_{22} \end{bmatrix}$	$\begin{bmatrix} I_1 \\ I_2 \end{bmatrix} = \begin{bmatrix} y_{11} & y_{12} \\ y_{21} & y_{22} \end{bmatrix} \begin{bmatrix} V_1 \\ V_2 \end{bmatrix}$ For reciprocal networks, $y_{12} = y_{21}$, and for symmetrical networks, in addition, $y_{11} = y_{22}$.
3.	Hybrid parameters $[h] = \begin{bmatrix} h_{11} & h_{12} \\ h_{21} & h_{22} \end{bmatrix}$	$\begin{bmatrix} V_1 \\ I_2 \end{bmatrix} = \begin{bmatrix} h_{11} & h_{12} \\ h_{21} & h_{22} \end{bmatrix} \begin{bmatrix} I_1 \\ V_2 \end{bmatrix}$ For reciprocal networks, $h_{12} = -\,h_{21}$, and for symmetrical networks, in addition $h_{11}h_{22} - h_{12}h_{21} = 1$.
4.	Transmission parameters $T = \begin{bmatrix} A & B \\ C & D \end{bmatrix}$	$\begin{bmatrix} V_1 \\ I_1 \end{bmatrix} = \begin{bmatrix} A & B \\ C & D \end{bmatrix} \begin{bmatrix} V_2 \\ -I_2 \end{bmatrix}$ For reciprocal networks, $AD - BC = 1$, and for symmetrical networks, in addition, $A = D$.

9.7 RELATIONSHIP BETWEEN TWO-PORT PARAMETERS

	$[z]$	$[y]$	$[h]$	$[T]$
$[z]$	$\begin{bmatrix} z_{11} & z_{12} \\ z_{21} & z_{22} \end{bmatrix}$	$\begin{bmatrix} \frac{y_{22}}{\Delta_y} & -\frac{y_{12}}{\Delta_y} \\ -\frac{y_{21}}{\Delta_y} & \frac{y_{11}}{\Delta_y} \end{bmatrix}$	$\begin{bmatrix} \frac{\Delta_h}{h_{22}} & \frac{h_{12}}{h_{22}} \\ -\frac{h_{21}}{h_{22}} & \frac{1}{h_{22}} \end{bmatrix}$	$\begin{bmatrix} \frac{A}{C} & \frac{\Delta_T}{C} \\ \frac{1}{C} & \frac{D}{C} \end{bmatrix}$
$[y]$	$\begin{bmatrix} \frac{z_{22}}{\Delta_z} & -\frac{z_{12}}{\Delta_z} \\ -\frac{z_{21}}{\Delta_z} & \frac{z_{11}}{\Delta_z} \end{bmatrix}$	$\begin{bmatrix} y_{11} & y_{12} \\ y_{21} & y_{22} \end{bmatrix}$	$\begin{bmatrix} \frac{1}{h_{11}} & -\frac{h_{12}}{h_{11}} \\ \frac{h_{21}}{h_{11}} & \frac{\Delta_h}{h_{11}} \end{bmatrix}$	$\begin{bmatrix} \frac{D}{B} & -\frac{\Delta_T}{B} \\ -\frac{1}{B} & \frac{A}{B} \end{bmatrix}$
$[h]$	$\begin{bmatrix} \frac{\Delta_z}{z_{22}} & \frac{z_{12}}{z_{22}} \\ -\frac{z_{21}}{z_{22}} & \frac{1}{z_{22}} \end{bmatrix}$	$\begin{bmatrix} \frac{1}{y_{11}} & -\frac{y_{12}}{y_{11}} \\ \frac{y_{21}}{y_{11}} & \frac{\Delta_y}{y_{11}} \end{bmatrix}$	$\begin{bmatrix} h_{11} & h_{12} \\ h_{21} & h_{22} \end{bmatrix}$	$\begin{bmatrix} \frac{B}{D} & \frac{\Delta_T}{D} \\ -\frac{1}{D} & \frac{C}{D} \end{bmatrix}$
$[T]$	$\begin{bmatrix} \frac{z_{11}}{z_{21}} & \frac{\Delta_z}{z_{21}} \\ \frac{1}{z_{21}} & \frac{z_{22}}{z_{21}} \end{bmatrix}$	$\begin{bmatrix} -\frac{y_{22}}{y_{21}} & -\frac{1}{y_{21}} \\ -\frac{\Delta_y}{y_{21}} & -\frac{y_{11}}{y_{21}} \end{bmatrix}$	$\begin{bmatrix} -\frac{\Delta_h}{h_{21}} & -\frac{h_{11}}{h_{21}} \\ -\frac{h_{22}}{h_{21}} & -\frac{1}{h_{21}} \end{bmatrix}$	$\begin{bmatrix} A & B \\ C & D \end{bmatrix}$

Δ is the determinant of the corresponding parameter matrix.

9.8 INTERCONNECTED TWO-PORT NETWORKS

The input ports of the two or more networks are in series if at the input ports, the current leaving network 1 enters network 2, that leaving network 2 enters network 3, and so on. Similar statements apply for the output ports in series.

Two or more networks are in parallel, if the corresponding terminals at the input ports of the networks are joined together, and the corresponding terminals at the output ports also are connected together.

In making series, or parallel connection, the original characteristic of the networks should not get altered; as such wherever necessary, isolating transformers can be interposed between the networks before any such connection is made.

When networks are in series, the z-matrix of the resulting network is obtained by taking the product of the z-matrices of the individual networks in the order of the networks.

Similarly, for networks in parallel, the y-matrix of the resulting network is obtained by taking the product of the y-matrices of the individual networks in the order of the networks.

Lastly, for networks in cascade (one after the other), the T-matrix of the resulting network is obtained by taking the product of the T-matrices of the individual networks in the order of the networks.

9.9 IMAGE IMPEDANCES

A reciprocal network can also be characterized by its image impedances Z_{1i}, and Z_{2i}; and the image transfer constant θ. Z_{1i} is the impedance seen at port 1 when Z_{2i} is connected at port 2. Similarly, Z_{2i} is the impedance seen at port 2 when Z_{1i} is connected at port 1.

The image transfer constant with Z_{2i} connected at port 2, $\theta = \frac{1}{2}\,\text{Ln}\left[\frac{V_1 I_1}{V_2 I_2}\right]$

$$Z_{1i} = \sqrt{Z_{1oc} Z_{1sc}}$$

$$Z_{2i} = \sqrt{Z_{2oc} Z_{2sc}}$$

$$\theta = \tanh^{-1}\sqrt{\frac{Z_{1sc}}{Z_{1oc}}} = \tanh^{-1}\sqrt{\frac{Z_{2sc}}{Z_{2oc}}}$$

In terms of A, B, C, D constants,

$$Z_{1oc} = A/C, \qquad Z_{1sc} = B/D$$

$$Z_{2oc} = D/C, \qquad Z_{2sc} = B/A$$

$$Z_{1i} = \sqrt{(AB/CD)} \qquad Z_{2i} = \sqrt{(DB/CA)}$$

$$\theta = \tanh^{-1}\sqrt{\frac{BC}{AD}}$$

For ***symmetrical networks***,

$$Z_{1oc} = Z_{2oc} = Z_{oc} \text{ (say); and } Z_{1sc} = Z_{2sc} = Z_{sc} \text{ (say)}$$

We find that $\quad Z_{1i} = Z_{1i} = Z_i \text{ (say)} = \sqrt{(Z_{oc} Z_{sc})}$

The image impedance Z_i for a symmetrical network is called the iterative, or the characteristic impedance; denoted by Z_o (or Z_c).

$$Z_o = Z_i = \sqrt{(B/C)}$$

θ is replaced by γ; which is called the propagation constant.

With $\quad V_2 = Z_o I_2,$ or $V_1 = Z_o I_1,$

$$\gamma = \text{Ln}(V_1/V_2) = \text{Ln}(I_1/I_2)$$

Also, $\quad \gamma = \tanh^{-1}\frac{\sqrt{A^2-1}}{A} = \cosh^{-1} A$

Thus, $\quad A = \cosh\gamma$

We can also find that

$$B = Z_o \sinh\gamma \quad \text{and} \quad C = \frac{\sinh\gamma}{Z_o}$$

For the symmetrical T-network shown (usual form),

$$Z_{oT} = \sqrt{\frac{Z_1^2}{4} + Z_1 Z_2} = \sqrt{Z_1 Z_2\left(1 + \frac{Z_1}{4Z_2}\right)} \quad \text{and}$$

For the symmetrical π-network shown (usual form),

$$Z_{o\pi} = \frac{Z_1 Z_2}{\sqrt{Z_1 Z_2\left(1 + \frac{Z_1}{4Z_2}\right)}}$$

We see that

$$Z_{oT}\, Z_{o\pi} = Z_1\, Z_2$$

It is again reminded that all the impedances, in general, are complex in nature; as such all mathematical operations are to be performed as per the rules of complex algebra.

9.10 FILTERS

Filters are the networks designed to allow (***pass***) signals of a certain range of frequencies through it, without any change in the magnitudes; and block(***stop***) the signals of the remaining range of frequencies.

Filters are symmetrical L-C networks. A filter circuit is assumed to be terminated in its characteristic impedance.

The voltages, or currents at the input and at the output ports differ by one neper if one of them is *e* times the other.

With termination in characteristic impedance Z_o, the propagation constant γ is given by $\frac{I_1}{I_2} = \frac{V_1}{V_2} = e^{\gamma} = e^{\alpha + j\beta}$, α is the attenuation constant, and β is the phase shift constant.

For a T-network, or a π- network,

$$\sinh\frac{\gamma}{2} = \sqrt{\frac{Z_1}{4Z_2}}$$

With reactances of the same type, the filter action is in the stop band with

$$\alpha = 2\sinh^{-1}\sqrt{\frac{Z_1}{4Z_2}}, \quad \text{and} \quad \beta = \pi.$$

With reactances of opposite sign in the range from – 1 to – ∞, it again is in the stop band with $\alpha = 2\cosh^{-1}\sqrt{\frac{Z_1}{4Z_2}}$, and $\beta = \pi$.

With reactances of opposite sign in the range from 0 to – 1,

the filter action is of pass band with $\alpha = 0$, and $\beta = 2\sin^{-1}\sqrt{\frac{Z_1}{4Z_2}}$

The cut-off frequencies are given by $Z_1 = 0$, and $Z_1 = -4Z_2$.

The constant-*k* ***low pass filter*** :

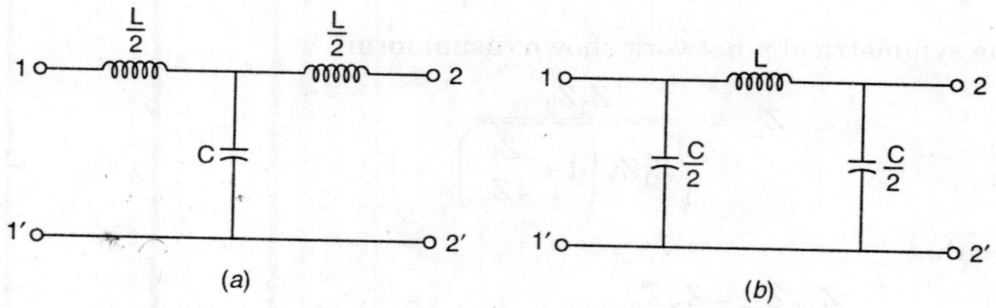

FIGURE 9.4. A low pass filter. (*a*) T-section, (*b*) π-section.

$$Z_1 = j\omega L,\quad Z_2 = -\frac{j}{\omega C},\qquad Z_1Z_2 = L/C = R^2$$

$$f_c = \frac{1}{\pi\sqrt{LC}},\ \mathrm{L} = \frac{R}{\pi f_c},\ C = \frac{1}{\pi f_c R}$$

$$\sinh\frac{\gamma}{2} = \frac{j\omega\sqrt{LC}}{2} = j\frac{f}{f_c}$$

For $\frac{f}{f_c} < 1,\quad \alpha = 0,\quad \beta = 2\sin^{-1}\frac{f}{f_c}$

For $\frac{f}{f_c} > 1,\quad \alpha = 2\cosh^{-1}\frac{f}{f_c},\quad \beta = \pi$

$$Z_{oT} = R\sqrt{1-\left(\frac{f}{f_c}\right)^2}$$

The constant-*k* ***high pass filter*** :

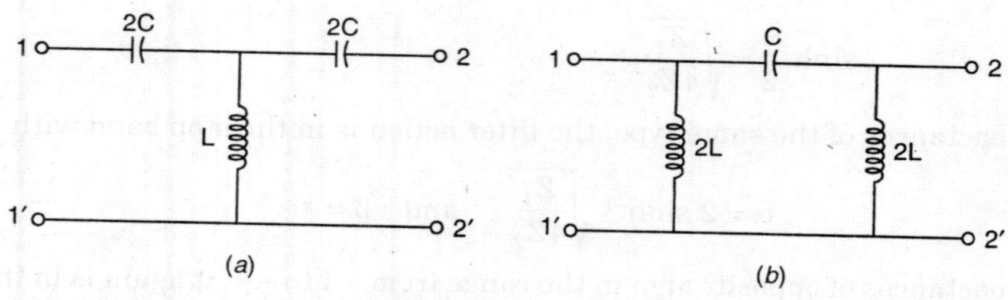

FIGURE 9.5. A high pass filter. (*a*) T-section, (*b*) π-section.

$$Z_1 = -\frac{j}{\omega C}, \qquad Z_2 = j\omega L, \quad Z_1Z_2 = L/C = R^2$$

$$f_c = \frac{1}{4\pi\sqrt{LC}}, \qquad L = \frac{R}{4\pi f_c}, \qquad C = \frac{1}{4\pi f_c R}$$

For $\frac{f_c}{f} < 1$, $\quad \alpha = 0, \quad \beta = -2\sin^{-1}\frac{f_c}{f}$

For $\frac{f_c}{f} > 1$, $\quad \alpha = 2\cosh^{-1}\frac{f_c}{f}, \quad \beta = -\pi$

$$Z_{oT} = \text{R}\sqrt{1-\left(\frac{f_c}{f}\right)^2}$$

9.11 BAND PASS FILTER

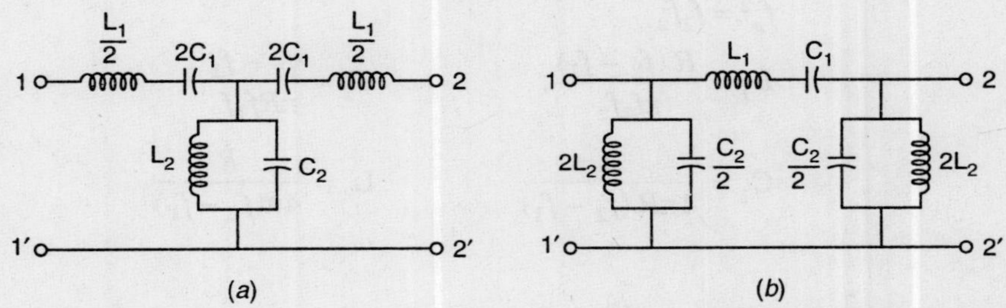

FIGURE 9.6. A band pass filter. (*a*) T-section, (*b*) π-section.

$$L_1C_1 = L_2C_2$$

$$Z_1Z_2 = \frac{L_1}{C_2} = \frac{L_2}{C_1} = R^2$$

$$\omega_o = \frac{1}{\sqrt{L_1C_1}}, \qquad f_o^2 = f_1f_2;$$

f_1 and f_2 are the two end frequencies of the pass band.

$$C_1 = \frac{f_2 - f_1}{4\pi R f_1 f_2}, \qquad L_2 = \frac{R(f_2 - f_1)}{4\pi f_1 f_2}$$

$$L_1 = \frac{R}{\pi(f_2 - f_1)}, \qquad C_2 = \frac{1}{\pi R(f_2 - f_1)}$$

9.12 BAND ELIMINATION FILTER

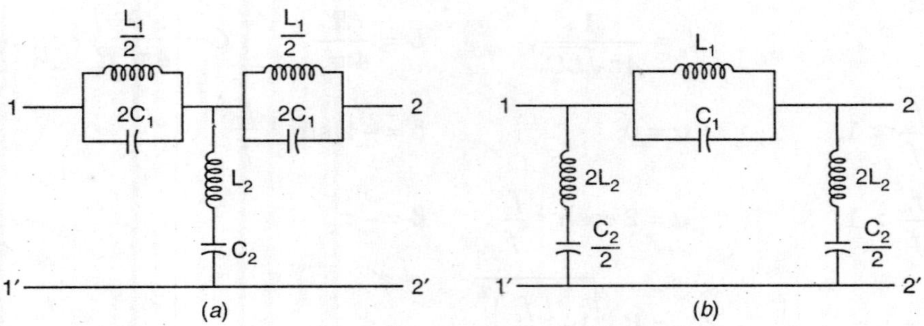

FIGURE 9.7. A band elimination filter. (*a*) T-section, (*b*) π-section.

$$L_1C_1 = L_2C_2$$

$$\frac{L_1}{C_2} = \frac{L_2}{C_1} = R^2$$

$$f_o^2 = f_1f_2$$

$$L_1 = \frac{R\,(f_2 - f_1)}{\pi f_1 f_2}; \qquad C_2 = \frac{f_2 - f_1}{\pi R f_1 f_2}$$

$$C_1 = \frac{1}{4\pi R\,(f_2 - f_1)} \qquad L_2 = \frac{R}{4\pi(f_2 - f_1)}$$

9.13 m-DERIVED FILTERS (0 < m < 1)

m-derived filters are used to achieve sharp cut-off in comparison to that in constant-*k* type filters.

T-*section* low pass filter

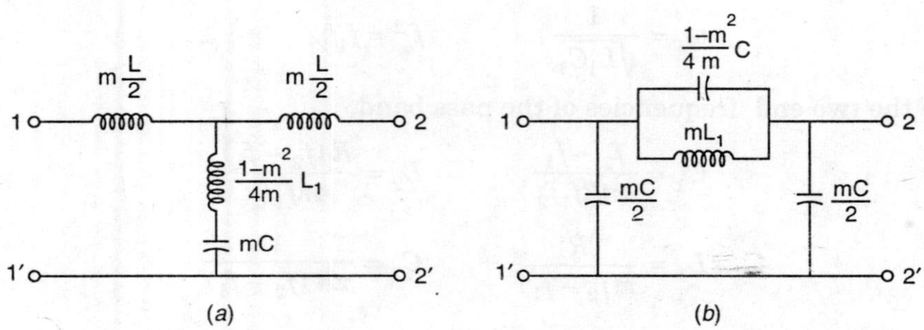

FIGURE 9.8. An *m*-derived low pass filter. (*a*) T-section, (*b*) *p*-section.

$$Z_1' = m\,Z_1, \qquad Z_2' = \frac{Z_2}{m} + \frac{1-m^2}{4m}\,Z_1$$

$$f_c = \frac{1}{\pi\sqrt{LC}}, \quad F_\infty = f_c\sqrt{1-m^2}, \quad m = \sqrt{1-\left(\frac{f_\infty}{f_c}\right)^2}$$

For $f_c < f < f_\infty$, $\alpha = 2\cosh^{-1}\dfrac{mf/f_c}{\sqrt{1-f^2/f_\infty^2}}$

For $f_\infty < f$ $\alpha = 2\sinh^{-1}\dfrac{mf/f_c}{\sqrt{f^2/f_\infty^2-1}}$

In the pass band, $\beta = 2\sin^{-1}\dfrac{mf/f_c}{\sqrt{1-(f/f_c)^2(1-m^2)}}$

From f_c to f_∞, $\beta = \pi$

For $f > f_\infty$ $\beta = 0$

m-derived π-***section***:

$$Z_2' = Z_2/m, \; Z_1' = \frac{1}{\dfrac{1}{mZ_1} + \dfrac{1}{\dfrac{4m}{1-m^2}Z_2}}$$

Z_{oT}, and $Z_{o\pi}$ vary considerably over the pass band. m-derived L-section termination can be used to improve upon this also.

Chapter **10**

ELEMENTRY NETWORK SYNTHESIS

10.1 POSITIVE REAL FUNCTIONS

For a driving point immitance function to be realizable, the function must be a positive real function. An immitance function denoted by $F(s) = \dfrac{N(s)}{D(s)}$, where s is the complex frequency $= \sigma + j\omega$, is positive real, if

(*i*) when s is real, $F(s)$ also is real, and

(*ii*) when real of $s \geq 0$, then real of $F(s)$ also ≥ 0.

Assuming that all the coefficients in the numerator polynomial $N(s)$ and in the denominator polynomial $D(s)$ are real; the above two conditions are equivalent to the following three simple conditions :

(*i*) $F(s)$ must have no poles in the right half of the complex s-plane,

(*ii*) $F(s)$ may have only simple poles on the $j\omega$-axis with real and positive residues, and

(*iii*) Real of $F(j\omega) \geq 0$ for all ω.

A polynomial in s with real coefficients, which has none of its roots in the right half of the complex s-plane, and the roots on the $j\omega$-axis are all simple; is called a Hurwitz polynomial. Thus, condition (*i*), and the first part of condition (*ii*) are satisfied if $D(s)$ is a Hurwitz polynomial. Therefore, for a function $F(s)$ to be a positive real function, its denominator polynomial $D(s)$ must be a Hurwitz polynomial, and the second part of condition (*ii*) along with condition (*iii*) should also be satisfied.

10.2 PROPERTIES OF L-C IMMITANCE FUNCTIONS

1. An *L-C* immitance function is the ratio of odd to even, or even to odd polynomials.
2. The poles and the zeros are simple, and lie on the $j\omega$-axis.
3. The poles and the zeros alternate on the $j\omega$-axis.

4. The highest powers of $N(s)$ and $D(s)$ must differ by 1; and the lowest powers too must differ by 1.
5. There must be either a pole, or a zero at the origin, as well as at infinity.

A term of the type $\frac{s^2 + s_1}{s_2 s}$ in an impedance function, or of the type $\frac{s_2 s}{s^2 + s_1}$ in an admittance function; is realized as a series L-C branch, with $L = \frac{1}{s_2}$, and $C = \frac{s_2}{s_1}$. Similarly, a term of the type $\frac{s_2 s}{s^2 + s_1}$ in an impedance function, or of the type $\frac{s^2 + s_1}{s_2 s}$ in an admittance function is realized as a parallel L-C branch, with $C = \frac{1}{s_2}$, and $L = \frac{s_2}{s_1}$.

10.3 PROPERTIES OF R-C IMPEDANCE, OR R-L ADMITTANCE FUNCTIONS

1. Poles and zeros lie on the negative real axis and these alternate.
2. The singularity nearest (or at) the origin must be a pole, and that nearest (or at) σ = –must be a zero.
3. The residues of the poles must be real and positive.

A term of the type $\frac{s_2}{s + s_1}$ in an R-C impedance function is realized as a parallel R-C branch, with $C = \frac{1}{s_2}$, and $R = \frac{s_2}{s_1}$; and in an R-L admittance function is realized as a series R-L branch, with $L = \frac{1}{s_2}$, and $R = \frac{s_1}{s_2}$.

10.4 PROPERTIES OF R-L IMPEDANCE, OR R-C ADMITTANCE FUNCTIONS

1. Poles and zeros lie on the negative real axis and these alternate.
2. The singularity nearest (or at) the origin must be a zero, and that nearest (or at) σ = – must be a pole.
3. The residues of the poles must be real and negative.

A term of the type $\frac{s_2 s}{s + s_1}$ in an R-L impedance function is realized as a parallel R-L branch, with $R = s_2$, and $L = \frac{s_2}{s_1}$; and in an R-C admittance function is realized as a series R-C branch, with $R = \frac{1}{s_2}$, and $C = \frac{s_2}{s_1}$.

Section 3 FIELDS AND WAVES

Chapter 11

VECTOR ANALYSIS

11.1 A GENERAL NOTE ON BOLD-FACED SYMBOLS

The bold-faced symbols are used to denote the following different quantities :

The vector quantities—like force **F**, electric field **E**, conduction current density **J**, unit vector **i** etc. These quantities are space directed.

The phasors to represent the quantities varying sinusoidally with time- like the current phasor **I**, the voltage phasor **V** etc.

The complex quantities—like impedance **Z**, admittance **Y**, and complex power **S** etc. These quantities are neither vectors, nor sinusoids.

If a vector phasor, and a scalar phasor are simultaneously occurring, then to denote these phasors, we use bar—like $\overline{\mathrm{E}}$, $\overline{H}_y$, $\overline{\rho}_v$ etc. In this situation, scalar phasors are not bold-faced, rather an overbar has been used.

The ordinary typed letters denote the magnitude of the corresponding vector, or the phasor—like A is the magnitude of **A**, Z is the magnitude of the impedance **Z** etc.

11.2 VECTOR PRODUCTS

$$\mathbf{A.B} = AB \text{ Cos } \alpha, \text{ where } \alpha \text{ is the angle between vectors } \mathbf{A} \text{ and } \mathbf{B}.$$

$$= \mathbf{B.A}$$

$\mathbf{A} \times \mathbf{B} = AB \sin \alpha \, \mathbf{i}_n$ where $\mathbf{i}_n$ is the unit vector normal to the plane of the vectors **A** and **B**, and in a direction in which the right hand screw advances when turned from **A** towards **B** through the angle α.

$$\mathbf{B} \times \mathbf{A} = -\mathbf{A} \times \mathbf{B}$$

$$\mathbf{A.(B{\times}C)} = \mathbf{B.(C{\times}A)} = \mathbf{C.(A{\times}B)} = \begin{vmatrix} A_1 & A_2 & A_3 \\ B_1 & B_2 & B_3 \\ C_1 & C_2 & C_3 \end{vmatrix}$$

11.3 UNIT VECTORS

A unit vector is a vector of unit magnitude and pointing in the required direction. If at a point there are three mutually perpendicular unit vectors $\mathbf{i}_1$, $\mathbf{i}_2$, and $\mathbf{i}_3$; then the vector **A** can be expressed as

$$\mathbf{A} = (\mathbf{A}.\mathbf{i}_1)\mathbf{i}_1 + (\mathbf{A}.\mathbf{i}_2)\mathbf{i}_2 + (\mathbf{A}.\mathbf{i}_3)\mathbf{i}_3$$

$$\mathbf{A} = A_1\mathbf{i}_1 + A_2\mathbf{i}_2 + A_3\mathbf{i}_3$$

$$A = \sqrt{A_1^2 + A_2^2 + A_3^2}$$

$$\mathbf{i}_A = \frac{A_1\mathbf{i}_1 + A_2\mathbf{i}_2 + A_3\mathbf{i}_3}{\sqrt{A_1^2 + A_2^2 + A_3^2}} = \cos\alpha\,\mathbf{i}_1 + \cos\beta\,\mathbf{i}_2 + \cos\gamma\,\mathbf{i}_3$$

Where α, β, γ are the angles which **A**, or $\mathbf{i}_A$ makes with $\mathbf{i}_1$, $\mathbf{i}_2$, and $\mathbf{i}_3$ respectively at that point.

$$\mathbf{A} \cdot \mathbf{B} = A_1B_1 + A_2B_2 + A_3B_3$$

$$\mathbf{A} \times \mathbf{B} = \begin{vmatrix} \mathbf{i}_1 & \mathbf{i}_2 & \mathbf{i}_3 \\ A_1 & A_2 & A_3 \\ B_1 & B_2 & B_3 \end{vmatrix}$$

11.4 COORDINATE SYSTEMS

In ***Cartesian*** system of coordinates, the variables are x, y, and z; the elemental lengths are dx, dy, and dz; the unit vectors are $\mathbf{i}_x$, $\mathbf{i}_y$, and $\mathbf{i}_z$ and are uniform throughout; the elemental displacement is $d\mathbf{l} = dx\,\mathbf{i}_x + dy\,\mathbf{i}_y + dz\,\mathbf{i}_z$.

In ***cylindrical*** system of coordinates, the variables are r, φ and z; the elemental lengths are dr, $rd\varphi$, and dz; the unit vectors are $\mathbf{i}_r$, $\mathbf{i}_\varphi$, and $\mathbf{i}_z$; $\mathbf{i}_z$ is uniform throughout but $\mathbf{i}_r$ and $\mathbf{i}_\varphi$ are point position dependent; the elemental displacement is $d\mathbf{l} = dr\,\mathbf{i}_r + rd\varphi\,\mathbf{i}_\varphi + dz\,\mathbf{i}_z$.

In ***spherical*** system of coordinates, the variables are r, θ, and φ; the elemental lengths are dr, $rd\theta$, and $r\sin\theta d\varphi$; the unit vectors are $\mathbf{i}_r$, $\mathbf{i}_\theta$ and $\mathbf{i}_\varphi$, and are point position dependent; the elemental displacement is $d\mathbf{l} = dr\,\mathbf{i}_r + rd\theta\,\mathbf{i}_\theta + r\sin\theta d\varphi\,\mathbf{i}_\varphi$.

r of cylindrical coordinates, and r of spherical coordinates are not the same.

z of Cartesian coordinates, and z of cylindrical coordinates are identical.

φ of cylindrical coordinates, and φ of spherical coordinates are also identical.

The coordinate systems have

$$\mathbf{i}_x \times \mathbf{i}_y = \mathbf{i}_z,\ \mathbf{i}_r \times \mathbf{i}_\varphi = \mathbf{i}_z, \text{ and } \mathbf{i}_r \times \mathbf{i}_\theta = \mathbf{i}_\varphi$$

The other similar relations can easily be written following the cyclic order.

The right hand screw rule is to be observed in all direction markings and the relations.

11.5 INTERRELATIONS BETWEEN VARIABLES

$$x = r_c\cos\varphi = r_s\sin\theta\cos\varphi$$

$$y = r_c \sin\varphi = r_s \sin\theta \sin\varphi$$
$$z = r_s \cos\theta$$
$$r_c = \sqrt{(x^2 + y^2)} = r_s \sin\theta$$
$$\varphi = \tan^{-1}(y/x)$$
$$r_s = \sqrt{(x^2 + y^2 + z^2)} = \sqrt{(r_c^2 + z^2)}$$
$$\theta = \tan^{-1}((x^2 + y^2)/z) = \tan^{-1}(r_c/z)$$

11.6 INTERRELATIONS BETWEEN UNIT VECTORS

$$\mathbf{i}_{rc} = \cos\varphi\, \mathbf{i}_x + \sin\varphi\, \mathbf{i}_y$$
$$\mathbf{i}_\varphi = -\sin\varphi\, \mathbf{i}_x + \cos\varphi\, \mathbf{i}_y$$
$$\mathbf{i}_{rs} = \sin\theta \cos\varphi\, \mathbf{i}_x + \sin\theta \sin\varphi\, \mathbf{i}_y + \cos\theta\, \mathbf{i}_z$$
$$\mathbf{i}_\theta = \cos\theta \cos\varphi\, \mathbf{i}_x + \cos\theta \sin\varphi\, \mathbf{i}_y - \sin\theta\, \mathbf{i}_z$$

11.7 DIFFERENTIATION

Differentiation (with respect to space variables) of a vector **A**

$$d\mathbf{A} = d(A_1\mathbf{i}_1) + d(A_2\mathbf{i}_2) + d(A_3\mathbf{i}_3)$$
$$= dA_1\mathbf{i}_1 + A_1 d\,\mathbf{i}_1 + dA_2\mathbf{i}_2 + A_2 d\,\mathbf{i}_2 + dA_3\mathbf{i}_3 + A_3 d\,\mathbf{i}_3$$
$$d(\mathbf{A}.\mathbf{B}) = d\mathbf{A}.\mathbf{B} + \mathbf{A}.d\mathbf{B}$$
$$d(\mathbf{A} \times \mathbf{B}) = d\mathbf{A} \times \mathbf{B} + \mathbf{A} \times d\mathbf{B}$$

11.8 THE DEL(∇) OPERATOR

The ***gradient*** of a function (a scalar function) at a point, ∇V, is the maximum value of the space rate of change of the function at that point, and is in the direction in which this maximum rate of change occurs.

$$\nabla V = \frac{\partial V}{\partial x}\mathbf{i}_x + \frac{\partial V}{\partial y}\mathbf{i}_y + \frac{\partial V}{\partial z}\mathbf{i}_z = \frac{\partial V}{\partial r_c}\mathbf{i}_{r_c} + \frac{1}{r_c}\frac{\partial V}{\partial \phi}\mathbf{i}_\phi + \frac{\partial V}{\partial z}\mathbf{i}_z$$
$$= \frac{\partial V}{\partial r_s}\mathbf{i}_{r_s} + \frac{1}{r_s}\frac{\partial V}{\partial \phi}\mathbf{i}_\theta + \frac{1}{r_s \sin\theta}\frac{\partial V}{\partial \phi}\mathbf{i}_\phi$$

The ***divergence*** of a function (a vector function) at a point, $\nabla \,.\, \mathbf{A}$ is the net outward flow(or flux) of **A** per unit area at that point.

$$\nabla \,.\, \mathbf{A} = \frac{\partial A_x}{\partial x} + \frac{\partial A_y}{\partial y} + \frac{\partial A_z}{\partial z}$$

$$= \frac{1}{r_c}\frac{\partial}{\partial r_c}(r_c A_{r_c}) + \frac{1}{r_c}\frac{\partial A_\phi}{\partial \phi} + \frac{\partial A_z}{\partial z}$$

$$= \frac{1}{r_s^2}\frac{\partial}{\partial r_s}(r_s^2 A_{r_s}) + \frac{1}{r_s \sin\theta}\frac{\partial}{\partial \theta}(\sin\theta A_\theta) + \frac{1}{r_s \sin\theta}\frac{\partial A_\phi}{\partial \phi}$$

If $\nabla \cdot \mathbf{A} = 0$, **A** is then called a solenoidal vector.

The ***curl*** of a function (a vector function) at a point, $\nabla \times \mathbf{A}$ is the maximum value of the circulation of **A** per unit area at that point, and its direction is the direction of that area which causes maximum value of the circulation.

$$\nabla \times \mathbf{A} = \begin{vmatrix} \mathbf{i}_x & \mathbf{i}_y & \mathbf{i}_z \\ \frac{\partial}{\partial x} & \frac{\partial}{\partial y} & \frac{\partial}{\partial z} \\ A_x & A_y & A_z \end{vmatrix}$$

$$\nabla \times \mathbf{A} = \begin{vmatrix} \frac{\mathbf{i}_{r_c}}{r_c} & \mathbf{i}_\phi & \frac{\mathbf{i}_z}{r_c} \\ \frac{\partial}{\partial r_c} & \frac{\partial}{\partial \phi} & \frac{\partial}{\partial z} \\ A_{r_c} & r_c A\phi & A_z \end{vmatrix}$$

$$\nabla \times \mathbf{A} = \begin{vmatrix} \frac{\mathbf{i}_{r_s}}{r_s^2 \sin\theta} & \frac{\mathbf{i}_\theta}{r_s \sin\theta} & \frac{\mathbf{i}_\phi}{r_s} \\ \frac{\partial}{\partial r_s} & \frac{\partial}{\partial \theta} & \frac{\partial}{\partial \phi} \\ A_{r_s} & r_s A_\theta & r_s \sin\theta A_\phi \end{vmatrix}$$

If $\nabla \times \mathbf{A} = 0$, **A** is then called an irrotational vector.

11.9 DIVERGENCE THEOREM

$$\int_v (\nabla \cdot \mathbf{A})\, dv = \oint_s \mathbf{A} \cdot d\mathbf{s}$$

S is the bounding surface of volume v, hence forming a closed surface.

11.10 STOKES' THEOREM

$$\int_s (\nabla \times \mathbf{A}) \cdot d\mathbf{s} = \oint_c \mathbf{A} \cdot d\mathbf{l}$$

C is the periphery of the open surface S, hence forming a contour.

11.11 HELMHOLTZ THEOREM

$$\mathbf{F} = -\nabla V + \nabla \times \mathbf{A}$$

11.12 LAPLACIAN OF A SCALAR

$$\nabla^2 V = \nabla.\nabla V$$

$$= \frac{\partial^2 V}{\partial x^2} + \frac{\partial^2 V}{\partial y^2} + \frac{\partial^2 V}{\partial z^2}$$

$$= \frac{1}{r_c}\frac{\partial}{\partial r_c}\left(r_c \frac{\partial V}{\partial r_c}\right) + \frac{1}{{r_c}^2}\frac{\partial^2 V}{\partial \phi^2} + \frac{\partial^2 V}{\partial z^2}$$

$$= \frac{1}{{r_s}^2}\frac{\partial}{\partial r_s}\left({r_s}^2 \frac{\partial V}{\partial r_s}\right) + \frac{1}{{r_s}^2 \sin\theta}\frac{\partial}{\partial \theta}\left(\sin\theta \frac{\partial V}{\partial \theta}\right) + \frac{1}{{r_s}^2 \sin^2\theta}\frac{\partial^2 V}{\partial \phi^2}$$

11.13 LAPLACIAN OF A VECTOR

$$\nabla^2 \mathbf{A} = (\nabla^2 A_x)\mathbf{i}_x + (\nabla^2 A_y)\mathbf{i}_y + (\nabla^2 A_\mathbf{z})\mathbf{i}_z$$

11.14 SOME VECTOR IDENTITIES

$$\nabla.\nabla \times \mathbf{A} = 0$$

$$\nabla \times \nabla V = 0$$

$$\nabla.(V\mathbf{A}) = \nabla V.\mathbf{A} + V\nabla.\mathbf{A}$$

$$\nabla.(\mathbf{A} \times \mathbf{B}) = \nabla \times \mathbf{A}.\mathbf{B} - \mathbf{A}.\nabla \times \mathbf{B}$$

$$\nabla \times \nabla \times \mathbf{A} = \nabla(\nabla.\mathbf{A}) - \nabla^2 \mathbf{A}$$

Chapter 12

ELECTROSTATICS

12.1 COULOMB'S LAW

This law is a force law and applies to point charges.

$$\mathbf{F}_{12} = \frac{Q_1Q_2}{4\pi\varepsilon\, r^2}\,\mathbf{i}_{21} = \frac{Q_1Q_2}{4\pi\varepsilon_r\varepsilon_o\, r^2}\,\mathbf{i}_{21} = \frac{9\times10^9\, Q_1Q_2}{\varepsilon_r r^2}\,\mathbf{i}_{21}$$

r is the distance between the charges Q_1 and Q_2.

$\mathbf{F}_{12}$ is the force which acts on charge Q_1 due to charge Q_2, $\mathbf{i}_{21}$ is the unit vector along the straight line joining charge Q_2 to Q_1 in the direction from Q_2 to Q_1.

$$\mathbf{F}_{21} = -\,\mathbf{F}_{12}$$

The ***electric field intensity* E** at a point P is the force per unit charge and its direction is the direction of the force. The test charge in determining **E** should be as small as possible so that it does not alter the distribution of charges, whose very electric field we are finding. If the source charges are held firmly in position, then the test charge may be of any value.

$$\mathbf{E} = \lim_{\Delta Q\to 0}\frac{\Delta \mathbf{F}}{\Delta Q}$$

The electric field lines in Cartesian coordinates are governed by the relation

$$\frac{dx}{E_x} = \frac{dy}{E_y} = \frac{dz}{E_z}$$

The ***electric potential V*** at a point P is the work done by the external agency per unit charge in bringing the small test charge from reference point (usually infinity) to the point P.

$$V = \lim_{\Delta Q\to 0}\frac{\Delta W}{\Delta Q}$$

$$V_P = -\int_{\infty}^{P} \mathbf{E}.d\mathbf{l} \quad \text{we can follow any path from } \infty \text{ to } P,$$

$$V_{AB} = -\int_{B}^{A} \mathbf{E}.d\mathbf{l} \text{ we can follow any path from } B \text{ to } A$$

$$\mathbf{E} = -\nabla V$$

The electric field is a vector quantity, while the electric potential is a scalar quantity. For static (*i.e.,* not varying with time) electric fields

$\oint \mathbf{E}.d\mathbf{l} = 0$, hence $\nabla \times \mathbf{E} = 0$. Electrostatic field is thus ***irrotational.***

12.2 ELECTRIC FIELD AND POTENTIAL AT POINT P FOR VARIOUS CHARGE CONFIGURATIONS

A point charge

$$\mathbf{E} = \frac{Q}{4\pi\varepsilon\, r^2}\mathbf{i}_{QP}, \qquad V = \frac{Q}{4\pi\varepsilon\, r}$$

n point charges

$$\mathbf{E} = \frac{1}{4\pi\varepsilon}\left[\frac{Q_1}{r_1^{\,2}}\mathbf{i}_{1P} + \frac{Q_2}{r_2^{\,2}}\mathbf{i}_{2P} + \ldots + \frac{Q_n}{r_1^{\,2}}\mathbf{i}_{nP}\right]$$

$$= \frac{1}{4\pi\varepsilon}\sum_{i=1}^{n}\left[\frac{Q_i}{r_i^{\,2}}\mathbf{i}_{iP}\right] \qquad V = \frac{1}{4\pi\varepsilon}\sum_{i=1}^{n}\left[\frac{Q_i}{r_i}\right]$$

r_i is the distance of the point P from the charge Q_i.

An electric dipole

The relations given below are in spherical coordinates, and are for points far away from the dipole. The origin is at the centre of the dipole, and (r, θ) are the coordinates of the point P. The dipole moment is $p = qd$; d being the distance between q and $-q$ of the dipole($d << r$).

$$\mathbf{E} = \frac{p}{4\pi\varepsilon\, r^3}\,(2\cos\theta\, \mathbf{i}_r + \sin\theta\, \mathbf{i}_\theta),\; V = \frac{p\cos\theta}{4\pi\varepsilon\, r^2}$$

A ***line of charge*** having uniform line charge density ρ.

The relations given are in cylindrical system of coordinates. The line is located along the z-axis. The point P has the coordinates (r, φ, z).

Finite line of charge extending from $z = z_1$ to $z = z_2$

$$E_r = \frac{\rho}{4\pi\varepsilon\, r}\left[\frac{z - z_1}{\sqrt{(z - z_1)^2 + r^2}} - \frac{z - z_2}{\sqrt{(z - z_2)^2 + r^2}}\right]$$

$$E_\phi = 0$$

$$E_z = \frac{\rho}{4\pi\varepsilon}\left[\frac{1}{\sqrt{(z-z_2)^2+r^2}} - \frac{1}{\sqrt{(z-z_1)^2+r^2}}\right]$$

$$V = \frac{\rho}{4\pi\varepsilon} Ln \frac{\left(\sqrt{((z-z_2)^2+r^2}\right)-(z-z_2)}{\left(\sqrt{((z-z_1)^2+r^2}\right)-(z-z_1)}$$

Infinitely long line

$$\mathbf{E} = \frac{\rho}{2\pi\varepsilon r}\mathbf{i}_r$$

$$V_{r-r_{\text{ref}}} = \frac{\rho}{2\pi\varepsilon} Ln \frac{r_{\text{ref}}}{r},$$

with respect to the coaxial equipotential cylinder of radius r_{ref}.

Uniformly charged ***circular ring***, ring being of radius R and co-axial with the z-axis. The point P being on the z-axis, distant h from the centre of the ring.

$$\mathbf{E} = \frac{\rho h R}{2\varepsilon(R^2+h^2)^{3/2}}\mathbf{i}_z$$

$$V = \frac{\rho R}{\varepsilon\sqrt{R^2+h^2}}$$

Disc of charge having uniform surface charge density ρ_s, the disc being of radius R and co-axial with the z-axis. The point P located on the z-axis, distant h from the centre of the ring.

$$\mathbf{E} = \frac{\rho_s}{2\varepsilon}\left[1-\frac{h}{\sqrt{R^2+h^2}}\right]\mathbf{i}_z$$

$$V = \frac{\rho_s h}{2\varepsilon}\left[1-\frac{h}{\sqrt{R^2+h^2}}\right]$$

An ***infinite sheet*** of charge

$$\mathbf{E} = \frac{\rho_s}{2\varepsilon}\mathbf{i}_n$$

$$V_{r\text{-}r\ \text{ref}} = \frac{\rho_s}{2\varepsilon}[r_{\text{ref}} - r]$$

with respect to a parallel equipotential surface distant r_{ref} from the sheet.

Two straight, infinitely long, ***equally and oppositely charged lines***; positively charged line passing through the point ($d/2$, 0, 0) and the negatively charged line through the point ($-$ $d/2$, 0, 0); and parallel to the z-axis. The coordinates of the point P are x, y. The z-ordinate may have any value.

$$\mathbf{E} = \frac{\rho}{2\pi\varepsilon}\left[\frac{1}{r_+}\mathbf{i}_+ - \frac{1}{r_-}\mathbf{i}_-\right]$$

where r_+ is the perpendicular distance of the point P from the positively charged wire and $\mathbf{i}_+$ is the unit vector along the perpendicular towards P; and r_- is its perpendicular distance from the negatively charged wire and $\mathbf{i}_-$ is the unit vector along the perpendicular towards P.

$$r_+^2 = \left(x - \frac{d}{2}\right)^2 + y^2, \qquad r_-^2 = \left(x + \frac{d}{2}\right)^2 + y^2$$

$$V = \frac{\rho}{2\pi\varepsilon} Ln \frac{r_-}{r_+}$$

For equipotential surfaces,

$$V = \frac{\rho}{2\pi\varepsilon} Ln\ k, \qquad k = \frac{r_-}{r_+}$$

In terms of potential, $k = e^{\frac{2\pi\varepsilon V}{\rho}}$

The equipotential surfaces are cylindrical surfaces whose radii are given by $\frac{kd}{k^2 - 1}$; the axes are parallel to the z-axis and pass through the points $\left(\frac{k^2+1}{k^2-1}\frac{d}{2}, 0\right)$.

For $V = 0, k = 1$, radius ∞ and the location of the axis also tends to ∞. This equipotential surface is, in fact, the y-z plane.

12.3 CONDUCTORS

On a conducting body, under electrostatic conditions, the charge, if any, must reside on the surface only. The charge density, and the electric field at a point on the surface are inversely proportional to the radius of curvature of the surface at that point. The conducting body is an equipotential body and the electric field is normal to the surface at every point.

The surface charge density $\rho_s = D = \varepsilon E$

The ***relaxation time*** T, the time by which any charge released in the interior of the conductor decays to e^{-1} (=37 %) of its initial value is ε/σ, σ being the conductivity of the conductor. The charge, almost instantaneously, appears on the surface.

12.4 GAUSS'S LAW

The net outward electric field flux through any closed surface S equals $1/\varepsilon$ times the net charge enclosed by that surface, or the net (outward) electric displacement through any closed surface S equals the net charge enclosed by that surface.

$$\oint_S \mathbf{E}.d\mathbf{s} = \frac{1}{\varepsilon}\int_v \rho_v \, dv = \text{the net charge enclosed by the closed surface } S.$$

S is the bounding surface of the volume v. The closed surface is also called the Gaussian surface.

The electric displacement density **D** is related to the electric field intensity **E** as

$$\mathbf{D} = \varepsilon\,\mathbf{E} = \varepsilon_r\,\varepsilon_o\,\mathbf{E}$$

$$\oint_S \mathbf{D}.d\mathbf{s} = \int_v \rho_v dv$$

Above is the integral form of the Gauss's law. The point form, or the differential form of the law is

$$\nabla.\mathbf{E} = \frac{1}{\varepsilon}\rho_v \quad \text{or} \quad \nabla.\mathbf{D} = \rho_v$$

Gauss's law constitutes one of the Maxwell's four equations.

Gauss's law can be used to determine the magnitude of electric field for situations where field pattern is known and field magnitude is uniform at symmetrical locations. Appropriate Gaussian surface is chosen to take advantage of the symmetry. A few relations found applying the Gauss's law are given next.

12.5 SPHERICAL VOLUME OF CHARGE

Uniform charge density ρ_v, inner radius R_1, and the outer radius R_2

Total charge $\quad Q_t = \rho_v \frac{4}{3}\pi(R_2^3 - R_1^3)$

For $r \geq R_2$

$$\mathbf{E} = \frac{Q_t}{4\pi\varepsilon\, r^2}\mathbf{i}_r, \quad V = \frac{Q_t}{4\pi\varepsilon\, r}$$

For $R_1 \leq r \leq R_2$

Charge contained in a sphere of radius r ($R_1 \leq r \leq R_2$)

$$Q_r = \rho_v \frac{4}{3}\pi(r^3 - R_1^3)$$

$$\mathbf{E} = \frac{Q_r}{4\pi\varepsilon r_2}\mathbf{i}_r = \frac{\rho_v}{3\varepsilon}(r - \frac{R_1^3}{r^2})\,\mathbf{i}_r$$

$$V = \frac{Q}{4\pi\varepsilon R_2} + \frac{\rho_v}{3\varepsilon}\left[\frac{1}{R_2}(R_1^3 + \frac{R_2^3}{2}) - \frac{1}{r}(R_1^3 + \frac{r^3}{2})\right]$$

For $r \leq R_1$

$$\mathbf{E} = 0$$

$$V = \frac{Q}{4\pi\varepsilon R_2} + \frac{\rho_V}{6\varepsilon}\left[R_1^3(\frac{2}{R_2} - \frac{3}{R_1}) + R_2^2\right]$$

A spherical shell of radius R and having a uniformly distributed charge Q

For $r \geq R$

$$\mathbf{E} = \frac{Q}{4\pi\varepsilon r^2}\mathbf{i}_r, \qquad V = \frac{Q}{4\pi\varepsilon r}$$

For $r < R$

$$\mathbf{E} = 0, \qquad V = \frac{Q}{4\pi\varepsilon R}$$

As the breakdown strength of air is approximately 30 kV/cm, the maximum potential to which a sphere of radius R cm can be charged is $30 \times R$ kV.

12.6 UNIFORM CYLINDRICAL CHARGE DISTRIBUTION

Inner radius R_1, outer radius R_2, Cylinder infinitely long, and Volume charge density ρ_v

Total charge per unit length $\rho_l = \rho_v \pi (R_2^2 - R_1^2)$

For $r \geq R_2$

$$\mathbf{E} = \frac{\rho_l}{2\pi\varepsilon r}\mathbf{i}_r, \qquad V = \frac{\rho_l}{2\pi\varepsilon} Ln\frac{r_{ref}}{r}$$

For $R_1 \leq r \leq R_2$

Charge per unit length in a cylinder of radius r ($R_1 \leq r \leq R_2$)

$$\rho_{1r} = \rho_v \pi (r^2 - R_1^2)$$

$$\mathbf{E} = \frac{\rho_{1r}}{2\pi\varepsilon r}\mathbf{i}_r = \frac{\rho_v}{2\varepsilon}\left(r - \frac{R_1^2}{r}\right)\mathbf{i}_r$$

$$V = \frac{\rho_l}{2\pi\varepsilon} Ln\frac{r_{ref}}{R_2} + \frac{\rho_v}{2\varepsilon}\left[\frac{1}{2}(R_2^2 - r^2) - R_1^2 Ln\frac{R_2}{r}\right]$$

For $r < R_1$

$$\mathbf{E} = 0,$$

$$V = \frac{\rho_l}{2\pi\varepsilon} Ln\frac{r_{ref}}{R_2} + \frac{\rho_v}{2\varepsilon}\left[\frac{1}{2}(R_2^2 - R_1^2) - R_1^2 Ln\frac{R_2}{R_1}\right]$$

A cylindrical tube of charge

Radius of tube R, uniformly distributed charge per unit length ρ_l

For $r \geq R$

$$\mathbf{E} = \frac{\rho_l}{2\pi\varepsilon r}\mathbf{i}_r, \quad V = \frac{\rho_l}{2\pi\varepsilon} Ln\frac{r_{ref}}{r}$$

For $r < R$

$$\mathbf{E} = 0, \qquad V = \frac{\rho_1}{2\pi\,\varepsilon} Ln \frac{r_{ref}}{R}$$

12.7 POISSON'S AND LAPLACE'S EQUATIONS

Knowing that $\mathbf{E} = -\nabla \mathbf{V}$, and $\nabla.\mathbf{E} = \frac{1}{\varepsilon}\rho_v$, we find $\nabla^2 V = -\frac{\rho_v}{\varepsilon}$.

This is the ***Poisson's equation.*** For a charge free region, *i.e.*, the region for which ρ_v is zero, the equation reduces to $\nabla^2 V = 0$. This is the ***Laplace's equation.***

For given boundary conditions, a solution of Poisson's equation, or Laplace's equation found by any method which satisfies the boundary conditions is the only solution of the equation.

The method of images is based on this fact.

12.8 THE METHOD OF IMAGES

The method of images substitutes the conducting body (boundary) by a set of charges, called the image charges. The image charges are the images of the given charges in the conducting body and are of opposite polarity to that of the given ones.

The images have to be of such magnitudes, and so located that the given charges and the image charges together produce the given potential at the location of the conducting body.

The electric field and potential are then found due to the given, and the image charges acting together; and this would be the same as due to the given charges in the presence of the conducting body.

The image charges in an infinite conducting plane surface are of the same magnitude as the original charges; and are located in the same way as optical mirror images.

The image of a point charge Q, in a grounded conducting sphere of radius R is $-QR/d$, and is located on the line joining Q and the centre of the sphere O, at a distance R^2/d from O; d being the distance between Q and O.

12.9 BOUNDARY CONDITIONS

Let $\mathbf{i}_n$ be the unit normal vector at the boundary directed from medium 2 to medium 1.

$\mathbf{i}_n$, $\mathbf{E}_1$, and $\mathbf{E}_2$ lie in the same normal plane.

$$D_{1n} - D_{2n} = \rho_s \qquad \text{or} \qquad \mathbf{i}_n.(\mathbf{D}_1 - \mathbf{D}_2) = \rho_s$$

$$E_{1t} - E_{2t} = 0 \qquad \text{or} \qquad \mathbf{i}_n \times (\mathbf{E}_1 - \mathbf{E}_2) = 0$$

In the absence of the surface charge,

$$\frac{\tan \alpha_1}{\tan \alpha_2} = \frac{\varepsilon_1}{\varepsilon_2},$$

where α_1 is the angle of incidence of **E** in medium 1, and α_2 is the angle of refraction of **E** in medium 2.

12.10 POLARIZATION

Polarization **P** is the dipole moment per unit volume.

$$\mathbf{P} = \chi_e \, \varepsilon_o \, \mathbf{E}, \quad \text{where } \chi_e \text{ is the electric susceptibility}$$

$$\chi_e = \varepsilon_r - 1$$

$$\mathbf{D} = \varepsilon_o \, \mathbf{E} + \mathbf{P}$$

12.11 ELECTROSTATIC ENERGY DENSITY

Electrostatic energy density = $\frac{1}{2} \, \varepsilon E^2$

Electrostatic energy stored in a system of ***n point charges*** is $W_E = \frac{1}{2}\sum_{i=1}^{n} Q_i V_i$,

where V_i is the electric potential at the location of charge Q_i due to all other charges.

12.12 GENERAL Q-V RELATIONS AMONGST CONDUCTORS

If n conducting bodies are holding the charges $Q_1, Q_2, \ldots, Q_n$; and are at potentials $V_1, V_2, \ldots, V_n$ respectively, we can write :

$$Q_1 = q_{11} V_1 + q_{12} V_2 + \ldots + q_{1n} V_n$$

$$Q_2 = q_{21} V_1 + q_{22} V_2 + \ldots + q_{2n} V_n$$

$$q_{ij} = q_{ji}$$

$$\vdots$$

$$Q_n = q_{n1} V_1 + q_{n2} V_2 + \ldots + q_{nn} V_n$$

and

$$V_1 = p_{11} Q_1 + p_{12} Q_2 + \ldots + p_{1n} Q_n$$

$$V_2 = p_{21} Q_1 + p_{22} Q_2 + \ldots + p_{2n} Q_n$$

$$p_{ij} = p_{ji}$$

$$\vdots$$

$$V_n = p_{n1} Q_1 + p_{n2} Q_2 + \ldots + p_{nn} Q_n$$

Chapter 13 MAGNETOSTATICS

13.1 MAGNETIC FIELD

A static charge, as well as a moving charge, experience force in electric field.

In magnetic field, a static charge experiences no force, but a moving charge experiences.

Thus, in the absence of electric field, if a moving charge experiences a force, then a magnetic field is present.

$$\mathbf{F} = q\, \mathbf{v} \times \mathbf{B}$$

F is the force experienced by charge q moving with velocity **v**, and **B** is the ***magnetic flux density*** which characterizes the magnetic field.

For the given q, **v**, and **B**; a particular direction of the velocity causes maximum force F_m.

Let the unit vector in that direction of the velocity be $\mathbf{i}_{vm}$.

Then $$\mathbf{B} = \frac{\mathbf{F}_m \times \mathbf{i}_{vm}}{qv}$$

It is assumed that the moving charge q does not disturb the source of original magnetic field **B**; else qv should be negligibly small.

$$\int_S \mathbf{B} . d\mathbf{s} = 0 \quad \text{or} \quad \nabla . \mathbf{B} = 0$$

Magnetic flux density vector is thus ***solenoidal***.

$\nabla . \mathbf{B} = 0$ constitutes one of the Maxwell's four equations.

Magnetic field, in addition to the magnetic flux density vector **B**, is also characterized by the ***magnetic field intensity*** vector **H**. **B** and **H** are related as

$$\mathbf{B} = \mu\, \mathbf{H} = \mu_r\, \mu_o\, \mathbf{H}$$

13.2 LORENTZ FORCE EQUATION

Lorentz force equation is

$$\mathbf{F} = q\,(\mathbf{E} + \mathbf{v} \times \mathbf{B})$$

Conduction current density at a point is

$$\mathbf{J} = \rho_v \mathbf{v}$$

ρ_v is the volume charge density at that point, and $\mathbf{v}$ is the velocity (drift) with which the charges at that point move.

Force at a point on the moving charges can also be expressed in terms of $\mathbf{J}$ as

$d\mathbf{F} = \mathbf{J} \times \mathbf{B}\, dv$, where dv is the differential of volume.

Force on a current-carrying element is

$$d\mathbf{F} = I\, d\mathbf{l} \times \mathbf{B}$$

Force on a straight current-carrying conductor of length l, the conductor being perpendicular to the direction of the magnetic field $\mathbf{B}$, is

$$F = B\,I\,l$$

The conductor, the magnetic field, and the force are in three mutually perpendicular directions.

13.3 AMPERE'S LAW OF FORCE

Force experienced by current element $I_1 d\mathbf{l}_1$ due to another current element $I_2 d\mathbf{l}_2$ is

$$d\mathbf{F}_{12} = \frac{\mu_0}{4\pi} \frac{I_1 d\mathbf{l}_1 \times (I_2 d\mathbf{l}_2 \times \mathbf{i}_{21})}{R_{21}^2},$$

where R_{21} is the distance of element 1 from element 2, and $\mathbf{i}_{21}$ is a unit vector in the direction from element 2 to element 1.

13.4 BIOT- SAVART'S LAW AND APPLICATIONS

The magnetic flux density at a point P due to a current element $Id\mathbf{l}$ is

$$d\mathbf{B} = \frac{\mu_0}{4\pi} \frac{Id\mathbf{l} \times \mathbf{i}_R}{R^2}$$

where R is the distance of the point P from the element $Id\mathbf{l}$, and $\mathbf{i}_R$ is a unit vector in the direction from the element to the point P.

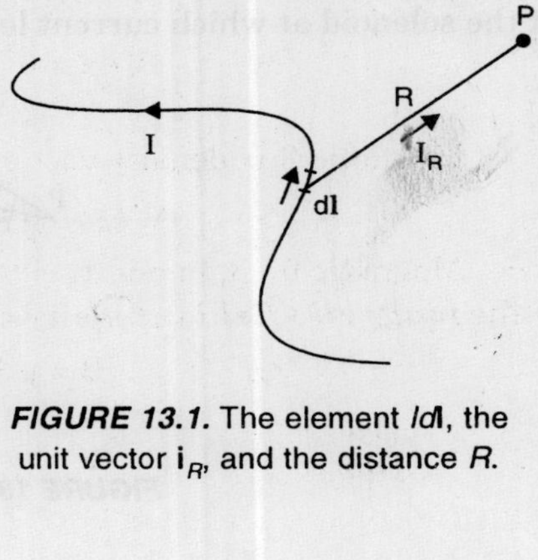

FIGURE 13.1. The element $Id\mathbf{l}$, the unit vector $\mathbf{i}_R$, and the distance R.

The magnetic flux density at a point P due to a ***straight section*** of a current-carrying wire is

$$\mathbf{B} = \frac{\mu_0 I}{4\pi R} (\cos\alpha_1 - \cos\alpha_2)\, \mathbf{i}_\phi$$

where R is the perpendicular distance of the point P from the section; α_1 is the angle which the line joining point P and the end of the section at which current enters, makes with the section; and α_2 is the angle which the line joining point P and the end of the section at which current leaves, makes with the straight section produced.

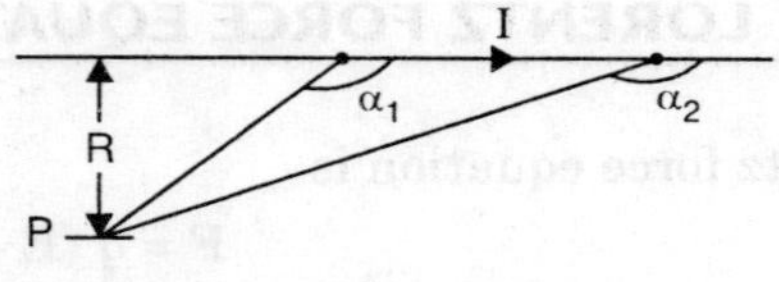

FIGURE 13.2. The angles α_1 and α_2, and the distance R for a straight section.

Field at a point P due to a straight, ***infinitely long***, current-carrying wire is

$$\mathbf{B} = \frac{\mu_0 I}{2\pi R}\,\mathbf{i}_\phi$$

where R is the perpendicular distance of the point P from the wire.

Force per unit length between two straight, parallel, infinitely long current-carrying conductors is

$$F = \frac{\mu_0 I_1 I_2}{2\pi d},$$

d being the perpendicular distance between the two wires.

The force is attractive if the two currents flow in the same direction, and repulsive if they flow in opposite directions.

Field at a point P located on the axis due to a current-carrying ***circular ring*** of radius R is

$$\mathbf{B} = \frac{\mu_0 I R^2}{2(R^2 + h^2)^{\frac{3}{2}}}\,\mathbf{i}_a$$

h is the distance of the point P from the centre of the ring, and $\mathbf{i}_a$ is the unit vector along the axis in accordance with the right hand screw rule.

Field at a point P located on the axis due to a current-carrying ***solenoid*** of radius R and having n turns per unit length is

$$\mathbf{B} = \frac{\mu_0 n I}{2}(\cos\alpha_2 - \cos\alpha_1)\,\mathbf{i}_a$$

α_1 is the angle which the line joining P to a point on the end of the solenoid at which current enters, makes with the axis; and α_2 is the angle which the line joining P to a point on the end of the solenoid at which current leaves it, makes with the axis.

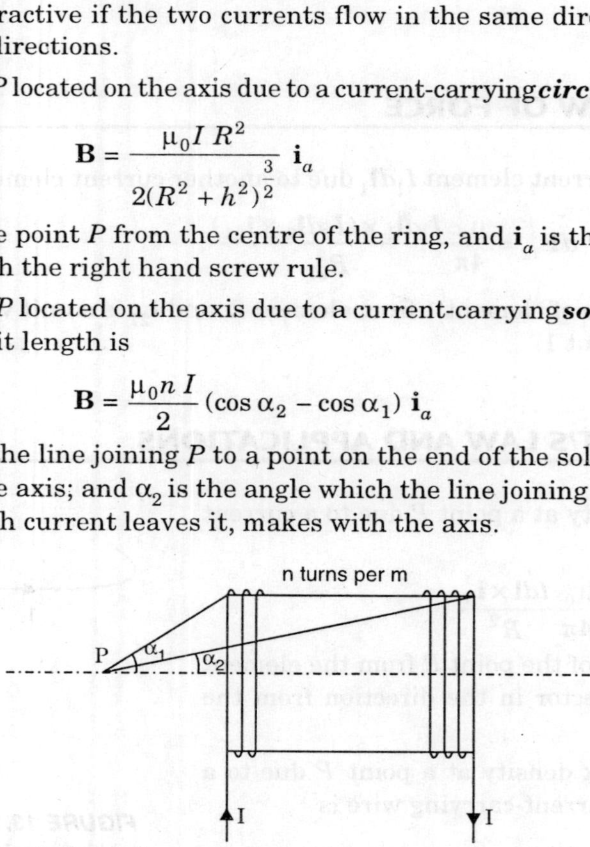

FIGURE 13.3. The angles α_1 and α_2 for a solenoid.

An ***infinite current sheet***, located in the *x-z* plane, having uniform surface current density of $J_S \mathbf{i}_z$ *A/m*, and extending from $x = x_1$ to $x = x_2$. Field at point $P(x, y)$ is

$$\mathbf{B} = \frac{\mu_0 J_s}{2\pi}\left[(\alpha_1 - \alpha_2)\mathbf{i}_x + Ln\left(\frac{\sin\alpha_2}{\sin\alpha_1}\right)\mathbf{i}_y\right]$$

If the width of the sheet also is infinite, *i.e.*, $x_1 \to -\infty$, and $x_2 \to \infty$, then

$$\mathbf{B} = -\frac{\mu_0 J_s}{2}\mathbf{i}_x \text{ for } y > 0, \text{ and } = \frac{\mu_0 J_s}{2}\mathbf{i}_x \text{ for } y < 0$$

α_1 is the angle which the perpendicular line joining P to the edge $x = x_1$, makes with the x-axis; and α_2 is similarly the angle which the perpendicular line joining P to the edge $x = x_2$, makes with the x-axis.

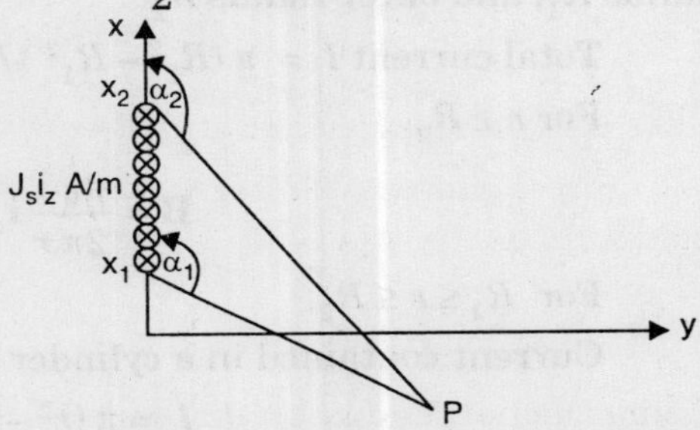

FIGURE 13.4. The angles α_1 and α_2 for a sheet of current.

Conduction current density at a point P,

$\mathbf{J} = \sigma \mathbf{E}$

Where σ is the conductivity of the medium, and $\mathbf{E}$ is the electric field acting on the medium at P.

13.5 CURRENT CONTINUITY EQUATION

Current continuity equation is

$$\oint_S \mathbf{J}.d\mathbf{s} = -\frac{\partial}{\partial t}\int_v \rho_v \, dv$$

$$\nabla.\mathbf{J} = -\frac{\partial}{\partial t}\rho_v$$

As $\nabla.\mathbf{D} = \rho_v$

$$\nabla.\mathbf{J} = -\nabla.\dot{\mathbf{D}} \quad \text{or} \quad \nabla.(\mathbf{J} + \dot{\mathbf{D}}) = 0$$

13.6 AMPERE'S CIRCUITAL LAW AND APPLICATIONS

Ampere's circuital law in the original form is

$$\oint_C \mathbf{H}.d\mathbf{l} = \oint_S \mathbf{J}.d\mathbf{s}$$

Since, it leads to $\nabla.\mathbf{J} = 0$; whereas if ρ_v is time-varying and hence the associated electric field and displacement too, then $\nabla.\mathbf{J} = -\nabla.\dot{\mathbf{D}}$. The law, therefore, gets modified as

$$\oint_C \mathbf{H}.d\mathbf{l} = \int_S \left(\mathbf{J} + \frac{\partial}{\partial t}\mathbf{D}\right).d\mathbf{s}$$

= The total current I piercing the surface S whose periphery is the contour C.

$\frac{\partial}{\partial t}$D is called the displacement current density.

We thus see that a ***time-varying electric field gives rise to magnetic field***.

Ampere's circuital law is analogous to Gauss's law in electrostatics, and can be applied to find B or H in situations where symmetry in **B** or **H** is known.

Ampere's circuital law also constitutes one of the Maxwell's four equations.

Cylindrical current of uniform current density $J\,\mathbf{i}_z$, axis of the cylinder is z-axis, inner radius R_1, and outer radius R_2.

Total current $I = \pi (R_2^2 - R_1^2)J$

For $r \geq R_2$

$$\mathbf{B} = \frac{\mu_0 \mathbf{I}}{2\pi r}\mathbf{i}_\varphi,$$

For $R_1 \leq r \leq R_2$

Current contained in a cylinder of radius r $(R_1 r R_2)$

$$I_r = \pi (r^2 - R_1^2)\mathbf{J}$$

$$\mathbf{B} = \frac{\mu_0 \mathbf{J}}{2}\left(r - \frac{R_1^2}{r}\right)\mathbf{i}_\varphi = \frac{\mu_0 I}{2\pi r}\frac{(r^2 - R_1^2)}{(R_2^2 - R_1^2)}\mathbf{i}_\varphi$$

For $r < R_1$ $\qquad \mathbf{B} = 0,$

13.7 MAGNETIC POTENTIAL

Magnetic scalar potential difference between points a and b is

$$U_a - U_b = -\int_b^a \mathbf{H}d\mathbf{l}$$

Magnetic ***vector potential A***, and magnetic flux density vector ***B*** are related as

$$\mathbf{B} = \nabla \times \mathbf{A}$$

For filamentary current, $\qquad \mathbf{A} = \frac{\mu_0}{4\pi}\int_l \frac{\mathbf{I}\,d\mathbf{l}}{R}$

For surface current, $\qquad \mathbf{A} = \frac{\mu_0}{4\pi}\int_s \frac{\mathbf{J}_S\,ds}{R}$

For volume current, $\qquad \mathbf{A} = \frac{\mu_0}{4\pi}\int_v \frac{\mathbf{J}\,dv}{R}$

R is the distance between the source element being considered and the point P at which **A** is to be found, and thus varies from element to element.

Analogous to Poisson's equation $\nabla^2 V = -\frac{\rho}{\varepsilon}$, we have $\nabla^2 \mathbf{A} = -\mu_0 \mathbf{J}$.

Analogous to Laplace's equation for charge-free region $\nabla^2 V = 0$, we have for current-free region $\nabla^2 \mathbf{A} = 0$

13.8 BOUNDARY CONDITIONS FOR MAGNETIC FIELDS

Let $\mathbf{i}_n$ be the unit normal vector at the boundary directed from medium 2 to medium 1.

$\mathbf{I}_n$, $\mathbf{H}_1$, and $\mathbf{H}_2$ lie in the same normal plane.

$$B_{1n} = B_{2n} \quad \text{or} \quad \mathbf{i}_n.(\mathbf{B}_1 - \mathbf{B}_2) = 0$$

$$H_{1t} - H_{2t} = J_s \quad \text{or} \quad \mathbf{i}_n \times (\mathbf{H}_1 - \mathbf{H}_2) = J_s$$

In the absence of the surface current,

$$\frac{\tan \alpha_1}{\tan \alpha_2} = \frac{\mu_1}{\mu_2},$$

where α_1 is the angle of incidence of **H** in medium 1, and α_2 is the angle of refraction of **H** in medium 2.

13.9 MAGNETIZATION AND SUSCEPTIBILITY

Magnetization, $\mathbf{M} = \chi_m \mathbf{H}$; where χ_m, the magnetic susceptibility $= \mu_r - 1$

$$\mathbf{B} = \mu_o [\mathbf{H} + \mathbf{M}]$$

For diamagnetic materials, $\mu_r \leq 1.0$, and $\chi_m < 0$

For paramagnetic materials, $\mu_r \geq 1.0$, and $\chi_m > 0$

For ferromagnetic materials, $\mu_r >> 1.0$, and $\chi_m >> 0$

Energy density in a magnetic field $= \dfrac{1}{2}\mu H^2$

13.10 BOUNDARY CONDITIONS FOR CONDUCTION CURRENTS

$$J_{1n} = J_{2n}$$

$$\frac{J_{1t}}{\sigma_1} = \frac{J_{2t}}{\sigma_2}$$

$$\frac{\tan \alpha_1}{\tan \alpha_2} = \frac{\sigma_1}{\sigma_2},$$

where α_1 is the angle of incidence of **J** in medium 1, and α_2 is the angle of refraction of **J** in medium 2.

For a medium of conductivity σ and permittivity ε, the conductance and the capacitance between any two arbitrarily shaped electrodes are related as

$$\frac{G}{C} = \frac{\sigma}{\varepsilon}$$

Chapter 14

ELECTROMAGNETIC FIELD AND MAXWELL'S EQUATIONS

14.1 FARADAY'S LAW

Integral form.

$$\oint_C \mathbf{E}.d\mathbf{l} = -\frac{d}{dt}\int_S \mathbf{B}.d\mathbf{s}$$

where C is the periphery of the surface S. The marking of the direction of the contour, and the orientation of $d\mathbf{s}$ has to be in accordance with the right hand screw rule.

The point, or the differential form is :

$$\nabla \times \mathbf{E} = -\frac{\partial \mathbf{B}}{\partial t} = -\dot{\mathbf{B}} = -\mu \dot{\mathbf{H}}$$

Faraday's law also constitutes one of the Maxwell's equations.

Faraday's law implies that a ***time-varying magnetic field gives rise to electric field.***

It may be recalled that the Ampere's law brought out the fact that a time-varying electric field gives rise to magnetic field.

Time-varying electric, and magnetic fields are coupled; that is, they are interdependent. Such coupled electric and magnetic field is ***electromagnetic*** field.

The ***constitutive relations*** are

$$\mathbf{D} = \varepsilon\, \mathbf{E}$$

$$\mathbf{B} = \mu\, \mathbf{H}$$

$$\mathbf{J} = \sigma\, \mathbf{E}$$

A medium for which ε is a scalar constant, is called an **isotropic** medium; the one for which it is a vector constant rather than a scalar constant, is called an **anisotropic** medium.

14.2 MAXWELL'S EQUATIONS

The point, or the differential form

$$\nabla \times \mathbf{H} = \mathbf{J} + \frac{\partial \mathbf{D}}{\partial t} = \mathbf{J} + \dot{\mathbf{D}} = \mathbf{J} + \varepsilon \dot{\mathbf{E}}$$

$$\nabla \times \mathbf{E} = -\frac{\partial \mathbf{B}}{\partial t} = -\dot{\mathbf{B}} = -\mu \dot{\mathbf{H}}$$ These are the two curl equations

$$\nabla . \mathbf{D} = \rho_v$$

$$\nabla . \mathbf{B} = 0$$ These are the two divergence equations

Integral form

$$\oint_C \mathbf{H}.d\mathbf{l} = \int_S \mathbf{J}.d\mathbf{s} + \frac{d}{dt}\int_S \mathbf{D}.d\mathbf{s}$$ S is any surface (open) whose periphery is the closed path, or the contour C.

$$\oint_C \mathbf{E}.d\mathbf{l} = -\frac{d}{dt}\int_S \mathbf{B}.d\mathbf{s}$$

$$\oint_S \mathbf{D}.d\mathbf{s} = \int_v \rho_v dv$$ v is the volume whose bounding surface is S, a closed surface.

$$\oint_S \mathbf{B}.d\mathbf{s} = 0$$

For ***static fields***, the time derivative terms vanish.

$$\nabla \times \mathbf{H} = \mathbf{J} \qquad \oint_C \mathbf{H}.d\mathbf{l} = \oint_S \mathbf{J}.d\mathbf{s}$$

$$\nabla \times \mathbf{E} = 0 \qquad \oint_C \mathbf{E}.d\mathbf{l} = 0$$

$$\nabla . \mathbf{D} = \rho_v \qquad \oint_S \mathbf{D}.d\mathbf{s} = \int_v \rho_v dv$$

$$\nabla . \mathbf{B} = 0 \qquad \oint_S \mathbf{B}.d\mathbf{s} = 0$$

For perfect dielectrics, $\sigma = 0$, $\rho_v = 0$, and $\mathbf{J} = 0$

For free space, in addition to the above conditions, $\varepsilon_r = 1$ and $\mu_r = 1$

For perfect conductors, $\sigma = \infty$, $\rho_v = 0$, and $\mathbf{E} = 0$

Only static magnetic field can exist in a perfectly conducting medium.

For good dielectrics, $|\sigma E| << \left|\frac{\partial D}{\partial t}\right|$; and therefore, the contribution due to the term $|\sigma E|$ can be neglected.

For good conductors, $|\sigma E| >> \left|\frac{\partial D}{\partial t}\right|$ and therefore the contribution due to the term $\left|\frac{\partial D}{\partial t}\right|$ can be neglected.

For all above cases, Maxwell's equations can be written keeping in mind the respective conditions.

The current continuity equation, though is not a Maxwell's equation, is grouped with Maxwell's four equations.

$$\nabla . \mathbf{J} = -\frac{\partial}{\partial t}\rho_v$$

$$\oint_S \mathbf{J}.d\mathbf{s} = -\frac{\partial}{\partial t}\int_v \rho_v dv$$

No relationship of field quantities can be in violation of any of the Maxwell's equations and the current continuity equation.

14.3 SINUSOIDALLY TIME-VARYING FIELDS

All the field quantities are, in general, functions of the space variables and time. If the variation with time is sinusoidal, we can suppress the time factor through the use of phasors, as we do in steady-state analysis of ac circuits. Then the vector field quantities are simultaneously vectors and phasors, and the scalar field quantities are merely phasors, just like current phasor (or voltage phasor) in ac circuit theory.

As an example, the electric field **E** having sinusoidal variation, may be expressed in Cartesian system of coordinates as :

$$\begin{aligned}
\mathbf{E}(x, y, z, t) &= E_x(x, y, z, t)\,\mathbf{i}_x + E_y(x, y, z, t)\,\mathbf{i}_y + E_z(x, y, z, t)\,\mathbf{i}_z \\
&= E_{xm}(x, y, z)\cos(\omega t + \varphi_x(x, y, z))\,\mathbf{i}_x \\
&\quad + E_{ym}(x, y, z)\cos(\omega t + \varphi_y(x, y, z))\,\mathbf{i}_y \\
&\quad + E_{zm}(x, y, z)\cos(\omega t + \varphi_z(x, y, z))\,\mathbf{i}_z \\
&= \text{Real of } [\{E_{xm}(x, y, z)\,e^{j\varphi_x(x, y, z)}\,\mathbf{i}_x \\
&\quad + E_{ym}(x, y, z)\,e^{j\varphi_y(x, y, z)}\,\mathbf{i}_y \\
&\quad + E_{zm}(x, y, z)\,e^{j\varphi_z(x, y, z)}\,\mathbf{i}_z\}e^{j\omega t}]
\end{aligned}$$

In terms of phasors,

$$\begin{aligned}
\overline{\mathbf{E}}(x,y,z) &= E_{xm}(x, y, z)\,e^{j\varphi_x(x, y, z)}\,\mathbf{i}_x \\
&\quad + E_{ym}(x, y, z)\,e^{j\varphi_y(x, y, z)}\,\mathbf{i}_y \\
&\quad + E_{zm}(x, y, z)\,e^{j\varphi_z(x, y, z)}\,\mathbf{i}_z \\
&= \overline{\mathbf{E}}_x(x, y, z)\,\mathbf{i}_x \\
&\quad + \overline{\mathbf{E}}_y(x, y, z)\,\mathbf{i}_y \\
&\quad + \overline{\mathbf{E}}_z(x, y, z)\,\mathbf{i}_z
\end{aligned}$$

Thus, $\overline{\mathbf{E}}\,(x,y,z) = \overline{\mathbf{E}}_x\,(x, y, z)\,\mathbf{i}_x + \overline{\mathbf{E}}_y\,(x, y, z)\,\mathbf{i}_y + \overline{\mathbf{E}}_z\,(x, y, z)\,\mathbf{i}_z$

Letters with ***bars*** are the ***phasors.***

Also, multiplying the original phasor by $j\omega$ implies taking first derivative in time domain; and dividing by $j\omega$ implies integration.

For sinusoidal time variations, Maxwell's equations, and the current continuity equation become

$$\nabla \times \overline{\mathbf{H}} = \overline{\mathbf{J}} + j\omega\overline{\mathbf{D}}$$

$$\nabla \times \overline{\mathbf{E}} = -\,j\omega\overline{\mathbf{B}}$$

$$\nabla.\overline{\mathbf{D}} = \overline{\rho}_v$$

$$\nabla.\overline{\mathbf{B}} = 0$$

$$\nabla.\overline{\mathbf{J}} = -\,j\,\omega\overline{\rho}_v$$

$$\oint_C \overline{\mathbf{H}}.d\mathbf{l} = \int_S (\overline{\mathbf{J}} + j\omega\overline{\mathbf{D}}).d\mathbf{s}$$

$$\oint_C \overline{\mathbf{E}}.d\mathbf{l} = -\int_S j\omega\overline{\mathbf{B}}.d\mathbf{s}$$

$$\int_S \overline{\mathbf{D}}.d\mathbf{s} = \int_v \overline{\rho}_v dv$$

$$\int_S \overline{\mathbf{B}}.d\mathbf{s} = 0$$

$$\int_S \overline{\mathbf{J}}.d\mathbf{s} = -\int_v J\,\omega\overline{\rho}_v dv$$

Chapter 15 ELECTROMAGNETIC WAVES

15.1 THE WAVE EQUATION

The wave equation for a conducting medium is

$$\nabla^2\mathbf{E} - \mu\varepsilon\ddot{\mathbf{E}} - \mu\sigma\dot{\mathbf{E}} = 0$$

For perfect dielectrics, it is

$$\nabla^2\mathbf{E} - \mu\varepsilon\ddot{\mathbf{E}} = 0$$

A uniform plane wave is one for which **E** (and **H** too) is same everywhere in a plane perpendicular to the direction of travel of the wave. If the direction of ***travel*** of the wave is ***x-direction***, then the uniform planes are parallel to the *y-z* plane. In such planes, there is no variation of **E**, or **H** with y, and z.

Also, there is no component of **E**, or **H** in the direction of travel.

15.2 THE RELATION BETWEEN E AND H IN A UPW

For a UPW in free space, traveling in x-direction, we have

$$\frac{E_y}{H_z} = \sqrt{\frac{\mu}{\varepsilon}}, \quad \frac{E_z}{H_y} = -\sqrt{\frac{\mu}{\varepsilon}}, \quad \text{and} \quad \frac{E}{H} = \sqrt{\frac{\mu}{\varepsilon}}$$

The ratio $\dfrac{E_y}{H_z}$ or $-\dfrac{E_z}{H_y}$, denoted by η, is the characteristic, or also called the intrinsic impedance of the medium. In general, it is a complex quantity.

For a medium for which σ is zero, it is simply a scalar quantity and equals $\sqrt{\dfrac{\mu}{\varepsilon}}$

For free space, $\sqrt{\dfrac{\mu}{\varepsilon}} = 120\pi = 377$ ohms

For a UPW, we see that

$$\mathbf{E.H} = 0,$$
$$\mathbf{E{\times}H} = \mathbf{i}_x\, \eta H^2$$

15.3 SINUSOIDAL TIME VARIATION OF FIELD QUANTITIES

With ***sinusoidal time variation***, the corresponding wave equations are

$$\nabla^2\overline{\mathbf{E}} + (\omega^2\mu\varepsilon - j\omega\mu\sigma)\,\overline{\mathbf{E}} = 0 \quad \text{and} \quad \nabla^2\overline{\mathbf{E}} + \omega^2\mu\varepsilon\overline{\mathbf{E}} = 0$$

For a ***lossless*** medium, the ***solution*** for $E_z(x, t)$, or $E_y(x, t)$ is

$$C_1 \cos(\omega t - \beta x) + C_2 \cos(\omega t + \beta x),$$

where $\beta = \omega\sqrt{(\mu\varepsilon)}$; and β is called the ***phase shift constant***.

A ***standing wave*** is one for which, in the direction of propagation, the points at which the electric field intensity E, or the magnetic field intensity H become maximum (or minimum), are fixed. For a traveling wave, it is not so; at all the points E, or H reach their peak values periodically.

15.4 PROPAGATION PARAMETERS

The ***general*** propagation parameters

Propagation constant, $\gamma = \sqrt{j\omega\mu(\sigma + j\omega\varepsilon)} = \alpha + j\beta$

Attenuation constant, $\alpha = \omega\sqrt{\dfrac{\mu\varepsilon}{2}\left(\sqrt{1+\dfrac{\sigma^2}{\omega^2\varepsilon^2}} - 1\right)}$

Phase constant, $\beta = \omega\sqrt{\dfrac{\mu\varepsilon}{2}\left(\sqrt{1+\dfrac{\sigma^2}{\omega^2\varepsilon^2}} + 1\right)}$

Intrinsic impedance, $\eta = \sqrt{\dfrac{j\omega\mu}{\sigma + j\omega\varepsilon}}$

velocity of propagation, $v = \dfrac{\omega}{\beta}$, wavelength, $\lambda = \dfrac{2\pi}{\beta}$

velocity of propagation, v equals $f\lambda$ also.

The solution for $E_z(x, t)$, or $E_y(x, t)$ will now carry attenuation terms $e^{-\alpha x}$, $e^{\alpha x}$.

$$C_1 e^{-\alpha x} \cos(\omega t - \beta x) + C_2 e^{\alpha x} \cos(\omega t + \beta x)$$

For ***good conductors***, $\frac{\sigma}{\omega \varepsilon} >> 1$

The expressions for approximate values of various parameters become

$$\alpha = \beta = \sqrt{\frac{\omega \mu \sigma}{2}}$$

$$v = \sqrt{\frac{2\omega}{\mu \sigma}}$$

$$\eta = \sqrt{\frac{\omega \mu}{\sigma}} \angle 45°$$

The depth of penetration, $\delta = \frac{1}{\alpha} = \sqrt{\frac{2}{\omega \mu \sigma}}$

For ***good dielectrics***, the dissipation factor, $\frac{\sigma}{\omega \varepsilon} << 1$

The expressions for approximate values of various parameters become

$$\alpha = \frac{\sigma}{2}\sqrt{\frac{\mu}{\varepsilon}}$$

$$\beta = \omega \sqrt{\mu \varepsilon}\left(1 + \frac{\sigma^2}{8\omega^2 \varepsilon^2}\right)$$

$$v = \frac{1}{\mu \varepsilon}\left(1 - \frac{\sigma^2}{8\omega^2 \varepsilon^2}\right)$$

$$\eta = \sqrt{\frac{\mu}{\varepsilon}}\left(1 + j\frac{\sigma}{2\omega \varepsilon}\right)$$

If ψ be the loss angle, then $\psi = \tan^{-1}\frac{\sigma}{\omega \varepsilon}$

And the power factor $= \sin \psi \approx \tan \psi = \frac{\sigma}{\omega \varepsilon}$

15.5 WAVE POLARIZATION

Wave polarization is the characteristic which describes how the magnitude and the direction of **E** vary with time. At a point in space, the locus traced by the tip of the electric field intensity vector specifies the type of polarization.

For a wave traveling in x- direction

If E_y is zero, the wave is polarized in the z-direction

If E_z is zero, the wave is polarized in the y-direction

If $\overline{E}_y$ and $\overline{E}_z$ are in phase, the wave is polarized in a direction which makes angle $\theta = \tan^{-1}\dfrac{E_z}{E_y}$ with the y-axis.

If $\overline{E}_y$ and $\overline{E}_z$ have a phase difference of 90°, the wave is elliptically polarized.

If, in addition, $E_y = E_z$; the wave is then circularly polarized.

15.6 REFLECTIONS AND REFRACTIONS AT THE INTERFACE OF TWO DIFFERENT MEDIA

Free space and perfect conductor – normal incidence

Let the interface be defined by the plane $x = 0$, and the positive direction of x be towards the perfect conductor.

At $x = 0$, the incident electric field strength : E_{io}, and the reflected field strength : E_{ro}.

$E_{ro} = -E_{io}$

The resultant electric field strength at any point $-x$ and at any time t is

$$2E_{io}\sin\beta x\sin\omega t$$

and the resultant magnetic field strength is

$$2H_{io}\cos\beta x\cos\omega t$$

Medium 1 of constants μ_1, ε_1, σ_1; and ***Medium 2 of constants*** μ_2, ε_2, σ_2 - ***normal*** incidence

At the interface ($x = 0$), the ratio of the amplitude of electric field in the reflected wave to that in the incident wave is the

reflection coefficient, $\dfrac{\mathbf{E}_r}{\mathbf{E}_i} = \dfrac{\eta_2 - \eta_1}{\eta_2 + \eta_1}$; and the

transmission coefficient, $\dfrac{\mathbf{E}_t}{\mathbf{E}_i} = \dfrac{2\eta_2}{\eta_2 + \eta_1}$

The reflection coefficient, and the transmission coefficient, for the magnetic field are

$$\frac{\mathbf{H}_r}{\mathbf{H}_i} = \frac{\eta_1 - \eta_2}{\eta_1 + \eta_2};$$

$$\frac{\mathbf{H}_t}{\mathbf{H}_i} = \frac{2\eta_1}{\eta_1 + \eta_2}$$

For ***perfect dielectrics***, the respective expressions (normal incidence) are

$$\frac{\mathbf{E}_r}{\mathbf{E}_i} = \frac{\sqrt{\varepsilon_1} - \sqrt{\varepsilon_2}}{\sqrt{\varepsilon_1} + \sqrt{\varepsilon_2}}$$

$$\frac{E_t}{E_i} = \frac{2\sqrt{\varepsilon_1}}{\sqrt{\varepsilon_1}+\sqrt{\varepsilon_2}}$$

$$\frac{H_r}{H_i} = \frac{\sqrt{\varepsilon_2}-\sqrt{\varepsilon_1}}{\sqrt{\varepsilon_2}+\sqrt{\varepsilon_1}}$$

$$\frac{H_t}{H_i} = \frac{2\sqrt{\varepsilon_2}}{\sqrt{\varepsilon_2}+\sqrt{\varepsilon_1}}$$

Perfect dielectrics, oblique incidence :

Plane of incidence is the plane containing the incident ray, and the normal to the interface.

Perpendicular (horizontal) polarization is one in which the electric field vector is perpendicular to the plane of incidence, or it is parallel to the boundary surface.

Parallel (vertical) polarization is one in which the electric field vector is parallel to the plane of incidence, or the magnetic field vector is parallel to the boundary surface.

Let the angle of incidence be denoted by θ_1, and that of transmission, by θ_2. Then

$$\frac{\sin\theta_1}{\sin\theta_2} = \sqrt{\frac{\varepsilon_2}{\varepsilon_1}} \qquad \text{(Snell's law)}$$

$$\frac{E_r^2}{E_i^2} = 1 - \frac{\sqrt{\varepsilon_2}\,E_t^2\cos\theta_2}{\sqrt{\varepsilon_1}\,E_i^2\cos\theta_1}$$

With ***perpendicular*** polarization

$$\frac{E_r}{E_i} = \frac{\cos\theta_1 - \sqrt{(\varepsilon_2/\varepsilon_1)-\sin^2\theta_1}}{\cos\theta_1 + \sqrt{(\varepsilon_2/\varepsilon_1)-\sin^2\theta_1}}$$

With ***parallel*** polarization

$$\frac{E_r}{E_i} = \frac{(\varepsilon_2/\varepsilon_1)\cos\theta_1 - \sqrt{(\varepsilon_2/\varepsilon_1)-\sin^2\theta_1}}{(\varepsilon_2/\varepsilon_1)\cos\theta_1 + \sqrt{(\varepsilon_2/\varepsilon_1)-\sin^2\theta_1}}$$

The condition for no reflection is given by

$$\tan\theta_1 = \sqrt{\frac{\varepsilon_2}{\varepsilon_1}};$$

this angle of incidence is called the ***Brewster angle***.

15.7 THE SURFACE IMPEDANCE

The ***surface impedance*** Z_s is given by $\dfrac{E_{tan}}{J_s} = \dfrac{\gamma}{\sigma}$

For good conductors, $\mathbf{Z}_s \approx \sqrt{\dfrac{j\omega\mu}{\sigma}} = \eta$

$$\mathbf{Z}_s = R_s + jX_s\,; \qquad R_s \approx \sqrt{\frac{\omega\mu}{2\sigma}} = \frac{1}{\sigma\delta}, \quad X_s \approx \sqrt{\frac{\omega\mu}{2\sigma}}$$

15.8 THE POYNTING VECTOR

Poynting Vector **P = E×H**

Poynting Vector at a point **P** is a measure of the time rate of energy flow per unit area at that point and points in the direction of energy flow.

For ***sinusoidal time variation***, the ***complex*** poynting vector $\overline{\mathbf{P}}$ in terms of the complex electric field vector $\overline{\mathbf{E}}$, and the complex magnetic field vector $\overline{\mathbf{H}}$ is

$$\overline{\mathbf{P}} = \frac{1}{2}\,\overline{\mathbf{E}} \times \overline{\mathbf{H}}^*$$

The complex power flow per unit area in the x-direction is

$$P_x = \frac{1}{2}(\overline{E}_y\overline{H}_z^* - \overline{E}_z\overline{H}_y^*);\ \overline{E}_y, \overline{E}_z, \overline{H}_y,$$

and $\overline{H}_z$ are all phasors; and P_x is a complex quantity. In this relation, $\overline{E}, \overline{H}$, and $\overline{P}$ are respectively analogous to ac circuit theory voltage phasor, current phasor, and complex power.

The expressions for y- and z- directions can be written following the cyclic order.

Chapter 16 GUIDED WAVES

16.1 GENERAL

The wave is assumed to be propagating in the z-direction.

The parallel conducting planes are situated parallel to the y-z plane for wave propagation between parallel conducting planes, and the distance between these planes is a. For rectangular waveguides, there are parallel plates which are here taken as parallel to the x-z plane, and the distance between these plates is being denoted by b.

Suffixes

o	for free space
p	for phase
g	for group
c	for cut-off, and also for conductor

m, and n are integers.

$$\gamma_p{}^2 = \omega^2\mu_o\varepsilon_o + h_p{}^2$$

For transverse electric waves (TE waves) $E_z = 0$, and H_z 0

For transverse magnetic waves (TM waves) $H_z = 0$, and E_z 0

For transverse electromagnetic waves(TEM waves) $E_z = 0$, and $H_z = 0$

While finding the solution for the various components of the field quantities for a wave propagating between parallel planes, h_p is constrained to equal $\frac{m\pi}{a}$ and the solution involves terms as $\sin \frac{m\pi}{a} x$, or $\cos \frac{m\pi}{a} x$.

For a rectangular waveguide, the solution, in addition, involves terms as sin $\frac{n\pi}{b}y$, or cos $\frac{n\pi}{b}y$.

Each value of m, and n specifies a particular field configuration, or a mode. We, accordingly, have TE_{mn}, or TM_{mn} waves. For wave propagation between parallel planes, $n = 0$.

The ***cut-off frequency***, f_c is that frequency below which there is only attenuation and $\beta_p = 0$, and above it, $\alpha_p = 0$, and β_p exists.

The ***dominant mode*** is a mode (specified by a value of m) which has the lowest cut-off frequency.

16.2 WAVE PROPAGATION BETWEEN PARALLEL PERFECTLY CONDUCTING PLATES

The field components of a ***TE*** wave are E_y, H_x, and H_z.

The field components of a ***TM*** wave are H_y, E_x, and E_z.

$$f_c = \frac{m}{2a\sqrt{\mu_o \varepsilon_o}}$$

$$\text{For } f > f_c, \quad \beta_p = \sqrt{\omega^2 \mu_o \varepsilon_o - \left(\frac{m\pi}{a}\right)^2}$$

$$\text{For } f < f_c, \quad \alpha_p = \sqrt{-\left[\omega^2 \mu_o \varepsilon_o - \left(\frac{m\pi}{a}\right)^2\right]}$$

$$\lambda_p = \frac{\lambda}{\sqrt{1 - \left(\frac{\lambda}{\lambda_c}\right)^2}}, \quad \lambda \text{ is the free space wavelength}$$

$$\text{Wave impedance } \eta_{mn}^{TE} = -\frac{E_y}{H_x} = \frac{\eta}{\sqrt{1 - \left(\frac{f_c}{f}\right)^2}}$$

$$\text{Wave impedance } \eta_{mn}^{TM} = \frac{E_x}{H_y} = \eta\sqrt{1 - \left(\frac{f_c}{f}\right)^2}$$

TEM waves :

A *TEM* wave is a TM_{oo} wave. The field components of a *TEM* wave are E_x, and H_y.

$$\alpha_p = 0$$

$$\beta_p = \omega\sqrt{\mu_o \varepsilon_o}$$

$$v_p = \frac{1}{\sqrt{\mu_o \varepsilon_o}}$$

$$f_c = 0$$

$$\eta_p = \left|\frac{E_x}{H_y}\right| = \sqrt{\frac{\mu_o}{\varepsilon_o}} = 120\pi$$

A *TEM* wave is also called the ***principal*** wave.

Attenuation in parallel plate(plates ***not perfectly conducting*** ones) guides is

$$\alpha = \frac{\text{power lost per unit length}}{2 \times \text{power transmitted}}$$

For ***TE*** waves, $$\alpha_p = \frac{2m^2\pi^2\sqrt{\omega\mu_c/2\sigma_c}}{\omega\mu a^2\sqrt{\omega^2\mu\varepsilon - \left(\frac{m\pi}{a}\right)^2}}$$

For ***TEM*** waves, $$\alpha_p = \frac{1}{\eta_p a}\sqrt{\frac{\omega\mu_c}{2\sigma_c}}$$

Free space velocity v_o is the velocity of propagation of the wave in free space.

Phase velocity v_p is the velocity of propagation of equiphase surfaces along the guide.

Group velocity v_g is the velocity with which the energy propagates in the waveguide.

$$v_o^2 = v_p\, v_g$$

16.3 RECTANGULAR WAVE GUIDES

$$\gamma_g = \sqrt{\left(\frac{m\pi}{a}\right)^2 + \left(\frac{n\pi}{b}\right)^2 - \omega^2\mu\varepsilon}$$

$$f_c = \frac{1}{2\pi\sqrt{\mu\varepsilon}}\sqrt{\left(\frac{m\pi}{a}\right)^2 + \left(\frac{n\pi}{b}\right)^2}$$

$$\lambda_c = \frac{2}{\sqrt{\left(\frac{m}{a}\right)^2 + \left(\frac{n}{b}\right)^2}}$$

$$v_p = \frac{\omega}{\sqrt{\omega^2\mu\varepsilon - \left(\frac{m\pi}{a}\right)^2 - \left(\frac{n\pi}{b}\right)^2}}$$

$$\lambda_g = \frac{2\pi}{\sqrt{\omega^2\mu\varepsilon - \left(\frac{m\pi}{a}\right)^2 - \left(\frac{n\pi}{b}\right)^2}} \quad \text{or} \quad \lambda_g = \frac{\lambda}{\sqrt{1 - \left(\frac{\lambda}{\lambda_c}\right)^2}}$$

16.4 CIRCULAR WAVE GUIDES

r is radius of the guide.

K_c is the cut-off wave number.

p_{nm} the roots of the Bessel function.

$J_n(K_c r)$ is the Bessel function of the first kind.

For a ***TE* wave,** the axial component of the field is

$$H_{z,nm} = J_n(K_c r)(A \cos n\varphi + B \sin n\varphi)$$

For a ***TM*** wave, the axial component of the field is

$$E_{z,nm} = J_n(K_c r)(A' \cos n\varphi + B' \sin n\varphi)$$

The cut-off wavelength for ***TM*** wave is

$$\lambda_{c,nm} = \frac{2\pi}{K_{c,nm}} = \frac{2\pi r}{P_{nm}}$$

$K_{c,nm}$, and p_{nm} are respectively the cut-off wave number and the roots of the Bessel function.

The cut-off wavelength for ***TE*** wave is

$$\lambda_{c,nm} = \frac{2\pi r}{P'_{nm}},$$

where p'_{nm} are the roots of derivative of the Bessel's function *w.r.t. r*.

The roots p_{nm} of the Bessel function

m	$n = 0$	$n = 1$	$n = 2$
1	2.40	3.83	5.14
2	5.52	7.02	8.42
3	8.65	10.17	11.62

The roots p'_{nm} of the derivative of Bessel function

m	$n = 0$	$n = 1$	$n = 2$
1	3.83	1.84	3.05
2	7.02	5.33	6.71
3	10.17	8.54	9.97

Guide wavelength, $$\lambda_{guide} = \frac{\lambda}{\sqrt{1 - \left(\frac{\lambda}{\lambda_{c,nm}}\right)^2}}$$

A ***TEM*** wave cannot exist in a hollow guide.

16.5 WAVEGUIDE RESONATOR

A rectangular waveguide with its open ends closed by shorts, becomes a waveguide resonator. The dimension of the resonator in the z-direction is taken as c.

For ***TM*** ($H_z = 0$) mode,

$$E_z = E_m \sin\left(\frac{m\pi x}{a}\right) \sin\left(\frac{n\pi y}{b}\right) \cos\left(\frac{l\pi z}{c}\right)$$

$m = 1, 2, 3,\ldots; n = 1, 2, 3,\ldots;$ and $l = 0, 1, 2,\ldots$

$$f_r = \frac{v}{2}\sqrt{\left(\frac{m}{a}\right)^2 + \left(\frac{n}{b}\right)^2 + \left(\frac{l}{c}\right)^2}$$

For ***TE*** ($E_z = 0$) mode,

$$H_z = H_m \cos\left(\frac{m\pi x}{a}\right) \cos\left(\frac{n\pi y}{b}\right) \sin\left(\frac{l\pi z}{c}\right)$$

$m = 0, 1, 2,\ldots; n = 0, 1, 2,\ldots;$ and $l = 1, 2, 3,\ldots$

The ***depth of penetration*** in the ***cavity*** walls,

$$\delta = \frac{1}{\sqrt{\pi f_{101} \mu_o \sigma_c}},$$

$f_{101,}$ the lowest order ***TE mode*** resonant frequency $= \frac{v}{2}\sqrt{\frac{1}{a^2} + \frac{1}{c^2}}$

$f_{110,}$ the lowest order ***TM mode*** resonant frequency $= \frac{v}{2}\sqrt{\frac{1}{a^2} + \frac{1}{b^2}}$

Chapter **17** TRANSMISSION LINES

17.1 GENERAL

A transmission line is a distributed parameter circuit.

The parameters used, *i.e*, R, L, $\mathbf{z}$, G, C, $\mathbf{y}$ are all line parameters per unit length.

At the receiving end we take $x = 0$, and at the sending end, $x = l$; thus x increases in the direction from R-end to S-end.

The relations given are for steady state ac conditions, and apply to single-phase lines, and also to the three-phase lines on per phase basis.

The two governing ***differential equations*** are :

$$\frac{d^2\mathbf{V}}{dx^2} = \mathbf{yzV} \quad \text{and} \quad \frac{d^2\mathbf{I}}{dx^2} = \mathbf{yzI}$$

$$\sqrt{\mathbf{yz}} = \gamma = \alpha + j\beta \quad \text{and} \quad \sqrt{\frac{\mathbf{z}}{\mathbf{y}}} = \mathbf{Z}_0$$

At the receiving end ($x = 0$), $\mathbf{V}_R = \mathbf{Z}_R\,\mathbf{I}_R$

At the sending end ($x = 1$), $\mathbf{V}_S = \mathbf{Z}_S\,\mathbf{I}_S$

17.2 LINE VOLTAGES AND CURRENTS

The ***solution*** for the voltage **V**, and the current **I**, at any point distant x from the R-end is

$$\mathbf{V} = \frac{\mathbf{V}_r + \mathbf{I}_R\mathbf{Z}_o}{2}e^{\alpha x}e^{j\beta x} + \frac{\mathbf{V}_R - \mathbf{I}_R\mathbf{Z}_o}{2}e^{-\alpha x}e^{-j\beta x}$$

$$\mathbf{I} = \frac{\mathbf{V}_r / \mathbf{Z}_o + \mathbf{I}_R}{2} e^{\alpha x} e^{j\beta x} - \frac{\mathbf{V}_R / \mathbf{Z}_o - \mathbf{I}_R}{2} e^{-\alpha x} e^{-j\beta x}$$

Following are ***other forms*** of the above solution :

$$\mathbf{V} = \frac{\mathbf{V}_R (\mathbf{Z}_R + \mathbf{Z}_o)}{2\mathbf{Z}_R} \left[e^{\gamma x} + \frac{\mathbf{Z}_R - \mathbf{Z}_o}{\mathbf{Z}_R + \mathbf{Z}_o} e^{-\gamma x} \right]$$

$$\mathbf{I} = \frac{\mathbf{I}_R (\mathbf{Z}_R + \mathbf{Z}_o)}{2\mathbf{Z}_o} \left[e^{\gamma x} - \frac{\mathbf{Z}_R - \mathbf{Z}_o}{\mathbf{Z}_R + \mathbf{Z}_o} e^{-\gamma x} \right]$$

$$\mathbf{V} = \mathbf{V}_R \cosh \gamma x + \mathbf{I}_R \mathbf{Z}_o \sinh \gamma x$$

$$\mathbf{I} = \mathbf{V}_R \frac{\sinh \gamma x}{z_0} + \mathbf{I}_R \cosh \gamma x$$

17.3 IMPEDANCES

The ***impedance***,

$$\mathbf{Z} = \frac{\mathbf{V}}{\mathbf{I}} = \mathbf{Z}_o \frac{\mathbf{Z}_R \cosh \gamma x + \mathbf{Z}_o \sinh \gamma x}{\mathbf{Z}_o \cosh \gamma x + \mathbf{Z}_R \sinh \gamma x}$$

$$= \mathbf{Z}_o \frac{e^{\gamma x} + \dfrac{\mathbf{Z}_R - \mathbf{Z}_o}{\mathbf{Z}_R + \mathbf{Z}_o} e^{-\gamma x}}{e^{\gamma x} - \dfrac{\mathbf{Z}_R - \mathbf{Z}_o}{\mathbf{Z}_R + \mathbf{Z}_o} e^{-\gamma x}}$$

The ***input impedance***,

$$\mathbf{Z}_s = \frac{\mathbf{V}_s}{\mathbf{I}_s} = \mathbf{Z}_o \frac{\mathbf{Z}_R \cosh \gamma l + \mathbf{Z}_o \sinh \gamma l}{\mathbf{Z}_o \cosh \gamma l + \mathbf{Z}_R \sinh \gamma l}$$

$$= \mathbf{Z}_o \frac{e^{\gamma l} + \dfrac{\mathbf{Z}_R - \mathbf{Z}_o}{\mathbf{Z}_R + \mathbf{Z}_o} e^{-\gamma l}}{e^{\gamma l} - \dfrac{\mathbf{Z}_R - \mathbf{Z}_o}{\mathbf{Z}_R + \mathbf{Z}_o} e^{-\gamma l}}$$

If $\mathbf{Z}_R = \mathbf{Z}_o$, then

$$\mathbf{V} = \mathbf{V}_R e^{\gamma x} = \mathbf{V}_S e^{-\gamma x'}; \qquad x' = l - x$$

$$\mathbf{I} = \mathbf{I}_R e^{\gamma x} = \mathbf{I}_S e^{-\gamma x'}$$

$$\mathbf{Z} = \mathbf{Z}_o$$

As in the case of electromagnetic waves, here also we have :

Wavelength $\lambda = \dfrac{2\pi}{\beta}$, velocity of propagation of energy $v = \dfrac{\omega}{\beta}$.

The velocity of propagation of signals is different for different frequencies. If the input comprises signals of several frequencies, the output is not the replica of the input signal. This is delay or phase distortion. To eliminate ***delay distortion*** in the line, β should be a direct function of frequency; which in turn leads to

$$\frac{L}{C} = \frac{R}{G}$$

Then $$\alpha = \sqrt{(RG)}$$

17.4 THE REFLECTION COEFFICIENT

The reflection coefficient at the receiving end,

$$\mathbf{K} = \frac{\text{Reflected voltage at the receiving end}}{\text{Incident voltage at the receiving end}} = \frac{\mathbf{Z}_R - \mathbf{Z}_o}{\mathbf{Z}_R + \mathbf{Z}_o}$$

Therefore,

$$\mathbf{V} = \frac{\mathbf{V}_R(\mathbf{Z}_R + \mathbf{Z}_o)}{2\mathbf{Z}_o}(e^{\gamma x} + \mathbf{K}e^{-\gamma x})$$

$$\mathbf{I} = \frac{\mathbf{I}_R(\mathbf{Z}_R + \mathbf{Z}_o)}{2\mathbf{Z}_o}(e^{\gamma x} - \mathbf{K}e^{-\gamma x})$$

$$\mathbf{Z} = \mathbf{Z}_o\frac{e^{\gamma x} + \mathbf{K}e^{-\gamma x}}{e^{\gamma x} - \mathbf{K}e^{-\gamma x}}$$

$$\mathbf{Z}_s = \mathbf{Z}_o\frac{e^{\gamma l} + \mathbf{K}e^{-\gamma l}}{e^{\gamma l} - \mathbf{K}e^{-\gamma l}}$$

$$\mathbf{Z}_{oc} = \mathbf{Z}_o \coth \gamma l$$

$$\mathbf{Z}_{sc} = \mathbf{Z}_o \tanh \gamma l$$

17.5 EQUIVALENT CIRCUITS

T-section equivalent circuit :

Arms $\mathbf{Z}_A = \mathbf{Z}_B = Z_o \tanh\frac{\gamma l}{2} = \frac{\mathbf{Z}_l}{2}\frac{\tanh\frac{\gamma l}{2}}{\frac{\gamma l}{2}}$; where $Z_l = zl$

Pillar $\mathbf{Z}_C = \mathbf{Z}_o/\sinh\gamma l$. The corresponding admittance is $\frac{\sinh\gamma l}{\mathbf{Z}_c} = \mathbf{Y}_l\frac{\sinh\gamma l}{\gamma l}$

Π-section equivalent circuit :

Beam $Z_C = Z_o \sinh \gamma l = Z_l \dfrac{\sinh \gamma l}{\gamma l}$

Pillars $Z_A = Z_B = Z_o / (\tanh \dfrac{\gamma l}{2})$. The corresponding admittance is $\dfrac{Yl}{2} \dfrac{\tanh \frac{\gamma l}{2}}{\frac{\gamma l}{2}}$

17.6 THE LOSSLESS LINE, OR THE LINE AT RADIO FREQUENCIES

$$\gamma = j\omega \sqrt{(LC)} \quad \text{Therefore,} \quad \alpha = 0, \qquad \beta = \omega \sqrt{(LC)}$$

$$Z_o = \sqrt{\frac{L}{C}} \qquad v = \frac{1}{\sqrt{LC}}$$

$$V = V_R \cos \beta x + jI_R R_o \sin \beta x$$

$$I = jV_R \frac{\sinh \beta x}{R_o} + I_R \cos \beta x$$

$$V_{oc} = V_R \cos \beta x$$

$$I_{oc} = jV_R \frac{\sinh \beta x}{R_o}$$

$$V_{sc} = jI_R R_o \sin \beta x$$

$$I_{sc} = I_R \cos \beta x$$

In all the relations, β may be substituted by $\dfrac{2\pi}{\lambda}$ as and when found useful.

17.7 STANDING WAVE

A ***standing wave*** is one for which the points of minimum voltage, and of maximum voltage are fixed. ***Nodes*** are points of zero voltage, and ***antinodes*** are points of maximum voltage.

Standing wave ratio, $S = \left|\dfrac{V_{max}}{V_{min}}\right| = \left|\dfrac{I_{max}}{I_{min}}\right|$

$$V_{max} = \frac{V_R (Z_R + Z_o)}{2Z_R} (1 + |K|)$$

$$V_{min} = \frac{V_R (Z_R + Z_o)}{2Z_R} (1 - |K|)$$

$$S = \frac{1 + |K|}{1 - |K|}$$

$$|\mathbf{K}| = \frac{S-1}{S+1} = \frac{|\mathbf{V}_{max}| - |\mathbf{V}_{min}|}{|\mathbf{V}_{max}| + |\mathbf{V}_{min}|}$$

$$S = \frac{R_R}{R_o} \text{ (for } R_R > R_o)$$

$$S = \frac{R_o}{R_R} \text{ (for } R_R < R_o)$$

$$\mathbf{Z} = R_o \frac{\mathbf{Z}_R + jR_o \tan \beta x}{R_o + j\mathbf{Z}_R \tan \beta x} = \frac{\frac{\mathbf{I}_R (\mathbf{Z}_R + R_o)}{2}(e^{j\beta x} + \mathbf{K}e^{-j\beta x})}{\frac{\mathbf{I}_R (\mathbf{Z}_R + R_o)}{2R_o}(e^{j\beta x} - \mathbf{K}e^{-j\beta x})}$$

$$= R_o \left(\frac{\angle \beta x + |\mathbf{K}| \angle(\psi - \beta x)}{\angle \beta x - |\mathbf{K}| \angle(\psi - \beta x)} \right) = R_o \left(\frac{1 + |\mathbf{K}| \angle(\psi - 2\beta x)}{1 - |\mathbf{K}| \angle(\psi - 2\beta x)} \right)$$

$$R_{max} = R_o \left(\frac{1+|\mathbf{K}|}{1-|\mathbf{K}|} \right) = SR_o$$

$$R_{min} = R_o \left(\frac{1-|\mathbf{K}|}{1+|\mathbf{K}|} \right) = R_o/S$$

$$Z_{sc} = jR_o \tan \beta x = jR_o \tan \frac{2\pi x}{\lambda}$$

$$Z_{oc} = -jR_o \cot \beta x = -jR_o \cot \frac{2\pi x}{\lambda}$$

$$\frac{\mathbf{E}_{max}}{\mathbf{I}_{max}} = R_o$$

$$\frac{\mathbf{E}_{min}}{\mathbf{I}_{min}} = R_o$$

$$\frac{\mathbf{E}_{max}}{\mathbf{I}_{min}} = SR_o = R_{max}$$

$$\frac{\mathbf{E}_{min}}{\mathbf{I}_{max}} = \frac{R_o}{S} = R_{min}$$

17.8 SPECIAL FRACTIONAL WAVELENGTH LINES

The ***eighth-wave*** ($\lambda/8$) line :

$$\mathbf{Z}_s = R_o \frac{\mathbf{Z}_R + jR_o}{R_o + j\mathbf{Z}_R}$$

If $\mathbf{Z_R} = R_R$, then

$$\mathbf{Z}_s = R_o \frac{R_R + jR_o}{R_o + jR_R}$$

$$Z_s = R_o$$

Thus, any resistance R_R gets transformed into an impedance whose magnitude is R_o.

The ***quarter-wave*** ($\lambda/4$) line :

$$\mathbf{Z}_s = \frac{R_o^{\ 2}}{\mathbf{Z}_R}$$

The line acts as an ***impedance invertor.***

The ***half-wave*** ($\lambda/2$) line :

$$\mathbf{Z}_s = \mathbf{Z}_R$$

It acts as one-to-one transformer.

17.9 STUB MATCHING

Single stub impedance matching on a line :

The length of the short-circuited stub is given by

$$l = \frac{\lambda}{2\pi} \tan^{-1} \frac{\sqrt{S}}{S-1}$$

The distance where the stub is to be connected, from a point at which a voltage minimum existed before connecting the stub, is given by

$$d = \frac{\lambda}{4\pi} \cos^{-1}\left(\frac{S-1}{S+1}\right)$$

The stub should be connected towards the load end.

17.10 BASIS OF SMITH CHART

The reflection coefficient at a distance x from the receiving end is

$$\mathbf{K}_x = \mathbf{K}_R\, e^{-2\gamma x} = u + j\, v$$

The normalized impedance at a distance x from the receiving end is

$$\mathbf{z} = \frac{\mathbf{Z}_x}{R_o} = r + j\, x$$

$$r = \frac{1-u^2-v^2}{(1-u)^2+v^2}$$

$$x = \frac{2v}{(1-u)^2+v^2}$$

Constant r circles are governed by

$$\left(u - \frac{r}{r+1}\right)^2 + (v-0)^2 = \left(\frac{1}{r+1}\right)^2$$

The radius of the circle is $\frac{1}{r+1}$, and the centre is located at $\left(\frac{r}{r+1}, 0\right)$

Constant x circles are governed by

$$(u-1)^2 + (v - \frac{1}{x})^2 = \frac{1}{x^2}$$

The radius of the circle is $1/x$, and the centre is located at $(1, 1/x)$

The constant r circles and the constant x circles drawn together on the same set of u-axis, and v-axis constitute Smith chart.

The Smith chart is used for various purposes-like finding the VSWR (denoted by S also), $\boldsymbol{K}, \boldsymbol{Z_s}$, locating the voltage maximum on the line, designing the stub for impedance matching etc.

Section 4

ELECTRICAL ENGINEERING MATERIALS

Chapter **18**

ELECTRICAL ENGINEERING MATERIALS

18.1 GENERAL

To understand the physical properties of a material and the effect of various external variable parameters on its properties, one need to examine the material on atomic scale. Simplified atomic models suffice for the purpose. Bohr's model is considered acceptable.

The properties of a material are not simply the sum total of the properties of the atoms which make the material, but are influenced by the strong interaction between the atoms. The properties of a substance may be completely different from those of its constituent separate atoms.

The subject of the science of materials, or the electrical engineering materials provides a platform for studying the materials and their properties on atomic scale. Given below are some of the relationships which belong to the subject.

18.2 CONDUCTIVITY IN METALS

$$\mathbf{J} = \sigma \mathbf{E}, \qquad \sigma = 1/\rho$$

J is the current density, **E** is the electric field, σ is the conductivity of the metal, and ρ, the resistivity.

Average drift velocity of N electrons in x-direction at time t, $< v_x >_t = \dfrac{1}{N} \sum_{i=1}^{N} V_{xi}(t)$

$$J(t) = -ne <v_x>_t;$$

n is the number of free electrons per m^3, and e is the electronic charge.

$<v_x>_t = <v_x>_0 \, e^{-\frac{t}{\tau}}$; τ is the relaxation time of the electrons.

$$\mu_e = e\tau/m;$$

μ_e is the mobility of the electrons, *i.e.*, the average drift velocity of the electrons per unit field, and m is the electronic mass.

$$J = (ne^2 \tau/m)\, E$$
$$\sigma = ne^2\tau/m$$
$$\sigma(\omega) = \sigma_{static} / (1+ \omega^2\tau^2);$$

$\sigma(\omega)$ is the conductivity at angular frequency ω.

$$v_x(t) = -\frac{e\tau}{m} E_0 \left[\frac{\cos \omega t}{1+\omega^2\tau^2} + \frac{\omega\tau \sin \omega t}{1+\omega^2\tau^2}\right]; \quad \text{with } \mathbf{E} = E_0 \cos \omega t\, \mathbf{i}_x$$

$$\tau = \frac{\tau_c}{1 - <\cos\theta>};$$

τ_c is the average time between collisions, and $<\cos\theta>$ is the average value of the cosine of the scattering angles upon collision.

When $<\cos\theta> = 0$, then $\tau = \tau_c$

Mean free path $\lambda = v\, t_c$, where v is the total velocity of an electron.

The probability of a state corresponding to energy W being occupied by an electron is

$$f(W) = \frac{1}{e^{(W - W_F)/kT} + 1};$$

W_F is the ***Fermi energy*** level of the electrons, k is the Boltzmann constant $= 1.380 \times 10^{-23}$ Joule degree^{-1}, and T is the temperature in degrees Kelvin. At room temperature, $kT = 0.025\, eV$, and for $W_F = 5eV$, $v = 4 \times 10^6\, ms^{-1}$.

The ***Debye temperature*** is the measure for the maximum frequency of the lattice vibrations.

At temperatures above Debye temperature, the energy of the lattice vibrations increases linearly with T, and consequently, the resistivity increases linearly with T.

$\therefore \quad \rho = \rho_i + aT;$ this is ***Matthiessen's rule***.

18.3 THERMAL CONDUCTIVITY K

$Q = -K\, dT/dx$, $\quad Q : W/m^2$, $\quad dT/dx : {}^0C/m$, $\quad K : Wm^{-1}\,{}^0C^{-1}$

$$Q = -\frac{1}{3}\frac{n\pi^2k^2T\tau}{m}\, dT/dx,$$ τ depends upon temperature.

$$K = \frac{1}{3}\frac{n\pi^2k^2T\tau}{m}$$

For $T = 300\, K$, $n \approx 5\times10^{28}$, $\tau \approx 10^{-14}$, $k = 1.38 \times 10^{-23}$, $m = 9\times10^{-31}$, we find $K \approx 100$.

Above Debye temperature, $\tau \propto T^{-1}$.

Combining $\sigma = ne^2\tau/m$ and the expression for K, we get

$$\frac{K}{\sigma} = \frac{\pi^2}{3}\frac{k^2}{e^2}T \quad \text{or} \quad \frac{K}{\sigma T} = \frac{\pi^2}{3}\frac{k^2}{e^2} \approx \mathrm{L}$$

This is ***Wiedemann-Franz*** law.

The Lorenz number, $L = 2.45 \times 10^{-8}$ WΩ degree^2.

The Wiedemann-Franz law holds above Debye temperature only.

18.4 SUPERCONDUCTIVITY

Superconductivity is generally observed at temperatures below $10K$. Every Superconductivity material has a transition temperature T_C. The compounds, alloys or metals which are good conductors at room temperatures do not necessarily exhibit superconductivity. The superconducting compounds and alloys do not necessarily have superconducting components. The application of magnetic field can reduce the transition temperature.

$$H_C(T) = H_0(1 - T^2/T_C^2),$$

T_C is the transition temperature without the application of any magnetic field, H_0 is the critical magnetic field required to return the material from superconducting state to normal state at temperature $T = 0$.

The superconductivity gets destroyed by self magnetic field also. The critical current I_C which a superconducting wire of radius r may carry is given by $I_C = 2\pi r H_C$. This is the ***Silsbee's rule.***

In a superconductor, $\mathbf{B} = 0$. Since, $\mathbf{B} = \mu_0 (\mathbf{H} + \mathbf{M})$, and $\mathbf{M} = (\mu_r - 1)\,\mathbf{H}$, it follows that $\mathbf{M} = -\mathbf{H}$ and $\mu_r = 0$.

With increase in the mass of various isotopes of an element, the transition temperature T_C decreases in a manner such that

$$M^{1/2}\, T_C = \text{a constant; where M is the isotopic mass.}$$

18.5 SEMICONDUCTORS

$$\sigma = \frac{n_e e^2 \tau_e}{m_e} + \frac{n_h e^2 \tau_h}{m_h};$$

τ_e and τ_h are the relaxation times of the electrons, and the holes respectively; m_e and m_h are the effective masses of the electrons and the holes.

For intrinsic semiconductors,

$$\sigma_{\text{intrinsic}} = ne^2\left(\frac{\tau_e}{m_e} + \frac{\tau_h}{m_h}\right)$$

$$P(E) = e^{-W_g/2kT}; \quad n = N\, e^{-W_g/2kT};$$

W_g is the difference between the conduction level energy W_c and the valence level energy W_v.

$$n = A \text{ constant} \times e^{-W_g/2kT}$$

$$n = \text{Avogadro number} \times \text{density } (\text{kg}/m^3)/\,(\text{Atomic mass in amu} \times 10^{-3})$$

$$n = 5 \times 10^{21} T^{3/2}\, e^{-W_g/2kT}$$

$$\sigma_{\text{intr}} = \text{constant} \times e^{-W_g/2kT}$$

At room temperature, $kT \approx 0.025$, and for $Wg = 1\ eV$, we find $n \approx 5 \times 10^{16}$

For germanium,

$$\mu_e(T) = 3.5 \times 10^3\, T^{-1.67}, = 0.38 \text{ at the room temperature}$$

$$\mu_h(T) = 9.1\times10^4\, T^{-2.3}, = 0.18 \text{ at the room temperature}$$

For silicon, $\mu_e(T) = 4.0\times10^5\, T^{-2.6}, = 0.17$ at the room temperature

$$\mu_h(T) = 2.5\times10^4\, T^{-2.3}, = 0.035 \text{ at the room temperature}$$

For doped semiconductors,

$$n_e n_h = n_{\text{intrinsic}}^2$$

18.6 ELECTRON ORBIT RADII AND ENERGY

$$\frac{mv^2}{r} = \frac{e^2}{4\pi\varepsilon_0 r^2}$$

$$W = \frac{1}{2}mv^2 - \frac{e^2}{4\pi\varepsilon_0 r} = -\frac{e^2}{8\pi\varepsilon_0 r}$$

$$mvr = nh/2\pi, \quad r_n = \frac{\varepsilon_0 h^2}{\pi m e^2} n^2 = 0.529\times10^{-10} n^2 \text{ metre}$$

$$W_n = -\frac{me^4}{8\varepsilon_0^2 h^2}\frac{1}{n^2} = -\frac{13.6}{n^2} eV$$

$$h\nu = |W_{n1} - W_{n2}|$$

18.7 HALL COEFFICIENT

$$R_H = \frac{E_y}{B_z J_x}$$

$R_H = -1/en_e$ for n-type semiconductors, and

$R_H = 1/en_h$ for p-type semiconductors.

In general, $$R_H = \frac{1}{e}\frac{n_h\mu_h^2 - n_e\mu_e^2}{(n_h\mu_h + n_e\mu_e)^2}$$

Large Hall coefficients correspond to low carrier density, and hence to purer samples.

18.8 DRIFT CURRENT AND DIFFUSION CURRENT

$$J_h = n_h e\,\mu_h E - eD_h \frac{dn_h}{dx};$$

E is the applied electric field, and D_h is the diffusion constant of the holes.

$J_e = n_e e\,\mu_e E + eD_e \frac{dn_e}{dx}$; D_e is the diffusion constant of the electrons.

Einstein relation $D_h = (kT/e)\,\mu_h, \quad D_e = (kT/e)\,\mu_e$

The minority carriers should satisfy the following continuity equations.

$$\frac{dn_e}{dt} = \frac{n_{e_0} - n_e}{\tau_e} + \frac{1}{e}\frac{d}{dx}J_e \ ;$$

n_{eo} is the thermal equilibrium density of electrons.

$$\frac{dn_h}{dt} = \frac{n_{h_0} - n_h}{\tau_h} + \frac{1}{e}\frac{d}{dx}J_h \ ;$$

n_{ho} is the thermal equilibrium density of holes.

The total current through a forward biased junction is given by

$$J = (J_{e_0} + J_{h_0})(e^{eV/kT} - 1)$$

18.9 DIELECTRICS

$$\mathbf{D} = \varepsilon_0\varepsilon_r E, \qquad \mathbf{E} = \mathbf{D}/\varepsilon_0\varepsilon_r, \quad \varepsilon_r = C/C_{vacuum},$$

ε_r is the relative permittivity of the dielectric, C and C_{vacuum} are respectively, the capacitances with, and without the dielectric.

The dipole moment, $\mu = \sum_i^n Q_i \ \mathbf{r}_i$; Q_i are the charges such that ΣQ_i is zero, $\mathbf{r}_i$ is the vector drawn from the origin to the charge Q_i. The dipole moment is independent of the choice of the origin.

The ***polarization***, *i.e.*, the electric dipole moment per unit volume is

$$\mathbf{P} = \varepsilon_o(\varepsilon_r - 1)\,\mathbf{E}$$

$$\mathbf{D} = \varepsilon_o\mathbf{E} + \mathbf{P}$$

The induced dipole moment in an atom

$$\mu_{\mathbf{ind}} = 4\pi\varepsilon_o R^3\mathbf{E} = \alpha_e\mathbf{E},$$

where R is the electron cloud radius-being of the order of 10^{-10} m.

α_e is the electronic polarizability.

For monoatomic, or rare gases

$$\mathbf{P} = N\alpha_e\mathbf{E}, \qquad \varepsilon_0(\varepsilon_r - 1) = N\alpha_e; \quad N \text{ is the number of atoms per } m^3.$$

$$(\varepsilon_r - 1) = 4\pi NR^3$$

For polyatomic gases

$$\mathbf{P} = \mathbf{P}_e + \mathbf{P}_i + \mathbf{P}_o$$

P is the total polarization,

$\mathbf{P}_e$ is the electronic polarization,

$\mathbf{P}_i$ is the ionic polarization,

$\mathbf{P}_o$ is the orientational polarization.

$$\mathbf{P}_e = N\alpha_e\mathbf{E}, \quad \mathbf{P}_i = N\alpha_i\mathbf{E}, \quad \mathbf{P}_o = N(\mu_p^2/3kT)E,$$

α_i is the ionic polarizability, defined in a way similar to α_e, μ_p is the permanent electric dipole moment of a molecule, and N is the number of molecules per m^3.

$$\varepsilon_0(\varepsilon_r - 1) = N(\alpha_e + \alpha_i + \mu_p^2/3kT)$$

18.10 THE INTERNAL FIELD IN SOLIDS AND LIQUIDS

$$E_i = \frac{E}{1 - 1.2\alpha_e/\pi\varepsilon_0 a^3} = \frac{E}{1-\beta};$$

a is the atom to atom distance in a string of atoms, E is the applied or the external electric field, and E_i is the internal field or the field seen by the atom.

$$E_i = E + (\gamma/E_o)P$$

γ is an internal field constant. $\gamma = 1.2/\pi$ for the linear chain of atoms.

$$E_{i\text{ Lorenz}} = E + P/3\varepsilon_o$$

Foe ***elemental dielectrics***, *i.e.*, the materials built up from only one kind of atoms,

$$P_i = P_0 = 0, \quad \text{and} \quad P = N\alpha_e E_i, \quad \text{or} \quad P = N\alpha_e[E + (\gamma/\varepsilon_o)P]; \quad P = \frac{N\alpha_e E}{1 - (\gamma N\alpha_e/\varepsilon_0)}$$

$$\varepsilon_0(\varepsilon_r - 1) = \frac{N\alpha_e}{1 - (\gamma N\alpha_e/\varepsilon_0)}$$

For $\gamma = 1/3$, $(\varepsilon_r - 1)/(\varepsilon_r + 2) = N\alpha_e/3\,\varepsilon_o$

For ***ionic dielectrics*** without permanent dipoles

$$P = P_e + P_i$$

$$\varepsilon_0(\varepsilon_{rs} - 1)\,\mathrm{E} = P_e + P_i, \quad \text{and} \quad \varepsilon_0(\varepsilon_{re} - 1)E = P_e.$$

ε_{rs} is the static (at zero frequency) value of the relative dielectric constant, and ε_{re} is the relative dielectric constant at optical frequencies.

ε_{re} may be taken equal to n^2; n is the index of refraction.

In the absence of permanent dipoles, ε_r is almost independent of temperature.

ε_r decreases with temperature due to reduction in orientational polarization (when materials have permanent dipoles).

Ferroelectric dielectrics are characterized by spontaneous polarization. The *P-E* curve exhibits hysteresis characteristic, identical to that in *B-H* curve for ferromagnetic materials. The spontaneous polarization vanishes above a certain temperature, called the ferroelectric Curie temperature, θ_f.

$$\varepsilon_0(\varepsilon_r - 1) = dP/dE \text{ below } \theta_f$$

$$\varepsilon = C/(T - \theta)$$

above θ_f; where C is a constant, and θ is the ch. temp., usually a few degrees smaller than the ferroelectric Curie temp. θ_f.

Above is the **Curie-Weiss** law.

18.11 SPONTANEOUS POLARIZATION

$$E_i = E + (\gamma/E_o)P \quad \text{(stated earlier too)}$$

$$P = N\alpha E_i, \quad \text{or} \quad P = N\alpha\,[E + (\gamma/\varepsilon_o)P];$$

α is the total polarizability per unit (molecule).

$$\therefore \quad P = \frac{N\alpha E}{1-(\gamma N\alpha/\varepsilon_0)}$$

$$\frac{P}{E} = \frac{A}{1-A\beta};\ \beta = \gamma/\varepsilon_0 \text{ and } A = N\alpha$$

$$\frac{1}{N}\frac{dN}{dT} = -\lambda, \quad \varepsilon_r = \frac{1/\lambda\gamma}{T-\theta_r},\ T > \theta_f$$

λ is the coefficient of volume expansion.

18.12 DIELECTRICS IN ALTERNATING FIELDS

The decay of polarization on switching of the applied dc field is described by

$$P_0(t) = P_0 e^{-t/\tau};\ \tau \text{ is the relaxation time.}$$

And the growth of polarization is described by

$$P_0(t) = P_0(1 - e^{-t/\tau})$$

$$P_0(t) = \varepsilon_o\, Re[(\varepsilon_{ro}{}^* - 1)E_o\, e^{j\omega t}]$$

$\varepsilon_{ro}{}^*$ is the complex orientational part of the relative dielectric constant, meaning that there can be a phase difference between $P_0(t)$ and $E(t)$.

$$\varepsilon_{ro}{}^* - 1 = (\varepsilon_{ro} - 1)/(1 + j\omega\tau);$$

ε_{ro} is static value of the orientational part of the relative dielectric constant.

$$P_0(t)\ \frac{\varepsilon_0\,(\varepsilon_{ro}-1)}{1+\omega^2\tau^2}\,E_0\cos\omega t + \frac{\varepsilon_0\,(\varepsilon_{ro}-1)\omega\tau}{1+\omega^2\tau^2}\,E_0\sin\omega t, \text{ with } E = E_o\cos\omega t.$$

18.13 MAGNETICS

As we have

$$\mathbf{P} = \varepsilon_o(\varepsilon_r - 1)\,\mathbf{E} \text{ and}$$

$$\mathbf{D} = \varepsilon_o\mathbf{E} + \mathbf{P} \text{ in dielectrics, similarly we have,}$$

$$\mathbf{M} = (\mu_r - 1)\,\mathbf{H} = \chi\,\mathbf{H} \quad \text{and} \quad \mathbf{B} = \mu_o(\mathbf{H} + \mathbf{M}).$$

M is the ***magnetization***, μ_r is the relative permeability of the material, **H** is the magnetic field intensity, χ is the ***magnetic susceptibility***, **B** is the magnetic flux density, and μ_o is the permeability of free space. The magnetization is the magnetic dipole moment per unit volume.

D and *H* are considered as corresponding quantities, rather than *D* and *B*, in spite of the similar names of the latter two quantities.

18.14 MAGNETIC MOMENT OF A CIRCULAR BOHR ORBIT

The magnitude of the orbital magnetic dipole moment, μ_m is given by

$$|\mu_m| = \pi R^2 e\omega/2\pi = \tfrac{1}{2} e\omega R^2,$$

where ω is the angular velocity of the electron in the circular orbit of radius R.

The angular momentum of the electron is

$\mathbf{M}_a = \mathbf{R} \times m\,\mathbf{v}$, $\mathbf{M}_a = m\,\omega R^2$. $\mathbf{M}_a$ and μ_m have opposite directions.

$\mu_m = -(e/2m)\,\mathbf{M}_a$

The atomic unit of magnetic moment is ***Bohr magneton***.

1 Bohr magneton = $(e/2m)(h/2\pi) = 9.27\times10^{-24}$ ampere m^2.

The orbital magnetic dipole moment of an electron in an atom is of the order of 1 Bohr magneton.

Magnetic moment of a spherical charge cloud of an electron

$$\mu_m = -(1/5)\,e\,\omega R^2$$

When the material is subjected to a magnetic field **B**, and considering circular electron orbit,

$$\omega = \omega_o + (e/2m)\,\mathbf{B} \approx \omega_o + \omega_L\,;$$ ω_L is called the Larmor angular frequency.

$$\mu_{m,\,induced} = \mu_m - \mu_{mo} = -e^2 R^2\,\mathbf{B}/4\text{m}.$$

Similarly, considering the electron cloud model,

$$\mu_{m,\,induced} = -e^2 R^2\,\mathbf{B}/10\text{m}$$

18.15 CLASSIFICATION OF MAGNETIC MATERIALS

Class	*Permanent dipoles*	*Interaction between neighboring dipoles*
Diamagnetic	No	—
Paramagnetic	Yes	Negligible
Ferromagnetic	Yes	Parallel orientation
Antiferromagnetic	Yes	Antiparallel orientation of equal moments
Ferrimagnetic	Yes	Antiparallel orientation of unequal moments

(*a*) paramagnetic (*b*) ferromagnetic (*c*) antiferromagnetic (*d*) ferrimagnetic

FIGURE 18.1. Relative dipole orientation in magnetic materials.

Examples

Diamagnetic materials, χ negative : water, copper, bismuth

Paramagnetic materials, χ positive and small : Aluminium, sidium

Ferromagnetic materials, χ positive and large : Iron, nickel, cobalt

For diamagnetic materials, $\mu_r \approx 1$ to 10^{-5}

For paramagnetic materials,

$$M = (N_p - N_a)\beta; \qquad \beta = eh/4\pi m$$

$$N = N_p + N_a \qquad N_a/N_p = \exp(-2\mu_o \beta H/kT)$$

N_p : Number of dipoles per m^3 parallel to the field at temperature T

N_a : Number of dipoles per m^3 antiparallel to the field at temperature T

N : Number of spins per m^3 in the material

$$M \approx N\mu_o \beta^2 H/kT, \qquad \text{for } \mu_o BH << kT$$

$$\chi = \mu_r - 1 = N\mu_o \beta^2/kT \approx C/T$$

This is the ***Curie law of paramagnetism***. $C = N\mu_o \beta^2/k$ is the Curie constant.

For ferromagnetic materials, and at $T > \theta_f$, the ferromagnetic Curie temperature, we have the ***Curie-Weiss*** law.

$$\chi = \mu_r - 1 = C/(T - \theta); \qquad T > \theta_f$$

θ is the paramagnetic Curie temperature.

$$M/N\beta = M/M_{sat} = \tanh x; \qquad \text{where } x = \gamma\mu_o \beta M/kT$$

$$M/M_{sat} = M/N\beta = (kT/N\gamma\mu_o \beta^2)\,x = (T/\theta)\,x$$

The temperature at which x has maximum value (for antiferromagnetic materials) is known as ***Neel temperature***, T_N.

$$\chi = C/(T + \theta)$$

For high temperature regions,

$\chi = C/T$ for paramagnetic materials,

$\chi = C/(T - \theta)$ for ferromagnetic materials for $T > \theta_f$

$\chi = C/(T + \theta)$ for antiferromagnetic materials for $T > T_N$.

Section 5

INSTRUMENTS AND MEASUREMENTS

Chapter **19** *INSTRUMENTS*

19.1 GENERAL

True value of a quantity is the average of a very large number of measurements made of that quantity, committing no gross errors and using instruments known to have no systematic errors.

The true value cannot be found exactly; we can only estimate it using a suitable error minimization criterion.

Error is the deviation in the measured value of the variable from its true value.

Accuracy of an instrument is the closeness with which the instrument reading approaches the true value of the variable being measured.

An accuracy of p percent of a meter implies that the maximum error which may be present in any reading of the meter is $\pm (p/100)$ times the marked full scale value.

Precision of an instrument is a measure of the reproducibility of the measurements. For a high precision meter, if a given fixed quantity is measured repeatedly, the readings would be very close to each other.

A result of greater precision shall contain relatively larger number of significant figures.

Precision is the necessary condition for an accurate measurement.

Sensitivity of an instrument is the ratio of the change in the output or the response of the meter to the change in input or the measured variable.

Resolution is the smallest change in the measured value to which the instrument responds.

19.2 ERROR CALCULATIONS

The ***mean*** of a large number of measurement observations (assuming all known causes of errors have been removed, or accounted for) of a given fixed quantity is taken as the most probable value of that quantity. The error or the ***deviation*** of any observation from the mean is $d_i = x_i - x_{mean}$. The average of d_i would be zero. The ***standard deviation*** is computed as σ

$$= \sqrt{\frac{\sum_{i=1}^{i=n} d_i^2}{n}}$$, where n, the number of observations, is assumed to be very large.

If n is small, n in the above expression for σ is replaced by $(n - 1)$.

$$\sigma = \sqrt{\frac{\sum_{i=1}^{i=n} d_i^2}{n-1}}$$

The ***variance*** is the square of the standard deviation. $V = \sigma^2$.

The deviations are known to follow ***normal distribution*** or also called Gaussian distribution. The area under a probability density curve is taken as unity.

One thing it means is that the probability that the error in a measurement is $+d$ is the same as the probability that the error is $-d$. As compared to large errors, small errors are more probable. The probability that the error is between $\pm\varepsilon$ is given by the area under the curve between $-\varepsilon$ and $+\varepsilon$. The error or the deviation can be expressed in terms of σ. It is found that the areas under the curve between $\mp 0.6745\sigma$, $\mp 1\sigma$, $\mp 2\sigma$ and $\mp 3\sigma$ are respectively 0.5000, 0.6828, 0.9546 and 0.9972. It, for example, means that the probability that the error in a measurement is between $\mp 2\sigma$ is 95.46%, and that the probability that the error in a measurement is between $\mp 0.6745\sigma$ is 50%. The latter is taken as the probable error, denoted by r. Thus, the ***probable error***

$$r = \pm 0.6745\ \sigma.$$

When a quantity is found in terms of several other quantities such that $A = f(x, y, z, \ldots)$, then the probable error is given by

$$r = \sqrt{\left(\frac{\partial f}{\partial x}\right)^2 r_x^{\,2} + \left(\frac{\partial f}{\partial y}\right)^2 r_y^{\,2} + \left(\frac{\partial f}{\partial z}\right)^2 r_z^{\,2} + \ldots}$$

The other errors which we speak about are the limiting error, and the relative error. The limiting error is the maximum error, as quoted by the manufacturer, which may be present in the component value, or a reading. And the relative error is the ratio of the limiting error and the nominal value. $\varepsilon_{rx} = \Delta x/x$. Again, if $A = f(x, y, z, \ldots)$, the relative error in A is

$$\frac{\Delta A}{A} = \frac{1}{A}\left(\frac{\partial A}{\partial x}\Delta x \pm \frac{\partial A}{\partial y}\Delta y \pm \frac{\partial A}{\partial z}\Delta z \pm \ldots\right)$$

In the expression on the RHS, out of + and –, that sign is to be used which makes ΔA maximum. All the deviations are regarded additive.

19.3 DEFLECTING INSTRUMENTS

In general, the deflecting instruments of various types are designed and constructed to develop ***deflecting torque*** which is a function of the variable being measured. Usually, it is either a linear function, or a quadratic function of the variable. The instruments are provided to develop ***control torque*** generally proportional to the deflection angle; and ***damping torque*** proportional to the velocity of the moving/deflecting system.

19.4 MOVING COIL INSTRUMENTS

Moving coil instruments are used for measuring *dc* quantities.

The deflecting torque, $T_{\text{def}} = G\,i$,

where G is the deflection constant of the meter and equals BaN, and i is the current in the coil of the meter. Further, B is the flux density present in the air gap in which the coil moves, a is the area of the coil, and N is the number of turns in the coil.

The control torque, $T_C = C\theta$,

where C is the control constant, and θ is the deflection, expressed usually in radians.

The damping torque, $T_D = D\dfrac{d\theta}{dt}$, D being the damping constant.

The inertia torque, $T_{\text{inertia}} = J\dfrac{d^2\theta}{dt^2}$, J being the moment of inertia of the movement.

The torque balance equation is

$$J\frac{d^2\theta}{dt^2} + D\frac{d\theta}{dt} + C\,\theta = G\,i$$

With $i = 0$, the general form of the solution is $\theta = A\,e^{\lambda 1t} + B\,e^{\lambda 2t}$

For the under damped case, *i.e.*, for $D^2 < 4CJ$, the form of the solution is

$$\theta = U\,e^{-(D/2J)\,t}\sin(\omega_o t + \varphi), \text{ where } \omega_o = \sqrt{\left(\frac{C}{J} - \frac{D^2}{4J^2}\right)}$$

If damping is negligible,

$$f_o = \frac{1}{2\pi}\sqrt{\frac{C}{J}}, \quad T_o = 1/f_o.$$

The steady state deflection $\theta_s = (G/C)\,I_s$, where $i = I_s$, the steady current.

The complete solution for the response of a steady state galvanometer is

$$\theta = \theta_s\,[1 - e^{-(D/2J)t}\,\mathrm{cosec}\varphi\,\sin(\omega_o t + \varphi)], \text{ with } \tan\varphi = 2J\omega_o/D.$$

Ballistic galvanometers are used for measuring the amount of charge which passes through the coil of the instrument. The charge must have passed through it in a very short time. The flow of charge imparts a torque impulse GQ to the movement. The inertia is kept relatively high. During the actual motion, no current is flowing through the coil, *i.e.*, there is no deflecting torque. The solution for the deflection is

$$\theta = U\,e^{-(D/2J)\,t}\,\sin\omega_o t$$

The ratio of the successive peaks (for instance second to third) of deflections is $-\,e^{(D/2J)(\pi/\omega_0)}$.

$$Q = (J\omega_o/G)\,U.$$

$$Q = \frac{T_0}{2\pi}\frac{C}{G}\left(1+\frac{\lambda}{2}\right)\theta_1 = \frac{T_0}{2\pi}\frac{I_s}{\theta_s}\left(1+\frac{\lambda}{2}\right)\theta_1 = K_1\Delta\,\theta_1 = K_1\,\theta_o,$$

where the logarithmic decrement

$$\lambda = \frac{1}{n-1}Ln\frac{\theta_1}{\theta_n}, \qquad \Delta\ 1 + \lambda/2, \quad \text{and} \quad K_1 = \frac{T_o}{2\pi}\frac{I_s}{\theta_s}.$$

Vibration galvanometers are used as null point detectors for alternating currents of frequencies upto 1000 *c/s*.

These have no iron core, high control constant, and low inertia.

These can be tuned to the supply frequency either by adjusting the tension in the taut suspension, or by adjusting the effective length of the suspension.

For a given angular frequency ω, the amplitude of the resulting oscillations for the peak value of the current I_m, is

$$A = \frac{GI_m}{\sqrt{[D^2\omega^2 + (C - J\omega^2)^2]}}$$

When the galvanometer is tuned, $\omega = \sqrt{(C/J)}$, and $A = (GI_m/D)\,(J/C)$.

The ***multiplying power*** of a shunt of an ammeter is

$I/i = 1 + r/S$, where I is the external current, i is the current through the coil of the meter, r its resistance, and S is the resistance of the shunt.

A ***fluxmeter*** has negligible damping and control torque. Inertia too is kept small.

The flux is measured as change in the flux linkages of a search coil of N_{sc} turns. The search coil is connected to the fluxmeter terminals. The current in the circuit comprising the search coil and the fluxmeter coil (in series) is given by

$$i = \frac{N_{sc}\dfrac{d\phi}{dt} - k\dfrac{d\theta}{dt} - L\dfrac{di}{dt}}{R};$$

L, and R are respectively the circuit inductance and resistance.

The equation of motion is

$$J\frac{d^2\theta}{dt^2} = G\,i$$

The flux change of the search coil is then found as

$$N_{sc}(\Phi_2 - \Phi_1) = K(\theta_2 - \theta_1)$$

If flux values are high, a shunt (a low resistance) can be used, which is connected across the meter. In that event,

$$N_{sc}(\Phi_2 - \Phi_1) = K(\theta_2 - \theta_1)(R_{sc} + R_{sh})/R_{sh}.$$

The ***ratiometers*** have two coils mounted on the same spindle, the angle between the two coils being 90°. There are no control springs. For the ratiometers, we have

$$G_1 I_1 \cos\theta = G_2 I_2 \cos(90° - \theta), \quad \text{or} \quad \tan\theta = G_1 I_1 / G_2 I_2,$$

thus the deflection θ is proportional to the ratio I_1 / I_2.

The ratiometer principle is used in some ***frequency meters*** wherein the deflection θ is proportional to the frequency f; and in ***meggers*** wherein θ is proportional to the unknown high resistance, R.

19.5 MOVING IRON INSTRUMENTS

Moving iron instruments are simple, cheap, robust, and can be used in *dc*, as well as *ac* circuits.

These can be either ***repulsion*** type, or ***attraction*** type. Both types have a stationary coil which carries the current proportional to the quantity being measured. There is a piece of iron (mumetal), suitably shaped, which is attached to one end of an arm, the other end of the arm being attached to the spindle. When the coil carries current, a force acts on this piece of iron, and it moves and causes the spindle to turn, and hence the pointer to deflect. Under the assumption of linear $B - H$ curve, the ***deflecting torque*** in the repulsion type (which has two pieces of iron, one being fixed) of instrument is proportional to

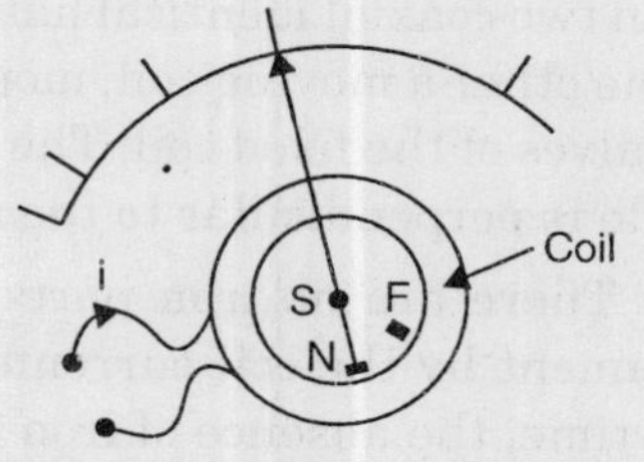

FIGURE 19.1. The angular displacement due to repulsion between the moving iron and the fixed iron.

$$\frac{i^2 \cos\frac{\phi}{2}}{\sin^2\frac{\phi}{2}}, \text{ where } \varphi \text{ is the angular displacement between the irons.}$$

The essential point is that the torque is proportional to i^2, and the mean torque is thus proportional to the square of the rms value of the current through the coil. The *MI* instruments can, therefore, be used in *dc*, as well as ac circuits.

The scale of the meter is not uniform and is exceedingly cramped near the zero marking. By shaping the moving iron suitably, near uniform scale is achieved.

With the movement of the moving iron, the inductance of the coil changes-though by a small amount. The ***alternative torque expression*** is

$$\frac{1}{2} i^2 \frac{dL}{d\theta}.$$

The $\frac{dL}{d\theta}$ itself is dependent on the deflection angle (θ).

Unlike the coils of PMMC instruments, the coils of *MI* instruments (being stationary) can be wound for currents upto as high as 50 A. All the same, if a shunt is required, it has to be ensured that the coil and the shunt have the same time constant, *i.e.*, the same *L*/*R* ratio so that the current division is independent of the frequency.

For voltmeters, resistance multipliers can be used. To make a voltmeter nearly frequency independent over a small range of frequency, we connect a capacitor of value $C = 0.41\ L/R^2$ across the series resistance *R*, where *L* is the inductance of the coil of the meter.

MI instruments are used largely as ammeters and voltmeters. The principle of MI instruments can, however, also be used to measure power factor in balanced three-phase circuits. MI type synchroscopes also are available.

19.6 ELECTODYNAMIC INSTRUMENTS

An electodynamic instrument essentially consists of two coils, one of them being a fixed coil, split in two coaxial identical halves which are parallel to each other and a short distance apart; and the other a moving coil, mounted on the spindle and symmetrically located in between the two halves of the fixed coil. The axis of the moving coil is perpendicular to the spindle, and the spindle is perpendicular to the axis of the fixed coil.

There are no iron parts in the magnetic circuit. As such, the flux produced in the instrument by the coil currents is quite small. This causes poor torque/weight ratio. At the same time, the absence of iron parts make these instruments relatively error free, and these instruments are considered as one of the most accurate class of instruments. Also, these instruments are most costly type of instruments.

When in use, both the coils carry current.

The ***deflecting torque*** developed is given by

$$i_1 i_2 \frac{\delta M}{\delta \theta},$$

where *M* is the mutual inductance between the moving coil and the fixed coil, and θ is the angle between the moving coil axis and the fixed coil axis. *M* can be expressed as $-M_m \cos\theta$.

As such, $T = i_1 i_2 M_m \sin\theta.$

Because of the presence of $\sin\theta$, the scale will be uniform in the middle, but will close up near the ends.

The *ED* instruments can be made ammeters, voltmeters, and wattmeters.

In ***ammeters***, the coils are connected in series, in ***voltmeters*** also both the coils are in series, but there is a high series resistance also. In ***wattmeters***, the fixed coil constitutes the current coil and is connected in series with the load, and the moving coil having a high resistance in series forms the voltage coil, and is connected across the load.

The *ED* instruments can be used in *dc* as well as *ac* circuits.

In wattmeters, the current coil terminals are usually marked as *M* and *L*; and the voltage coil terminals are marked as Com and *V*. *M* is for the 'mains', and *L* for the 'load'. Com is the voltage coil terminal which is made common with one of the current coil terminals. Com can be made common with either *M* or *L*. If Com is connected to *M*, the wattmeter reading is higher than the load power by an amount equal to the loss in the current coil, and if it is connected to *L*, the reading is higher than the load power by an amount equal to the loss in the voltage coil.

It is assumed that the current in the voltage coil is in phase with the voltage across it; however, the coil inductance *L* causes this current to lag the voltage by a small angle α. This necessitates a correction factor to be applied to the wattmeter reading at very low power factor measurements.

The correction factor is $\dfrac{\cos\phi}{\cos\alpha\cos(\phi-\alpha)}$. At poor load power factors, the correction factor assumes significance.

In wattmeters meant for low power factor use, a capacitor of value $\boldsymbol{C = L/R^2}$ is connected across the high series resistance R. This makes the voltage coil circuit practically non-inductive.

Under *ac* conditions, the wattmeter reading is

W = (*rms* value of the voltage across the voltage coil) × (*rms* value of the current in the current coil) × cosine of the angle φ between the voltage phasor $V_{Com,\,V}$ and the current phasor I_{ML}.

If a wattmeter deflects negative, one can make it deflect up the scale by reversing the connections of either the current coil or the voltage coil. By doing so, we make angle φ (which was earlier > 90°) 180° – φ (which is now < 90°).

The ***two-wattmeter method*** is a common method of measuring power in three-phase circuits—for balanced, as well as for unbalanced circuits. The current coils of the wattmeters carry currents of any of the two lines, and the '*V*' terminals of the two voltage coils are connected to the third (leftover) line. The total three-phase power to the circuit is the sum of the readings of the two wattmeters, W_1 and W_2. For balanced conditions, the treatment is simple. One of the wattmeters reads proportional to cos (30° – φ), and the other reads proportional to cos (30° + φ), φ now is the load power factor angle. The sum of the readings is $\sqrt{3}V_L I_L \cos\varphi$. If the load power factor is 0.5, *i.e.*, φ = 60°, one wattmeter (say, W_2) reads zero. And if φ > 60°, the same wattmeter tends to read negative. We can make it read positive by reversing the connections of the voltage coil (or the current coil). Then the total three-phase power is $W_1 - W_2$.

We also see that the difference of the readings (normal connection) $W_1 - W_2$ is $V_L I_L \sin\varphi$.

The two wattmeters can be combined into one to yield a ***three-phase wattmeter***, in which there is a common control spring, a common spindle having two electrodynamic elements. The torque on the spindle is the algebraic sum of the torques developed by the two elements.

The electrodynamic principle can also be applied to develop single-phase, or three-phase ***power factor meters***.

In these power factor meters, there are two moving coils mounted on the same spindle, the coils being at an angle of nearly 90° to each other in single-phase meter, and at an angle of nearly 120° to each other in three-phase meters. There are no control springs. The fixed coil carries the load current.

In the single-phase power factor meter, one moving coil has a high series resistance and the other has a high series inductance. Both the coils are connected across the supply terminals, which also supplies the load. The resistance and the inductance are so chosen that the two currents have the same magnitude. Also, the currents have a phase difference of nearly 90°.

The theory shows that the deflection of the moving coils of this arrangement is the load power factor angle φ.

In the three-phase power factor meter, the two moving coils have identical high series resistances in their circuits. If the fixed coil carries the current of line R, then the moving coils (along with their high series resistances) are connected across R-Y, and across R-B. The two moving coil currents thus have a phase difference of nearly 120°. Again, the theory shows that the deflection of the movement is a measure of the load power factor angle φ. The load is assumed to be balanced.

19.7 ELECTROSTATIC INSTRUMENTS

Electrostatic instruments depend on the force of repulsion or attraction exerted between the charged surfaces. These forces are much smaller in comparison to the magnetic forces. To develop any significant force, high voltages are necessary. As such, these instruments are used for measurements at high voltages, of the order of kilo-volts. We have electrostatic ***voltmeters*** and electrostatic ***wattmeters***.

The force $F = \frac{1}{2} V^2 \frac{\delta C}{\delta d} = -\frac{a\varepsilon_0 V^2}{2d^2}$; C is the capacitance, d is the distance between the parallel surfaces, and a is the area of the plates. The – sign implies that the force is in a direction to decrease d.

If we have a turning arrangement, then

$$\text{The Torque} = \frac{1}{2} V^2 \frac{\delta C}{\delta \theta}.$$

To develop increased deflecting torque, usually, several double-ended vanes mounted on the common spindle, and capable of rotation in an equal number of quadrants are used. Such a construction is called multi-cellular construction. Each double-ended vane and the corresponding pair of quadrants give rise to four capacitances. If there are n vanes, there shall be $4 \times n$ capacitances, and the deflecting torque is increased in the same ratio.

From protection point of view, a high resistance is usually connected in series with the meter.If the voltages to be measured are high, we may use a resistance or a capacitance potential divider.

For high voltages, suitably shielded ***resistance potential divider*** can be used. The multiplying factor becomes R/r, where R is the resistance of the whole resistance chain, and r is the resistance of the portion of the divider across which the meter is connected. The power loss becomes high at high voltages.

Under *ac* conditions, the accuracy of the multiplier is affected due to stray capacitances.

For measuring high *ac* voltages, a ***capacitance potential divider*** can also be used. The multiplying factor is C_{meter}/C_{chain}, C_{meter} being the capacitance of the section across which the meter is connected, and C_{chain} the capacitance of the whole capacitor chain. Here too, stray capacitances and leakage resistances are present and introduce error in the multiplier. A capacitance potential divider should not be used for *dc* measurement as then the potential division is governed by the leakage resistance chain, rather than by the capacitor chain.

The electrostatic instruments are free from all the troublesome magnetic errors. They may be used on both *dc* and *ac* circuits. The indication is the rms value, whatever be the wave form. These draw very little power from the source.

These, however, have low torque/weight ratio, are expensive, and are not very robust.

19.8 QUADRANT ELECTROMETER

In a ***quadrant electrometer***, there are four hollow quadrants Q_1, Q_2, Q_3, and Q_4, within the hollow of which a suspended double-ended vane, positioned symmetrically, can turn. The quadrants are connected together in diagonally opposite pairs (Q_1 to Q_3, and Q_2 to Q_4). One of the pairs (say $Q_1 - Q_3$) is at ground potential. In one type of connection, called ***idiostatic connection***, the vane is connected to the pair $Q_2 - Q_4$, and the voltage V to be measured is applied across the quadrants Q_2 and Q_1. The vane and the quadrants Q_2 and Q_4 are at the same potential. There is, therefore, no interaction between the two. There is an interaction between the vane and the quadrants Q_1 and Q_3. The double-ended vane and the quadrants Q_1 and Q_3 form four capacitors. The deflecting torque is given by

$$T = \frac{1}{2}V^2 \frac{\partial}{\partial\theta}\left(\frac{2\varepsilon_0 r^2\theta}{d}\right) = \frac{\varepsilon_0 r^2}{d}V^2;$$

where θ is the angle of overlap between the vane and the quadrant Q_1, and r is the radius of the vane.

T is seen to be proportional to V^2.

In another connection, called ***heterostatic connection***, instead of connecting the vane to the pair $Q_2 - Q_4$, it is kept at a high *dc* potential V_1. The voltage to be measured is again to be applied across the quadrants Q_2 and Q_1.

The net deflecting torque is

$$T = \frac{\varepsilon_0 r^2}{d}V(2V_1 - V) \approx 2\frac{\varepsilon_0 r^2}{d}V_1 V.$$

T is seen to be proportional to V.

The quadrant electrometer is a laboratory type of meter, with the help of which other electrostatic meters can be calibrated.

19.9 ELECTROSTATIC WATTMETER

The power at high voltages can be measured with the help of the following circuit which uses a quadrant electrometer and a resistance R. The pair $Q_1 - Q_3$ is connected to A, and the pair $Q_2 - Q_4$ to point B. A voltage v_1, equal to $1/n$ times the supply voltage, is tapped from the supply at point N on the potential divider, and is applied to the vanes. There is a small resistance r in series with the circuit, whose value is $(n - 2)\ R/2$. If $n = 2$, r becomes zero, *i.e.*, r is not needed. Also, n should not be less than 2. The load voltage and the load current are v_L and i_L respectively.

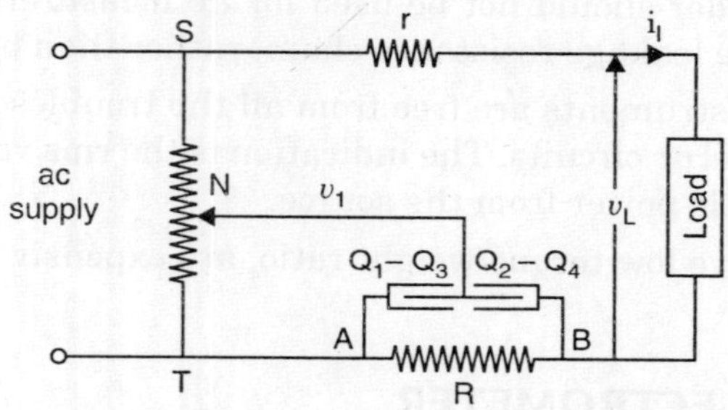

FIGURE 19.2. Measuring load power using quadrant electrometer.

Applying the theory of the electrostatic instruments, we find that the instantaneous torque is proportional to $(2R/n)\ v_L i_L$. Thus, the mean torque is proportional to the load power.

The arrangement is particularly suitable for measuring power loss in dielectrics at high voltages, and for measuring iron losses.

19.10 INDUCTION INSTRUMENTS

The action of induction instruments depends on the production of induced alternating currents by alternating fluxes and the interaction between the currents and the fluxes. The instantaneous torque acting on the disc in which the *ac* magnets 1 and 2 induce currents is given by

$$T = (Kr/R)\ \omega\Phi_{1m}\ \Phi_{2m}\ \sin\alpha\ ;$$

K is a constant, r is the perpendicular distance between the spindle of the disc and the line joining the two magnets, R is the resistance of the induced current path, Φ_{1m} and Φ_{2m} are the peak values of the fluxes of magnets 1 and 2 linking circular areas centered at the magnets and of radius equal to the distance between the two magnets, ω is the angular frequency of pulsation of the flux, and α is the angle by which the flux

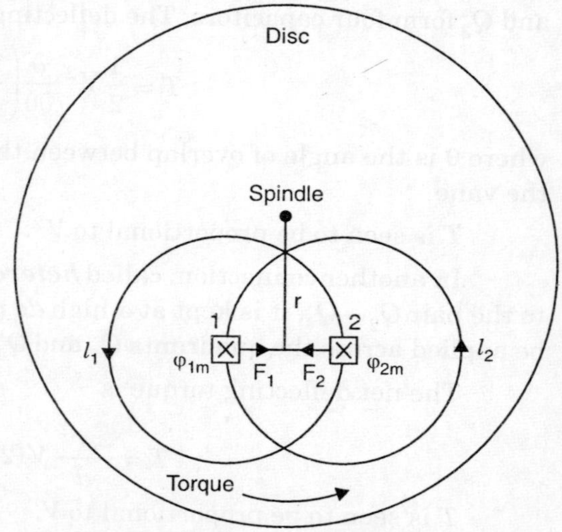

FIGURE 19.3. Fluxes, currents, forces, and the torque in a disc type induction instrument.

Φ_1 leads the flux Φ_2. It is assumed that the two *ac* magnets have equal radial distance from the spindle of the disc. The torque tends to turn the disc from magnet of leading flux toward the magnet of lagging flux.

In this class of instruments, we have ammeter, voltmeter, and wattmeter.

In all the induction instruments some arrangement to cause a phase difference between the two fluxes has to be incorporated. This may be achieved by ***phase splitting***, or by ***shading*** the pole.

In an ***ammeter***, the coils of the two magnets are connected in series; and the coil of magnet 2 is shunted by a resistance thereby introducing phase difference between the two coil currents, and hence between the fluxes produced by them.

In the ***shaded pole*** type, there is actually only one magnet which is split into two parts, one part having a shading ring (a short-circuited few-turn coil). The phase difference between the flux Φ_1 under the unshaded part of the pole and the flux Φ_2 under the shaded part, can be seen in the phasor diagram.

Due to the reluctance of the portion between B and B', there is a magnetic potential drop of $\Phi_2 R_2$ which lies along Φ_2 itself. The current I_2 which tends to set up a flux in the same direction as the flux causing it (Φ_2), lags the flux Φ_2 by an angle somewhat greater than 90°. The magnetic potential difference between B and B' is the phasor sum of $\Phi_2 R_2$ and $-N_2 I_2$. Let it be OE. The portion between A and A', and that between B and B' are in parallel. Therefore, the magnetic potential difference across $A - A'$ is the same as that across $B - B'$. The flux Φ_1, therefore, lies along OE. The total flux Φ is the phasor sum of Φ_1 and Φ_2. The net mmf of the main coil, NI, is the phasor sum of OE and the magnetic potential drop ΦR, R being the reluctance of the rest of the pole body in which flux Φ exists.

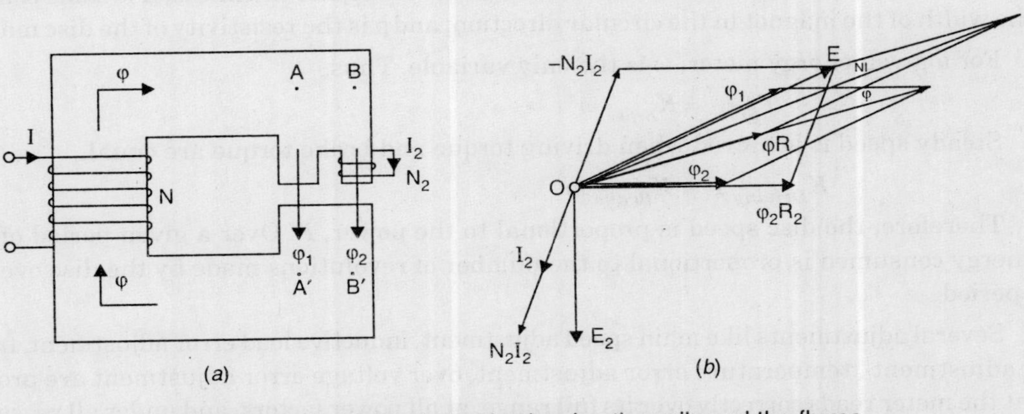

FIGURE 19.4. *(a)* Shaded-pole showing the coils and the fluxes.
(b) The phasor diagram of fluxes and currents.

The ***voltmeter*** magnet coils have relatively large number of turns of fine wire. The phase splitting is achieved by connecting a high series resistance in the circuit of one magnet, and an inductive coil in series in the circuit of the other magnet. The two circuits are then put across the voltage to be measured.

In the induction ***wattmeter***, the coil of one magnet carries the load current I and produces flux Φ_1 which is in phase with the current, and the circuit of the other magnet is made highly inductive to which is applied the load voltage. The flux Φ_2 produced by this magnet, therefore, lags the voltage by nearly 90°. The angle between the two fluxes is thus $(90° - \varphi)$, φ being the load power factor angle. Thus, the torque is proportional to $\omega(V/L\omega)I \sin(90° - \varphi)$, *i.e.*, $\propto VI \cos \varphi$, the load power. The wattmeter deflection is independent of frequency.

As in the case of electrodynamic wattmeters, we have three-element or two-element three-phase wattmeters, here also we have three-element or ***two-element three-phase wattmeters***.

The induction instruments have long scales. Full scale deflection is about 300°. These are, however, prone to frequency and temperature errors.

Energy meter, though not an indicating meter, is also an induction type meter. It is an integrating meter that integrates power wrt time, and hence measures energy (usually in *kWh*) over a period of time. It develops the ***driving torque*** in the same way as an induction wattmeter develops the deflecting torque. Since, it is an integrating meter, the disc rotates, and not merely turns. There are no control springs, but there is an arrangement to provide ***brake torque***. The brake torque is developed with a horse-shoe magnet or a pair of magnets, within the field of which the disc rotates.

The brake torque is given by

$$T_{Brake} = \frac{\Phi^2 r^2 \omega t}{K b^2 \rho},$$

Φ is the flux of the brake magnet, r is the radial distance of the magnet from the axis of the disc, ω is the angular speed of rotation of the disc, t is the disc thickness, K is some constant, b is the width of the magnet in the circular direction, and ρ is the resistivity of the disc material.

For a given energy meter, ω is the only variable. Thus,

$$T_{Brake} = K_{Brake}\ \omega$$

Steady speed is achieved when driving torque and brake torque are equal.

$$\boldsymbol{K_{Driving}\ P = K_{Brake}\ \omega}$$

Therefore, the disc speed is proportional to the power, P. Over a given period of time, the energy consumed is proportional to the number of revolutions made by the disc over that time period.

Several adjustments like main speed adjustment, inductive load error adjustment, friction error adjustment , temperature error adjustment, over voltage error adjustment are provided so that the meter reads correctly over its full range, at all power factors, and under all reasonable variations of temperature and voltage.

As we have two-element three-phase wattmeter, we also have two-element three-phase energy meter to record three-phase energy consumption.

As has already been noted, the difference of the two wattmeter readings under balanced load condition is $V_L I_L \sin \varphi$. Thus, by reversing the connection of the current coil of the second element of the two-element three-phase wattmeter and allowing for a factor of 3, we can arrange to measure the reactive power in a balanced three-phase circuit (= $\sqrt{3}\ V_L I_L \sin \varphi$). If instead of the control torque, we provide the brake torque, we then can record reactive energy consumed, kV Arh.

An energy meter should be tested for the following conditions :

1. at the highest current at which the accuracy is guaranteed,
2. at one intermediate current,
3. at the lowest current at which the accuracy is guaranteed, usually 5% of marked current,
4. at 0.5% of marked current to ensure that the meter runs continuously at the lowest load,
5. at the marked current at 0.5 lagging power factor, and
6. at zero current, with 10% excess voltage applied to the voltage magnet to ensure that there is no creep.

Tests 1–4 are performed at unity power factor.

All the tests, except test 6, are performed at the rated voltage.

Chapter 20

MEASUREMENT METHODS

20.1 GENERAL

We are required to measure several electrical parameters like resistance, inductance, quality factor, capacitance, loss angle, frequency etc. The ranges of their values are also very wide. One may estimate the values of impedances by simply applying the Ohm's law. However, the loading effects of the instruments, the instrument errors may cause the measurement to be inaccurate. The comparison method and the ***bridge methods*** on the other hand are free from instrument errors and are considered superior. These give more accurate results provided the components which are used in the measurement are of high quality and their values are accurately known.

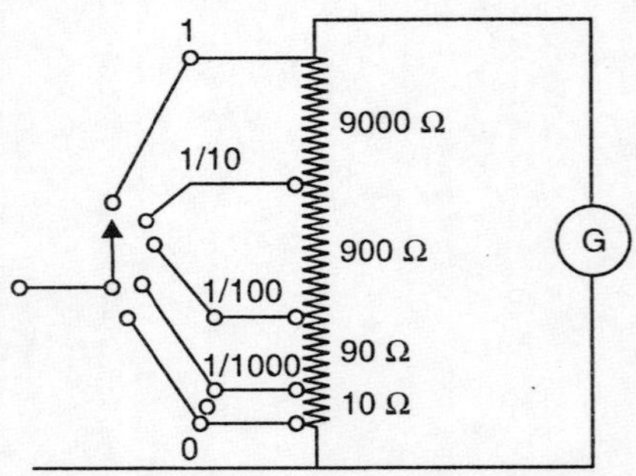

FIGURE 20.1. Universal shunt.

In ***dc bridge methods***, a sensitive galvanometer, in conjunction with a universal shunt is used as a null point detector. A ***universal shunt*** enables one to use low or high sensitivity in the null detection circuit. In the initial stages of balancing the sensitivity should be set low so that the galvanometer may not get damaged by likely high out of balance current.

The sensitivity of the circuit increases as the tap position is changed from the bottom most stud to the top most stud. In this action, the resistance in the *G* branch is successively decreased.

20.2 DC POTENTIOMETER

A ***dc potentiometer*** is a very reliable and accurate method of measuring small voltages and currents (indirectly, using a low standard resistance). Higher voltages can also be measured using a resistance potential divider or a volt-ratio box. The potentiometer essentially compares the unknown voltage with known voltage drop in the potentiometer resistance coils and the slide wire (which are in series). The coils and the wire carry a predetermined fixed current so that the meter is direct reading. This current has to be set first, *i.e.*, the meter is standardized, with the help of a standard emf cell.

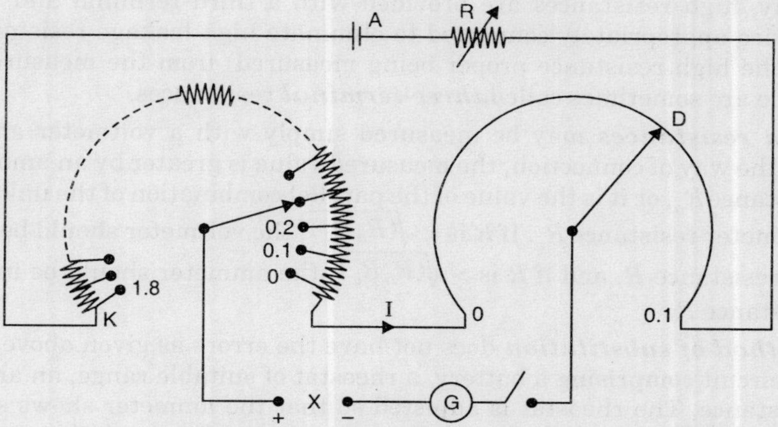

FIGURE 20.2. dc potentiometer circuit.

To improve the resolution and to do away with the possibility of non-uniformity, wear and tear of the slide wire, a noble arrangement of resistance coils, known as Kelvin-Varley arrangement can be used. The rotary switches which change the stud positions must be of very good quality. If I = 4 mA, then for the contacts positions shown the unknown X is 0.428 V.

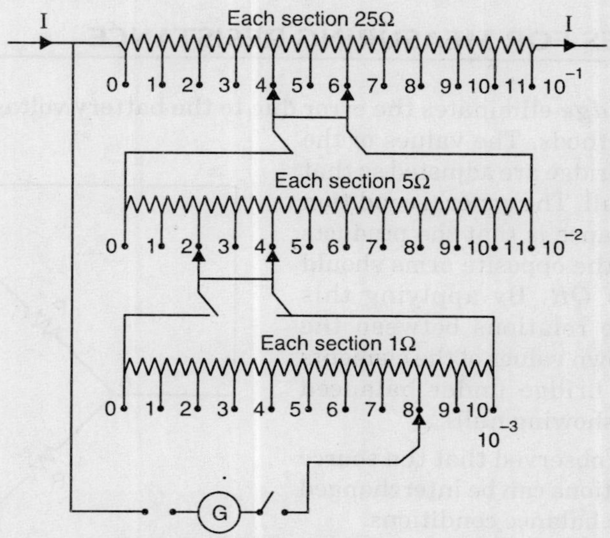

FIGURE 20.3. Kelvin-Varley arrangement.

20.3 MEASUREMENT OF RESISTANCE

Resistances may be classed as ***low*** (< 1 ohm), ***medium***, or ***high*** (> 10^6 ohm). Low resistances meant for use in measurement circuits have ***four terminals*** - two for current, and two for potential. By doing so, the resistance between the potential terminals is rendered definite and the small resistance which may be introduced in the potential circuit at the point of connection of the current leads with the resistance proper is then absent.

Similarly, high resistances are provided with a third terminal and when these are measured, a wire appropriately connected to eliminate high leakage resistance (which is in parallel with the high resistance proper being measured) from the measurement circuit is used; and hence are sometimes called ***three-terminal*** resistances.

Medium resistances may be measured simply with a voltmeter and an ammeter. Depending on the way of connection, the measured value is greater by an amount equal to the ammeter resistance R_a, or it is the value of the parallel combination of the unknown resistance R and the voltmeter resistance R_v. If R is $< \sqrt{(R_a R_v)}$, the voltmeter should be in parallel with the unknown resistance R, and if R is $> \sqrt{(R_a R_v)}$, the ammeter should be in series with the unknown resistance R.

The ***method of substitution*** does not have the errors as given above. The set-up is a simple series circuit comprising a battery, a rheostat of suitable range, an ammeter, and the unknown resistance. The rheostat is adjusted so that the ammeter shows some convenient reading. Next, the unknown resistance R is substituted by a suitable variable standard resistance S, and now this variable standard resistance is adjusted so that the ammeter shows the same reading. The unknown resistance R then equals this value of the standard resistance. It is assumed that in the course of measurement, the battery emf and the rheostat setting have not changed.

20.4 DC BRIDGES FOR MEASURING RESISTANCE

The ***Wheatstone bridge*** eliminates the error due to the battery voltage variation. All bridge methods are null methods. The values of the variable arms of the bridge are adjusted so that the detector shows null. The general condition for achieving the balance is that the products of the resistances of the opposite arms should be equal, *i.e.*, $PS = QR$. By applying this condition, we derive relations between the unknown and the known values of the elements in the arms of the bridge under balanced conditions (detector showing null).

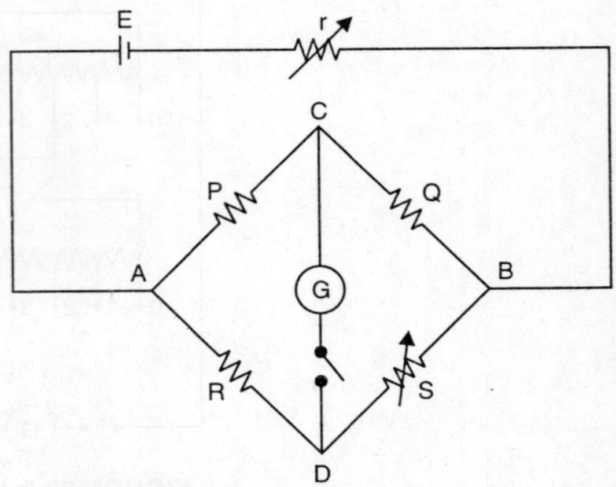

FIGURE 20.4. Wheatstone bridge.

It may also be observed that the source and the detector positions can be interchanged without affecting the balance conditions.

For the Wheatstone bridge, the unknown resistance is given by $R = (P/Q)\,S$. P and Q are known as the ratio arms. The bridge sensitivity is high if P and Q are of the same order as R and S. However, P and Q may have to be set quite different from this condition in order to increase the bridge measurement range and the precision. The thermoelectric *emf* errors can be eliminated by reversing the battery connection and taking the mean of the two measurements.

The ***Carey-Foster bridge*** measures the difference between two nominally equal resistances, and the unknown resistance is then determined very accurately. P and Q are set at equal values.

Two measurements are carried out, one as shown, and the other with the positions of R and S interchanged. P and Q remain at their equal values. The two null point positions are at L, and at M respectively. Then

$R–S = r_2 – r_1$. This requires the knowledge of resistance per unit length of the slide wire.

$R–S = r(l_2 – l_1)$, where r is the resistance per unit length of the slide wire.

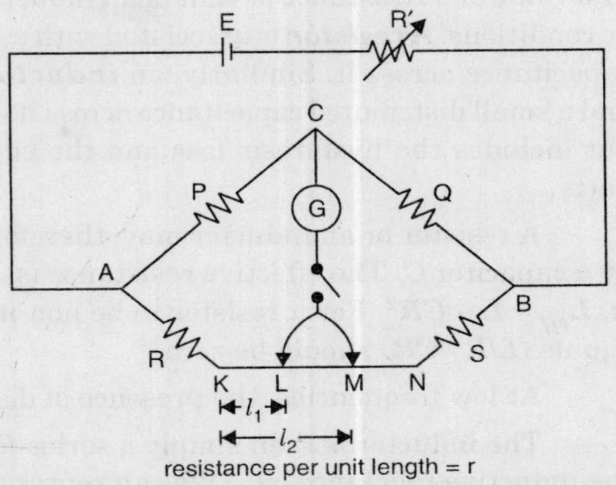

FIGURE 20.5. Carey-Foster bridge.

For measuring very low resistances, ***Kelvin's double bridge*** is used, which is free from errors due to contact resistance, and lead resistances. The circuit diagram is shown in Fig. 20.6.

We call it double bridge as there are two sets of ratio arms – P and Q and p and q. We set $P = p$, and $Q = q$, and to achieve the balance conditions, P and p are varied simultaneously, always keeping $P = p$. The resistance r of the connecting link is kept very small. In general, the value of the unknown resistance is given by

$$R = \frac{P}{Q}S + \frac{qr}{p+q}\left(\frac{P}{Q} - \frac{p}{q}\right)$$

If r is nearly zero, or $\frac{P}{Q} = \frac{p}{q}$, the second term disappears.

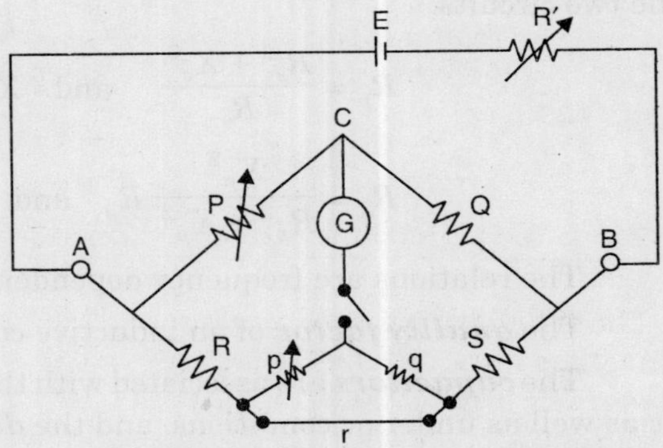

FIGURE 20.6. Kelvin's double bridge.

20.5 LOSS OF CHARGE METHOD FOR MEASURING HIGH RESISTANCE

For measuring very high resistances, ***loss of charge*** method can be used, the circuit diagram for which is shown in Fig. 20.7.

It is based on the fact that the charge on the capacitor plates, or the voltage across the capacitor plates decays as $v(t) = V_o e^{-t/RC}$.

The capacitor voltages are noted at suitable points of time after disconnecting the dc supply, and $v(t)$ is plotted to find RC.

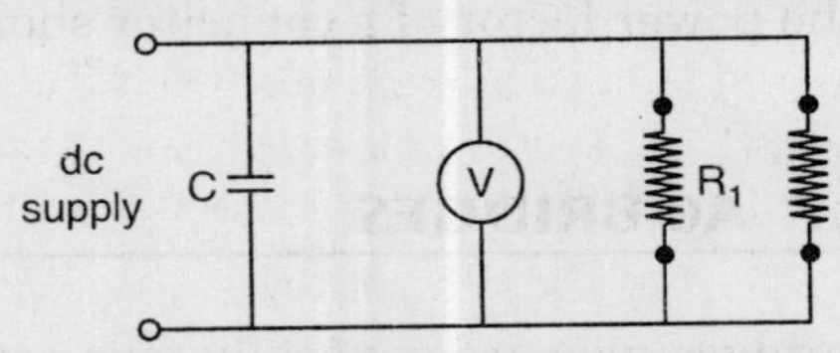

FIGURE 20.7. Loss of charge method.

The test is performed twice; once with R in position and second with R removed.

The resistances in the two tests are calculated. Let these be R_1 and R_2. The unknown resistance R is then found from the equation

$$R_1 = RR_2/(R + R_2).$$

20.6 AC MEASUREMENTS

The value of a resistance or that of an inductance under ac conditions is not the same as under dc conditions. A ***resistor*** is associated with a small inductance in series, and a small distributed capacitance across it. Similarly, an ***inductor*** is associated with a small resistance in series, and a small distributed capacitance across it. This resistance is not merely the ohmic resistance, but includes the hysteresis loss and the eddy current loss effects in the neighboring metal parts.

A resistor or an inductor may, therefore, be represented as a series *R-L* circuit shunted by a capacitor *C*. The effective resistance is, $R_{eff} \approx R\,(1+\omega^2 LC)$, and the effective inductance is, $L_{eff} \approx L - CR^2$. For a resistor to be non-inductive, the time constant of the resistor which equals $(L/R - CR)$ should be zero.

At low frequencies, the presence of distributed capacitance can be ignored.

The inductor is then simply a series $R_s - L_s$ circuit, where R_s is small in comparison to the inductive reactance ωL_s. We can represent the inductor by a parallel $R_p - L_p$ circuit also, where R_p is >> R_s and $L_p \approx L_s$. At any angular frequency ω, we have the following relations of the two circuits.

$$R_p = \frac{R_s^2 + X_s^2}{R_s} \quad \text{and} \quad X_p = \frac{R_s^2 + X_s^2}{X_s} \quad \text{where } X_s = \omega L_s, \ \text{and } X_p = \omega L_p.$$

$$R_s = \frac{X_p^2}{R_p^2 + X_p^2} R_p \quad \text{and} \quad X_s = \frac{R_p^2}{R_p^2 + X_p^2} X_p$$

The relations are frequency dependent.

The ***quality factor*** of an inductive coil is $Q = \omega L_s / R_s$.

The ***capacitors*** are associated with the leakage resistance (insulation resistance) under dc as well as under ac conditions, and the ***dielectric absorption*** losses under ac conditions. The total loss effect is shown by the resistance. A capacitor can thus be represented either by a parallel, or a series $R - C$ circuit, the interrelation between the two being similar to those for an inductor. C_p and C_s are nearly equal. We then have a parallel R_p–C circuit, or a series R_s–C circuit. R_p is >> R_s and R_s is quite small in comparison to the capacitive reactance. The ***loss angle*** δ in radians $\tan \delta = 1/(R_p \omega C)$. Also, $\tan \delta \approx \sin \delta$ = the power factor $\cos \varphi$. Similar relation holds for the series representation. δ in radians $\tan \delta = R_s \omega C \approx \cos \varphi$. The loss angle, or the power factor of a capacitor should be as small as possible.

20.7 AC BRIDGES

ac bridges also are generally four-arm bridges. Suitable sine-wave ac sources and detectors are used. The detector can be a vibration galvanometer, a pair of head phones, or a CRO.

The general balance condition remains the same, *i.e.*, the product of the impedances (Complex) of the opposite arms should be equal. $\mathbf{Z_1 Z_4 = Z_2 Z_3}$. This provides us with two scalar conditions to be satisfied. One condition is obtained by equating the reals, and the other by equating the imaginaries on the two sides of the complex equation. Or, in other words, we are to fulfill the magnitude equality condition as well as the phase equality condition. $|\mathbf{Z}_1\mathbf{Z}_4| = |\mathbf{Z}_2\mathbf{Z}_3|$ and $\theta_1 + \theta_4 = \theta_2 + \theta_3$. The angle equation resticts the choice of elements for various brannches. For instance, if $\mathbf{Z}_1\mathbf{Z}_4$ is inductive, $\mathbf{Z}_2\mathbf{Z}_3$ also will have to be inductive.

By adjusting the value of one variable element, we satisfy one condition, and by adjusting the value of the other variable element, we satisfy the other condition. The adjustments have to be made alternately. Thus, it is a little more difficult to balance an ac bridge.

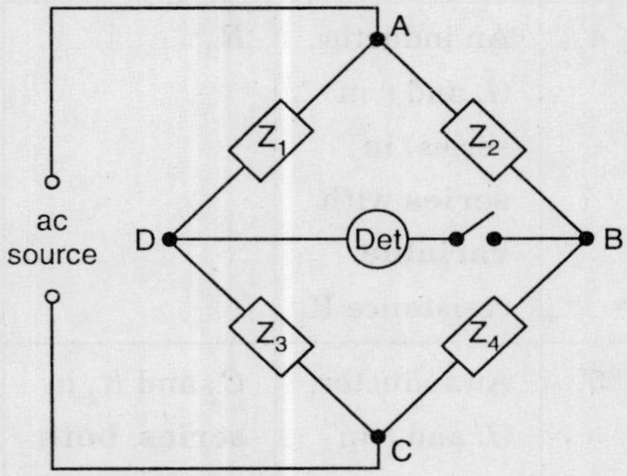

FIGURE 20.8. A general four-arm ac bridge.

It is simpler to have a variable capacitor than a variable inductor. Also, to measure a particular parameter over a range of values, the bridge should not require a very wide range of the variable elements in its arms.

There are several bridge configurations which one can think of. Some of the useful bridges which can be used for measuring inductance, capacitance, and frequency are given below.

The Various Bridges

S. No.	Z_1 *(unknown)*	Z_2	Z_3	Z_4	*Name of the bridge*	*Measurement*
1.	An inductor, (Land r in series)	An inductor, (L_2 and r_2 in series), in **series** with R_2, **both variable**	R_3	R_4	**Maxwell's** inductance bridge	Inductance $L = \frac{R_3}{R_4} L_2$ $r = \frac{R_3}{R_4}(R_2 + r_2)$
2.	An inductor, (L and r in series)	R_2	R_3	C_4 and R_4 in **parallel, both variable**	**Maxwell's** inductance capacitance bridge	Inductance $L = R_2 R_3 C_4$ $r = R_2 R_3 / R_4$
3.	An inductor, (L and r in series)	R_2	R_3	C_4 and R_4 in **series, both variable**	**Hay's** bridge	High inductances $L = \frac{R_2 R_3 C_4}{1 + \omega^2 C_4^2 R_4^2}$ $r = \frac{\omega^2 R_2 R_3 R_4 C_4^2}{1 + \omega^2 C_4^2 R_4^2}$

4.	An inductor, (L and r in series) in **series** with **variable** resistance R_1	R_2	R_3	* see below	**Anderson's** bridge	Small inductances $L = R_3C_4[R_2+R_5(1+R_2/R_4)]$ $r = R_2R_3/R_4$
5.	An inductor, (L and r in series)	C_2 and R_2 in **series, both variable**	R_3	C_4	**Owen's** bridge	Inductance $L= R_2R_3C_4$ $r = R_3C_4/C_2$
6.	A capacitor, (C and r in series)	A capacitor, (C_2 and r_2 in series) in **series with** R_2, **both variable**	R_3	R_4	**De Sauty's** bridge	Capacitance $C = C_2R_4/R_3$ $r = (R_2+r_2)\, R_3/R_4$
7.	A capacitor, (C and r in series)	C_2	R_3	C_4 in **parallel** with R_4, **both variable**	**Schering** bridge	Capacitance, a high voltage bridge $C = C_2R_4/R_3$ $r = R_3C_4/C_2$
8.	C_1 and R_1 in series (It is not unknown), $\mathbf{R_1}$ **variable**	C_2 and R_2 in parallel, $\mathbf{R_2}$ **variable** $C_1 = 2C_2$, $R_2 = 2R_1$	$R_3 = R_4$	R_4	**Wien's** frequency bridge	frequency $\omega = 1/(2\, R_1C_2)$

*Anderson's bridge, – one terminal of the detector is connected to the bridge junction D ; and from its second terminal a capacitor C_4 and a variable resistor R_5 are connected respectively to the junction C and B.

20.8 SCREENING OF AC BRIDGES

Errors may be introduced in bridge measurements due to imperfect insulation and residuals in the components, due to magnetic couplings between the components, due to electrostatic couplings between the components, and between the components and the earth. Under high frequency and high voltage conditions, the errors become significant.

The errors due to first three causes can be reduced by using high quality components, ***co-axial*** connecting leads, and laying out the bridge in such a manner that the least possible loop areas are formed.

The electrostatic couplings introduce several stray capacitances. To reduce the errors caused by these capacitances, we use ***screened components***, screen the whole bridge, and provide ***Wagner Earth***. Screening renders the capacitances definite in value. Further, screening also puts several of the stray capacitances effectively across the source or across the detector whence no error can be caused. Screening can place some of the capacitances across the standard components whence these can be accounted for in calibration.

When the capacitances are so reduced that these appear between the corner points A, B, C, and D of the bridge and the earth, and if one terminal of the detector branch is earthed; the capacitances C_A and C_C only can cause error. The errors due to this also can be eliminated by using Wagner earth, wherein the auxiliary arms Z_5 and Z_6, of the same type as Z_2 and Z_4, are provided. Instead of earthing point B, the point E, the junction point of Z_5 and Z_6 is earthed.

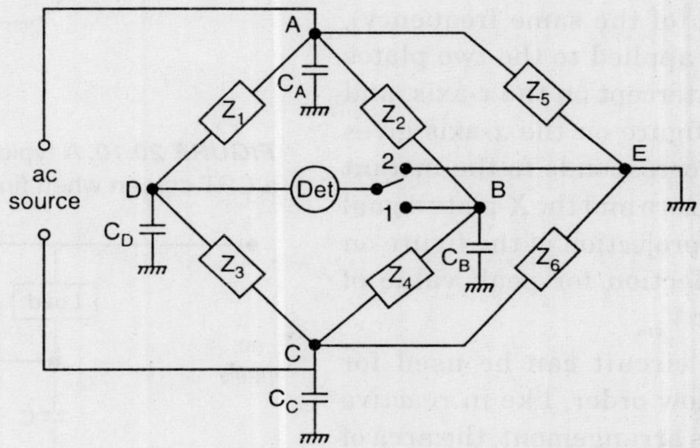

FIGURE 20.9. Earth capacitances and Wagner earth.

The bridge is balanced with $Z_2 - Z_4$, and $Z_5 - Z_6$ in circuit alternately, *i.e.*, the detector connected to B, or E alternately. This way, the points B and D are brought to earth potential without actually connecting either of these to earth. The stray earth capacitances C_A and C_C are now across Z_5, and Z_6; and not across Z_2, and Z_4.

The errors can also be eliminated from the measurement by applying the ***difference method***, or the ***substitution method***. In the difference method, two measurements are taken-one with the unknown in position, and the other with unknown removed and the terminals left open if the unknown is a capacitor and shorted if the unknown is an inductor or a resistor. The difference of the two measurements taken is the error-free value of the unknown.

In the substitution method the second measurement is taken by substituting the unknown by a variable standard of the same type and the bridge is balanced by varying the value of this standard component. The balance value of the standard is the value of the unknown.

20.9 MEASUREMENTS USING CRT

As already stated, a CRT can be employed as a null detector in ac bridges. A CRT can also be used in limited cases for finding frequency ratio of two sinusoidal voltages, phase difference between two voltages, and power of a circuit.

The ***frequency ratio*** should be an integer and also should not be large. On applying the two signals to the X-, and Y- plates, the familiar pattern known as Lissajous figure is obtained, the shape depending on the frequency ratio and the phase difference between the two signals. The frequency ratio f_y/f_x is given by the ratio of the number of intersections made on a horizontal line to the number of intersections made on a vertical line by the figure.

Using a double-beam CRO with the two signals applied to the two Y-plates and the time base synchronized to the lower frequency, the frequency ratio is simply the number of waves contained in one wave of lower frequency signal.

For finding the ***phase difference*** between two voltages (of the same frequency), again the signals are applied to the two plates and the ratio of the intercept on the x-axis, and the projection of the figure on the x-axis gives sin φ. The intercept corresponds to the instant when Y-plate signal is zero and the X-plate signal is $\mathbf{V_{xm}}$ **sin φ,** and the projection of the figure on the x-axis is the deflection for peak value of X-plate signal, *i.e.*, for V_{xm}.

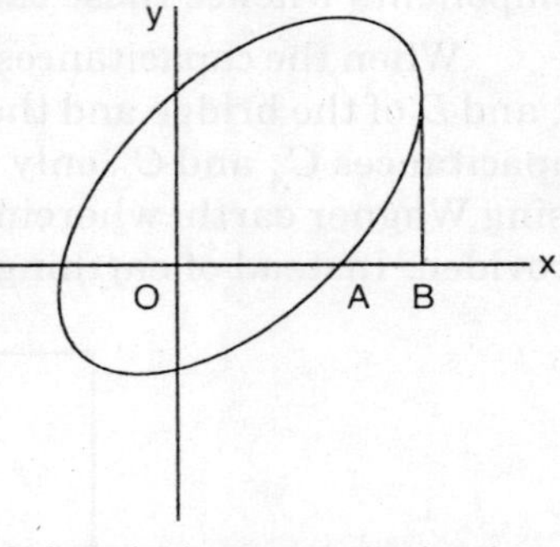

FIGURE 20.10. A typical pattern obtained on a CRT screen when finding phase difference.

The following circuit can be used for measuring ***losses*** of low order, like in reactive components. With this arrangement, the area of the loop obtained is proportional to the energy loss occurring in the component or the load under test in one cycle. The power is then given by

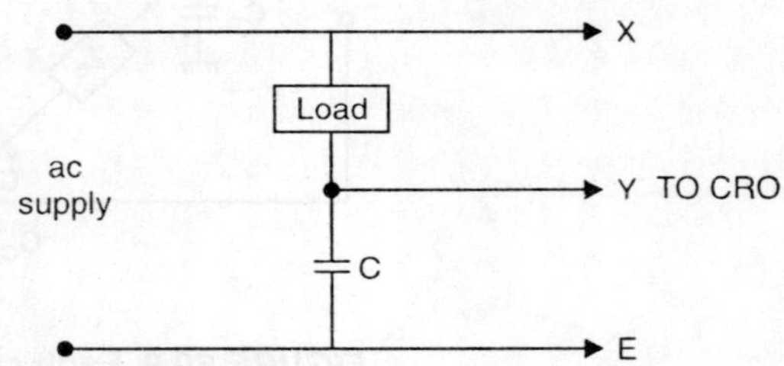

FIGURE 20.11. The circuit for measuring load power using CRT.

$$P = \frac{Cf}{S_x S_y} \times \text{area of the loop},$$

where S_x and S_y are the X-plate and the Y-plate sensitivities respectively.

20.10 HIGH FREQUENCY MEASUREMENTS

The measurements of residuals, the component values, the quality factors at frequencies of the order of Mc/s are largely based on series resonance methods. An oscillator of frequency range from 50 kHz to 50 MHz is employed which delivers current to a low resistance of the order of 20 m Ω which acts as a source to the measurement circuit, or if the element to be measured is of low impedance, the oscillator may be directly connected in the circuit. One may have to vary resistance, reactance, capacitance, or frequency to achieve resonance condition.

At series resonance, the current in the circuit becomes maximum; however, it will be assumed that when the voltage across the series capacitor becomes maximum, the resonance condition exists.

(*a*) In the ***direct measurement connection*** to measure Q of a coil, the ratio of the voltages V_C and V_S at resonance gives the quality factor of the coil. The resonance condition is achieved by varying either the frequency of the oscillator, or the capacitance C.

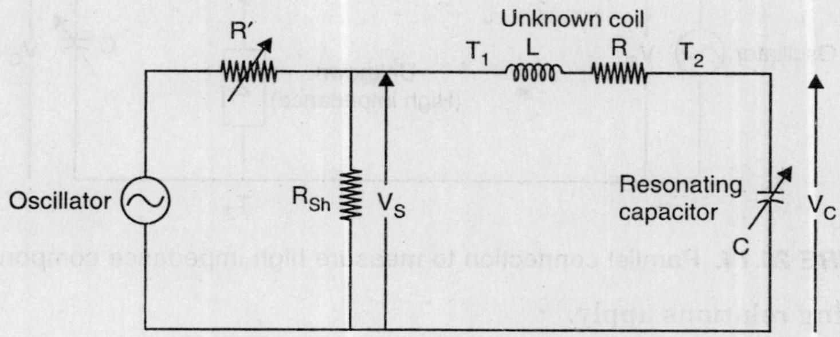

FIGURE 20.12. Direct connection to measure Q of a coil.

(*b*) In the ***series connection***, suitable for low impedance elements which may be a resistor, a coil, or a capacitor; the unknown is connected in series with the 'work coil' as shown. Two measurements are taken by tuning the capacitor C; one with the element shorted when the capacitance is C_1 and the circuit quality factor is Q_1, and the other with the element in the circuit when the capacitance is C_2 and the circuit quality factor is Q_2. The following relations which can be derived easily, apply.

$$L_x = \frac{C_1 - C_2}{\omega^2 C_1 C_2}$$

$$R_x = \frac{C_1 Q_1 - C_2 Q_2}{\omega C_1 C_2 \; Q_1 Q_2}$$

$$Q_x = \frac{(C_1 - C_2)\, Q_1 Q_2}{C_1 Q_1 - C_2 Q_2}$$

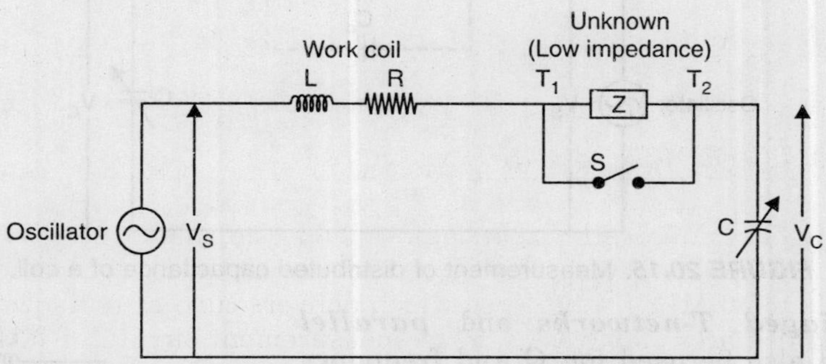

FIGURE 20.13. Series connection to measure low impedance components.

(*c*) In the ***parallel connection***, suitable for high impedance elements, the unknown is represented by a parallel combination of R_p and X_p, and is connected across the tuning capacitor. Two measurements are taken by tuning the capacitor; one with the unknown removed when the capacitance is C_1, and the other when the unknown is in the circuit when the tuning occurs at capacitance C_2.

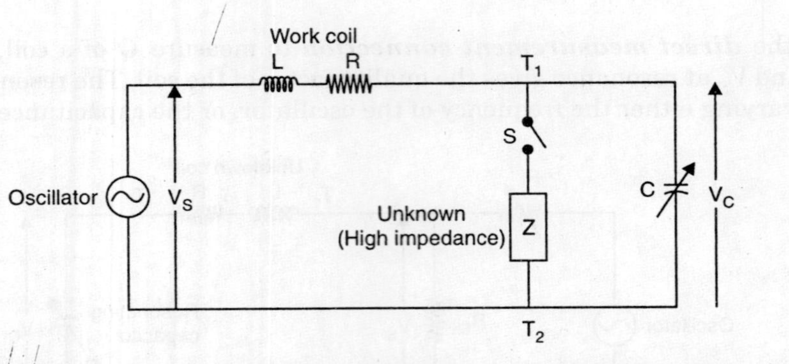

FIGURE 20.14. Parallel connection to measure high impedance components.

The following relations apply.

$$X_P = \frac{1}{\omega(C_1 - C_2)}$$

$$R_P = \frac{Q_1 Q_2}{\omega C_1 \Delta Q}$$

$$Q_P = \frac{(C_1 - C_2)\, Q_1 Q_2}{C_1 \Delta Q}, \quad \text{where } \Delta Q = Q_1 - Q_2.$$

(*d*) The ***distributed capacitance C_d*** of a coil which shunts the coil, can be measured by connecting it in the direct mode. Again two resonance measurements are taken, one at frequency f when the resonating capacitance is C_1, and the other at frequency $2f$ when the capacitance is C_2. The distributed capacitance of the coil is given by

$$C_d = (C_1 - 4C_2)/3.$$

It will be seen that C_2 has to be less than $C_1/4$.

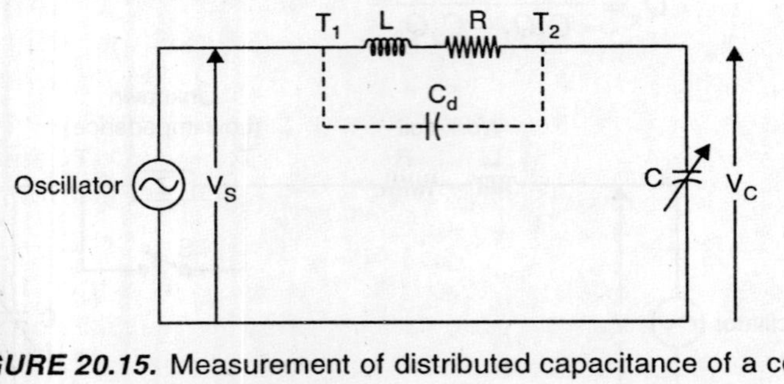

FIGURE 20.15. Measurement of distributed capacitance of a coil.

Some ***bridged T-networks*** and ***parallel T-networks*** can also be used for Q and frequency measurements.

For the bridged *T*-network shown above, the following equations hold under balanced conditions (no signal at the output port).

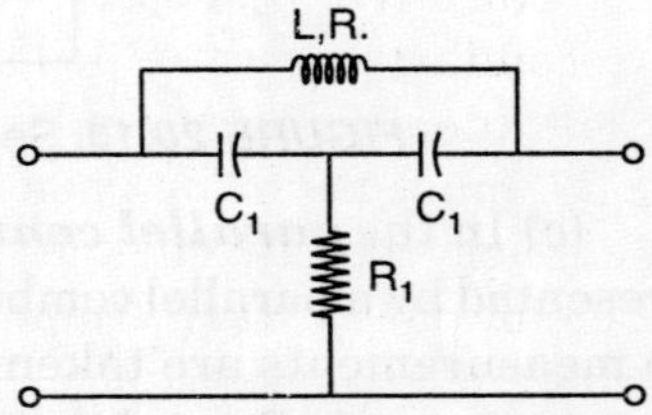

FIGURE 20.16. A bridged *T*-network for measurement of Q of a coil.

$$R = \frac{1}{\omega^2 C_1^{\ 2} R_1}$$

$$L = \frac{2}{\omega^2 C_1}$$

$$Q = 2\omega C_1 R_1$$

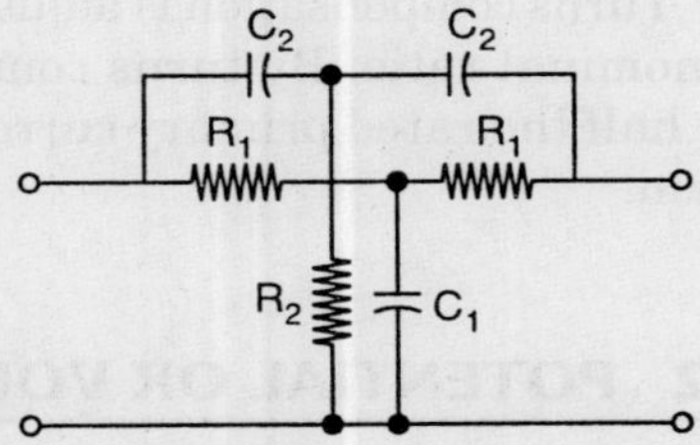

FIGURE 20.17. A parallel T-network for measurement of frequency.

For the parallel T-network shown above, the following relations hold under balanced conditions.

$$\omega^2 = \frac{1}{2R_1 R_2 C_2^{\ 2}}$$

$$\omega^2 = \frac{2}{R_1^{\ 2} C_1 C_2}$$

The balance is possible only if we maintain $R_1 = 4R_2$, and $C_1 = C_2$.

20.11 CURRENT TRANSFORMER

A current transformer (CT) is expected to transform high currents to low currents with fixed transformation ratio and exact phase reversal. The core of a CT is made up of low loss silicon steel stampings, or ***mumetal*** which has relatively high permeability and low 'iron' losses. The flux density too is kept low- of the order of **0.1 T**. The core size is small as the voltages across the windings are small. The primary conductor, of course, has to be insulated for full conductor voltage.

The nominal ratio k_n of a CT is the ratio of the rated primary current to the nominal corresponding secondary current. This should be the secondary to primary turns ratio (assuming no turns compensation). The actual ratio k is the ratio of the primary current I_p to the corresponding secondary current I_s. The ***ratio error*** σ is $(k_n - k)/k$; which is equivalent to the ratio of the difference between the primary current read and the true primary current to the true primary current. k_n is not equal to k because of the presence of small exciting current in the primary circuit. It can be shown that $\boldsymbol{\sigma = -I_w/I_p}$, where I_w is the iron loss component of the exciting current I_o in the CT.

The ***phase angle error*** β is defined as the angle by which the reversed secondary current phasor leads the actual primary current phasor I_p. It can be shown that $\boldsymbol{\beta = I_m/I_p}$, where I_m is the flux producing component of the exciting current I_o in the CT.

The ***burden*** of a CT is stated in terms of VA at the ***rated*** secondary current. Thus, more the secondary circuit impedance, more the burden; irrespective of the value of the secondary current. Ideally, the secondary circuit of a CT should have zero impedance. Also, the secondary of a CT must never be opened when primary is carrying current. Open-circuited secondary implies very high flux density in the core, consequently high core heating and very high induced voltage of peaky wave shape in the open-circuited secondary, and possibly the core may get permanently magnetized.

The ratio error and the phase angle error decrease with the increase in the primary current. An increase in the burden causes an increase in both the errors.

Turns compensation is adjusting number of secondary turns a little less than that required with nominal ratio. By turns compensation, the ratio error is usually adjusted to be zero at about half the rated primary current. This has the effect of making the ratio error more or less uniform.

20.12 POTENTIAL OR VOLTAGE TRANSFORMER

A potential transformer (*PT*) is expected to transform high voltages to low voltages with fixed transformation ratio and exact phase reversal. The core of a *PT* also is made up of low loss silicon steel stampings, the flux density being of the order of **0.8 T**. The primary consists of a relatively large number of turns of thin wire. The core size and insulation are comparable to those of a power transformer of similar voltage rating.

The nominal ratio, the ratio error, the phase angle error and the burden for a voltage transformer are defined in the same way as are these defined for a *CT*. The phase angle error in this case carries a negative sign, as the reversed secondary voltage phasor does not lead, but lags the primary voltage phasor. The external admittance in the secondary circuit should be as small as possible, *i.e.*, the burden should be small.

Here too, in order that the errors are small, I_w and I_m are required to be small. In addition, resistance drops and the reactance drops also should be small. The leakage reactances are, therefore, designed to be low.

The errors increase with the increase in the burden.

The ratio error can be reduced by turns compensation, which means keeping number of primary turns a little less than that required with nominal ratio.

20.13 MAGNETIC MEASUREMENTS

In magnetic measurements, one is required to measure magnetic flux density B, magnetic field intensity H, and magnetic losses in the materials. Such measurements are necessary in order to know the magnetic properties of the materials.

Inherent inaccuracies are present in magnetic measurements because of non-uniformities in the material, different characteristics in different directions, and difficulties in ensuring uniform B, or H.

Only in a finely and uniformly wound toroid, or a long straight solenoid we are able to achieve practically uniform field. In most of the experimental set ups, the specimen are chosen of such shape, and the windings are wound in a way such that as uniform B in the specimen as possible is achieved.

The average value of H over a magnetic circuit of length l and having N-turn winding carrying current I is NI/l. This is by definition.

Any B or H measurement involves changing the field suddenly in the coil connected to the meter.

H can be measured with the help of an H-coil, or a so-called magnetic potentiometer, both of which are simply coils of a large number of turns of fine wire wound on thin non-magnetic flexible strips. The coil is connected to the meter.

As already stated, sudden flux reversal or sudden flux change through the coil is needed to effect measurement.

A fluxmeter, or a ballistic galvanometer is used for measuring flux-turns change in a coil. The terminals of the coil are connected to the instrument. The instrument constant has to be first found by noting the deflection caused by a known (how accurately known!) change in flux-turns of the coil.

Having found the flux-turns change, and knowing the number of turns of the coil, average value of the flux linking the coil can be found. Then knowing the flux path cross-sectional area, average value of B is determined.

The flux-turn change may be caused by suddenly reorienting the coil in the magnetic field, or due to a sudden change in the current in the coil.

Once we are in a position to determine B, we can obtain the hysteresis loop and the $B-H$ curve of a magnetic specimen.

20.14 IRON LOSS MEASUREMENT TESTS

The iron loss P_i comprises hysteresis loss P_h and eddy-current loss P_e. It is known that

$$P_h = k_h f B_m^{\ x}, \quad \text{and} \quad P_e = k_e K_f^2 f^2 B_m^{\ 2},$$

where K_f is the form factor of the flux-time wave, or the induced emf-time wave.

We then have two ways of separating the hysteresis loss from the eddy-current loss.

(*a*) ***By varying the frequency***

If other parameters are held fixed, we see that

$$P_i = k_1 f + k_2 f^2$$
$$P_i/f = k_1 + k_2 f$$

P_i is measured at several frequencies keeping the applied voltage, and hence the flux density constant.

Thus, a plot of P_i/f against frequency, gives a straight line with intercept equal to k_1, and the slope of the straight line equal to k_2. P_h and P_e at the desired frequency can therefore be found out.

(*b*) ***By varying the form factor :*** The form factor variation affects P_e only. Keeping everything else constant, we measure P_i at several form factors of the applied voltage, and hence approximately of the induced emf. Thus, a plot of P_i against K_f^2, gives a straight line with intercept equal to P_h. The slope of the straight line multiplied by the given K_f^2 gives P_e.

The following experimental set up (Fig. 20.18) can be used for the test.

The voltage coil of the wattmeter has been connected across the coil C_1 to exclude the ohmic loss in the coil M from the wattmeter reading. The wattmeter reading should be multiplied by T_M/T_{C1}, where T_M is the number of turns on M and T_{C1} is the number of turns on C_1. V_1 is a true rms reading voltmeter, and V_2 is a PMMC voltmeter. K_f is given by the ratio of the readings of V_1 and of V_2.

The form factor is varied by including a resistor and a variable inductor in the supply circuit. The frequency f, and B_m are held constant. The series resistance and the inductance are so varied that the PMMC voltmeter V_2 reading stays constant, while that of V_1 varies considerably.

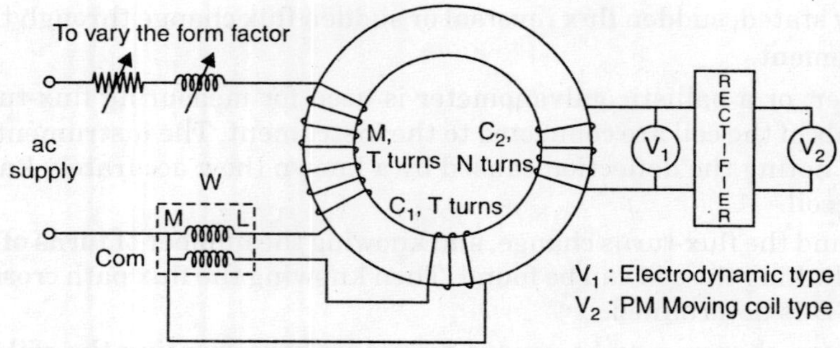

FIGURE 20.18. Circuit to separate hysteresis loss and eddy-current loss using form factor variation method.

For measuring iron losses of strip specimen the toroid or the ring specimen is replaced by ***Lloyd-Fisher Square.*** The square has four windings on its four sides. The sample strips of specified size can be inserted into one of the windings. All the windings should be excited for uniformity of the magnetic flux density in the strips. The losses are determined in the same way as for the ring specimen. The losses in the sample are the total losses measured less the losses in the other three sides and the core of the square.

(c) ***Oscillographic method :*** An oscillographic method to measure power has already been given at fig.20.11. With that arrangement the ohmic loss of the coil will also be included in the loop area. An improved version is given at Fig. 20.19.

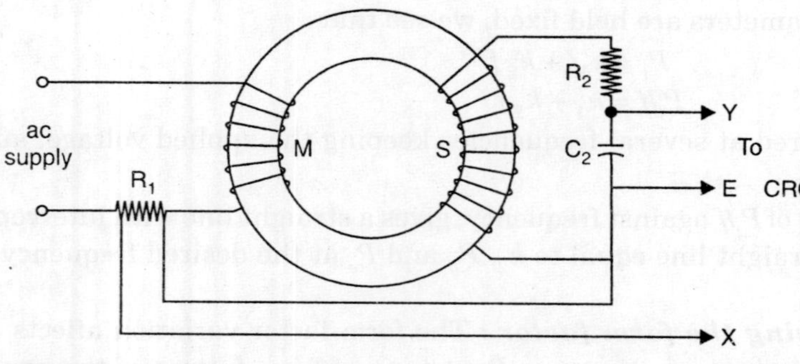

FIGURE 20.19. Measuring iron loss in the toroid using CRT oscilloscope.

R_1 is a low resistance. With this arrangement the X-plate deflection is proportional to the primary winding current, which is largely magnetizing but also includes the current component corresponding to the losses. In the secondary circuit, R_2 and C_2 are large in values. The secondary current i is then e/R_2. The voltage across the capacitor C_2 is $v = \dfrac{Na}{C_2R_2}\int dB$, *i.e.*, v is directly proportional to B. Therefore, the Y-plate deflection is proportional to B. The area of the loop traced on the screen is proportional to the loss in the specimen.

If the current in the primary winding is gradually reduced to zero, the locus of the tips of the loops can be taken as the B-H curve of the specimen.

Section 6 CONTROL SYSTEMS

Chapter 21 CONTROL SYSTEMS BASICS

21.1 GENERAL

In all walks of life, we always wish to keep one or more quantities or items of our interest at the desired levels. Because of various disturbances caused by society, environment, associated mechanisms etc., the quantities of our interest, may not be at the levels desired. We have to build something into the system so that the quantities in question stay at the desired levels, or within a narrow range of the desired levels. This 'building something into the system' is controlling the quantities of interest (henceforth called controlled quantities or variables) in a manner to keep them at the intended levels.

In context with the physical systems, the word 'control' is a term to mean manual or automatic adjustment of some part of the system so as to develop adequate controlling power which drives the controlled variable to the 'set' value, as and when it deviates or tends to deviate from the set value.

In devising an electrical control scheme, an electrical quantity must be selected whose value is a measure of the actual performance feature to be controlled, and this quantity or its deviation from an established ideal must actuate the controlling mechanism in a prescribed fashion.

The control power acts on a part of the system which directly controls the output variable, the control power is usually derived from the output of a power amplifying device, the input to the amplifying device being obtained from some kind of signal processing and conditioning devices to which the input is a weak signal, usually the error signal. All the components, devices and equipment appropriately connected/assembled to function so as to produce the intended result constitute the control system.

The control systems are categorized as—the open loop control systems, and the closed loop control systems.

In an open loop system, there is no link between the actual value of the performance, or the output variable and the 'set', or the 'reference' value of the variable.

For instance, if in a heating system our quantity of interest is the temperature attained, and the system is equipped merely with a timing control device wherein we can set the time for which the heating system remains on, the temperature attained may not be the desired one. Though the relationship between the heating time and the temperature attained can be found under the condition that everything else remains the same, and one can act accordingly to achieve the desired temperature; but if the conditions change, or/and disturbances are there, the time-temperature relation will fail, and the desired temperature will not be attained.

If the heating system is equipped with some thermostat device, and we set the desired temperature on the thermostat, the system will function to raise the temperature to the set level even in the face of the changing conditions and disturbances. There is a link between the desired temperature and the 'on'- 'off ' status of the heating circuit through the bimetallic strip.

The former is an open loop system, while the latter is the so called 'closed loop system'.

A closed loop control system is thus one in which the controlled quantity is compared with a standard or reference representing the desired performance and any deviation from the standard is fed back into the control system in such a sense as to reduce the deviation. A closed loop control system is also called the feedback control system.

When feedback is deliberately introduced for the purpose of control, its existence and function are easily identified. However, the existence of feedback in many physical systems is quite often obscure. Even if there is no visual feedback, the system may have feedback.

From a general point of view, feedback is said to exist whenever a closed sequence of cause and effect relationship exists among the variables of a system.

In contrast, an open loop system does not have any feedback.

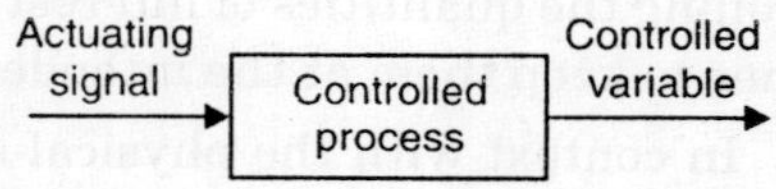

FIGURE 21.1. Representation of an open loop control system.

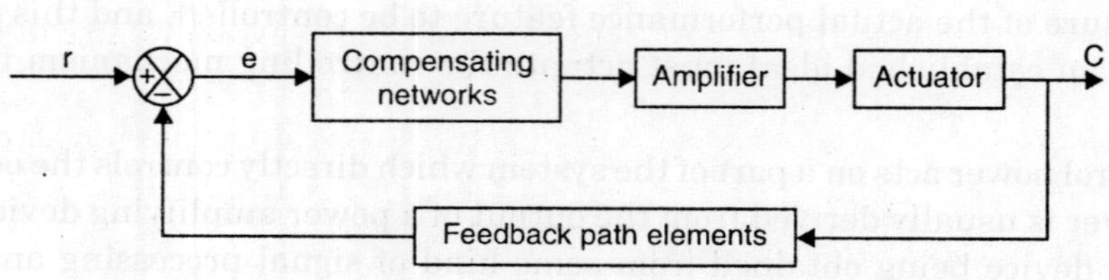

FIGURE 21.2. Representation of a closed loop control system.

We shall consider only continuous data feedback control systems, wherein the signals at various parts of the system are all functions of the continuous time variable *t*. Further, the signals at various points are considered to be small so that for any component a linear relation between cause and effect can be assumed. We shall consider linear systems only.

21.2 LINEAR SYSTEMS MODEL

To study any system, its appropriate model should be developed; and a model is nothing but a means of representing the links between the various components of the system. Every component /device is modeled by its governing algebraic equations, differential equations, state equations etc., which relate output and the input. From the knowledge of the interrelationships of the components, the system model is obtained from the components models.

A model can be a graphical model also, which is in the form of signal flow graphs, or block diagrams.

A single input and single output system can be described by an nth order differential equation relating the output variable $c(t)$ and the input variable $r(t)$.

$$a_0 \frac{d^n c(t)}{dt^n} + a_1 \frac{d^{n-1} c(t)}{dt^{n-1}} + \dots + a_{n-1} \frac{dc(t)}{dt} + a_n\, c(t) = b_0 \frac{d^m r(t)}{dt^m} + b_1 \frac{d^{m-1} r(t)}{dt^{m-1}} + \dots + b_{m-1} \frac{dr(t)}{dt} + b_m\, r(t)$$

The a's and the b's are constants.

21.3 TRANSFER FUNCTION

The above equation is cumbersome to solve and of little use for design purposes. From this differential equation model, we can, however, obtain the transfer function model of the system which is the ratio of the Laplace transform $C(s)$ of the output variable $c(t)$ to the Laplace transform $R(s)$ of the input variable $r(t)$, under the assumption of zero initial conditions. s is the Laplace transform variable which actually is the same as the complex frequency, $s = \sigma + j\omega$. We find from above equation

$$G(s) = \frac{C(s)}{R(s)} = \frac{b_0 s^m + b_1 s^{m-1} + \dots + b_{m-1} s + b_m}{a_0 s^n + a_1 s^{n-1} + \dots + a_{n-1} s + a_n}$$

The numerator polynomial in s, and the denominator polynomial in s in this expression should not be interpreted as equal to $C(s)$ and $R(s)$ respectively.

The denominator polynomial, when equated to zero is called the characteristic equation of the system.

We also see that $C(s) = G(s)\, R(s)$.

If there is a p input, q output linear system, then we can develop such equations as

$$C_i(s) = \sum_{j=1}^{j=p} G_{ij}(s) R_j(s); \quad (i = 1, 2, \dots, q)$$

The transfer function for nonlinear systems is not defined.

In the analysis problems, the transfer function of the system and the input are known, and the output $C(s)$ is found out. By finding the Laplace inverse of $C(s)$, the response $c(t)$ in time domain is obtained. However, the solution thus found does not include the effects of the non-zero initial conditions, if any.

As Laplace transform of a unit impulse function is 1, and as $C(s) = G(s)\,R(s)$, the Laplace inverse of $G(s)$ is the unit impulse response of the system. Thus, the function $G(s)$ and the impulse response are the two ways of describing a system.

The evaluation of the impulse response of the controlled process or system is also an important step in a class of systems called "adaptive control systems".

The control systems are composed of many noninteracting elements whose transfer functions can be determined independently. Each element of the control system can thus be represented by its own transfer function, which pictorially can be represented by a block whose value is the transfer function of the element.

21.4 BLOCK DIAGRAM OF A FEEDBACK CONTROL SYSTEM AND THE ASSOCIATED TERMINOLOGY

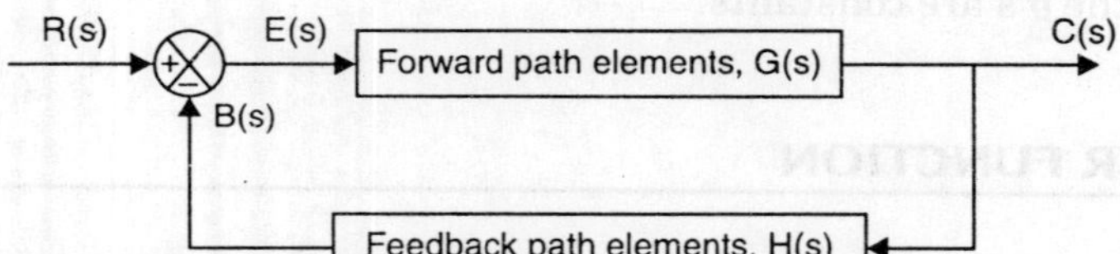

FIGURE 21.3. Basic block diagram of a feedback control system, with symbolic transfer functions of the blocks shown.

$R(s)$ – Reference input
$C(s)$ – Output signal, or the controlled variable
$B(s)$ – Feedback signal
$E(s)$ – Error signal
$G(s)$ – Forward path transfer function, or open-loop transfer function = $C(s)/E(s)$
$M(s)$ – Closed loop transfer function = $C(s)/R(s)$
$H(s)$ – Feedback path transfer function
$L(s)$ – Loop transfer function = $G(s)\,H(s)$

It can easily be found that for a negative feedback control system

$$M(s) = \frac{C(s)}{R(s)} = \frac{G(s)}{1 + G(s)\,H(s)},$$ a very frequently used expression.

21.5 COMPLEX BLOCK DIAGRAMS AND BLOCK DIAGRAMS REDUCTION

The blocks of the various elements are joined with arrowed lines in accordance with the manner of functonal linking of various elements, to yield the block diagram of the entire control system.

The block output signal value is the block input signal value multiplied by the block transfer function. The value of the signal at any point on the diagram is the sum of all the signals arriving there. The block diagram provides detailed cause and effect relationship between the input and the output of the system.

The block diagram comprising several blocks interconnected in complex way can be reduced to a single equivalent block with the same input and output as the original system had.

Several blocks in series are reduced to a single block whose transfer function is simply the product of the individual transfer functions.

Similarly, several blocks whose inputs and outputs are common, *i.e.*, which are in parallel can be reduced to a single block whose transfer function is simply the sum of the individual transfer functions.

The tapping off of a branch can be shifted to the other side of the block by appropriately multiplying /dividing the branch path transfer function by the transfer function of the block.

To summarize, the reduction should be carried out in a manner that the value of the signal at any input point remains unaltered.

21.6 SIGNAL FLOW GRAPHS

The alternative way of depicting graphically the flow of signals and the values of the signals at various points in the control system is the "Signal Flow Graph".

A signal flow graph is simply a graphical representation of the relationships between the variables. We can obtain the signal flow graph from the governing algebraic equations, or from the block diagram itself.

The block is replaced by an arrow marked line (*i.e.*, an oriented line) which now is called a branch, with ***TF*** of the block labeled along the line, which now is termed as branch gain.

Signals travel in the marked direction only. The two ends of the branch, *i.e.*, the nodes represent variables of the system. For an isolated branch, the branch out variable = the branch in variable × the branch gain.

For the interconnected branches, the value of the variable at a node is the sum of all the incoming signals at that node. The input or the source node, and the output or the sink node have to be specified by the system description. The input node has no incoming dangling branches incident to it and has at least one outgoing branch. The output node has no outgoing dangling branches incident to it and has at least one incoming branch.

A path on a signal flow graph is a continuous, unidirectional succession of branches traversed in the branch directions. The starting and the terminating nodes may be the same, or different; in the former case it is a feedback path or a loop. The loop is a self loop if no other node is encountered on the way. A path from the input node to the output node on which no node is encountered more than once is called a forward path. The gain of a path, or that of a loop is simply the product of all the branch gains encountered in traversing the path, or the loop.

21.7 FORMULA FOR GAIN BETWEEN THE INPUT AND THE OUTPUT NODES

The formula is known as Mason's gain formula.

$$M = \frac{X_{out}}{X_{in}} = \sum_k \frac{M_k \Delta_k}{\Delta} \quad \text{where}$$

x_{out} – output node variable

x_{in} – input node variable

k – forward path number

M_k – gain of the kth forward path

Δ – 1 – (sum of all individual loop gains) + (sum of gain products of all possible combinations of two non-touching loops) – (sum of gain products of all possible combinations of three non-touching loops) + ...

Δ_k – the value of Δ for that part of the graph not touching the kth forward path.

The two parts of a graph are said to be non-touching if no nodes are common to them.

21.8 TRANSFER FUNCTIONS PLOTS

The performance of a control system can be studied in the time domain, or in the complex frequency domain. We may wish to find the complete response, or only the steady state response. Further, if our interest is in the steady state frequency response to sinusoidally varying input signals, it can be found from the transfer function simply by replacing s in it by $j\omega$.

In control systems studies, the plots of the open loop transfer function $G(j\omega)$, or that of the loop transfer function $G(j\omega)\,H(j\omega)$ are often made and the performance of the closed loop system is predicted from these plots.

1. ***Polar plot.*** A polar plot is the locus of the tip of $G(j\omega)$ in the $G(j\omega)$ plane as ω is varied from zero to infinity.

Obtaining a polar plot is generally a tedious process. For many purposes, only the general shape of the plot suffices. The sketching of the polar plot is facilitated by the following information :

(*a*) the magnitude and the phase of the function at $\omega = 0$, and at $\omega = \infty$,

(*b*) the possible points of intersections of the polar plot with the *Re* $G(j\omega)$ axis and the *Im* $G(j\omega)$ axis; and the values of ω at these intersections.

The point, at which the magnitude of $G(j\omega)$ is unity is called the gain crossover point, and the corresponding frequency is the gain crossover frequency; and the point at which the phase of $G(j\omega)$ is 180° is the phase crossover point, and the corresponding frequency is the phase crossover frequency. Thus, the gain crossover point is the point where the polar plot intersects a circle centered at the origin and of radius unity; and the phase crossover point is the point where the polar plot cuts the negative real axis. The gain crossover point or the frequency, and the phase crossover point or the frequency are important in graphical stability study of feedback control systems.

2. ***Bode plot.*** The Bode plot is a twin plot—one, the magnitude DB of $G(j\omega)$ in decibels plotted as a function of log ω; and the other, φ, the phase of $G(j\omega)$ in degrees, also plotted as a function of log ω; as ω is varied from zero to infinity.

The characteristics of the Bode plot are that the product factors in the expression for $G(j\omega)$ become additive terms, as logarithms are used, and that the shape of the plot is approximated by straight line segments.

The magnitude and phase of $G(j\omega)$ are composed of simple factors of four types :

(*a*) Constant K.

$$DB = 20 \log |K|, \text{ which represents a horizontal line.}$$

$$\varphi = 0° \text{ or } 180°.$$

(*b*) Poles or zeros at the origin, $(j\omega)^{\pm n}$, $n = 1, 2, 3,...$

$$DB = \pm 20\, n \log \omega$$

which is a straight line passing through $\omega = 1$, and having a slope of $\pm 20n$ *db* per log ω.

$$\varphi = \pm n \times 90°.$$

(*c*) Simple pole or zero, $(1 + j\omega T)^{\pm 1}$.

For $\omega T << 1$,

$DB \approx 0$ *db*, and

For $\omega T >> 1$,

$DB \approx \pm 20 \log \omega T$, which is a straight line having value zero at $\omega = 1/T$, and of slope of $\pm$ 20 *db* per decade of ωT.

The value of ω at which $\omega T = 1$, is called the corner frequency.

For both the cases,

$$\varphi = \tan^{-1} \pm \omega T.$$

(*d*) Quadratic pole or zero, $\left(\dfrac{\omega_n^2}{s^2 + 2\delta\omega_n s + \omega_n^2}\right)^{\pm 1}$

At very low frequencies, *i.e.*, for $\omega/\omega_n << 1$,

$$DB \approx \mp 20 \log 1 = 0.$$

At very high frequencies, *i.e.*, for $\omega/\omega_n >> 1$,

$$DB \approx \mp 40 \log (\omega/\omega_n)\ db$$

For both the cases,

$$\varphi = \mp \tan^{-1} \left(\frac{\dfrac{2\delta\omega}{\omega_n}}{1 - \left(\dfrac{\omega}{\omega_n}\right)^2} \right)$$

On the DB plot one can easily locate gain crossover frequency ($DB = 0$), and on the φ plot can be located phase crossover frequency ($\varphi = -180°$).

3. ***The magnitude versus phase plot*:** It is a plot of the magnitude of the transfer function $G(j\omega)$ in decibels against its phase in degrees, as frequency is varied from zero to infinity.

On this plot also, one can identify the gain crossover point and the phase crossover point and hence the corresponding frequencies.

Such plots, in conjunction with the Nichols chart, are used to determine the relative stability and the frequency response of the closed loop system.

Chapter 22

State Variable Characterization of Linear Systems

22.1 GENERAL

The transfer function description of a control system is useful for frequency domain analysis of the system. The time domain response of the system as found using transfer function does not include the effect of initial conditions as the transfer function is defined with zero initial conditions.

The other way of describing a system, which takes into account the initial condition as well, is to use state variables and state equations. The 'state' of the system is described through suitable system variables. Knowing the state of the system is knowing the behavior of the system.

The state variables of a linear system is a minimal set of variables $x_1(t)$, $x_2(t)$,..., $x_n(t)$ such that the knowledge of these variables at any time t_o plus information on the input excitation subsequently applied is sufficient to determine the state of the system at any time $t \geq t_o$. There can be more than one such set for a given system. Thus, the state variable characterization is not unique.

22.2 SINGLE INPUT, SINGLE OUTPUT SYSTEMS

In general, for a single input, single output system described by an nth order differential equation

$$\frac{d^n c(t)}{dt^n} + a_1 \frac{d^{n-1}c(t)}{dt^{n-1}} + a_2 \frac{d^{n-2}c(t)}{dt^{n-2}} + \ldots + a_{n-1}\frac{dc(t)}{dt} + a_n c(t),$$

$$= b_0 \frac{d^n r(t)}{dt^n} + b_1 \frac{d^{n-1}r(t)}{dt^{n-1}} + \ldots + b_{n-1}\frac{dr(t)}{dt} + b_n r(t)$$

the state variables are defined as

$$x_1(t) = c(t) - b_o\, r(t)$$

$$x_2(t) = \frac{dx_1(t)}{dt} - h_1\, r(t)$$

$$x_3(t) = \frac{dx_2(t)}{dt} - h_2\, r(t)$$

$$\vdots$$

$$x_k(t) = \frac{dx_{k-1}(t)}{dt} - h_{k-1}\, r(t) \qquad (k = 2, 3, \ldots, n)$$

where h's are expressed in terms of a's and b's as

$$h_k = (b_k - a_k\, b_o) - a_{k-1}\, h_1 - a_{k-2}\, h_2 - \ldots - a_2\, h_{k-2} - a_1\, h_{k-1};$$

every h_k expression is to have k terms.

For this system, the n state equations are

$$\frac{dx_{k-1}(t)}{dt} = x_k(t) + h_{k-1}\, r(t) \qquad (k = 2, 3, \ldots, n)$$

and
$$\frac{dx_n(t)}{dt} = -a_n\, x_1(t) - a_{n-1}\, x_2(t) - a_{n-2}\, x_3(t) - \ldots - a_1\, x_n(t) + h_n\, r(t)$$

This set of ***n state equations*** is expressed in matrix form as

$$\dot{\mathbf{x}}(t) = \mathbf{A}\mathbf{x}(t) + \mathbf{B}\, r(t); \qquad \dot{\mathbf{x}}(t) \text{ being } \frac{d\mathbf{x}(t)}{dt}.$$

$\mathbf{x}(t)$ is an $n \times 1$ matrix of the state variables, usually called the state vector; $\mathbf{A}$ is an $n \times n$ matrix and

B is an n×1 matrix.

For such a system, the matrix **A** is

$$\begin{bmatrix} 0 & 1 & 0 & 0\ldots & 0 \\ 0 & 0 & 1 & 0\ldots & 0 \\ 0 & 0 & 0 & 1.. & 0 \\ \cdots & & & & \\ -a_n & -a_{n-1}\cdots & & & -a_1 \end{bmatrix}$$

The output equation, which is the very first state variable equation, is

$$c(t) = x_1(t) + b_o r(t)$$

The state equations and the output equations together form the dynamic equations of the system.

22.3 MULTIVARIABLE SYSTEMS

For a linear system, with p inputs and q outputs, the state equations are a set of n first-order differential equations.

$$\frac{dx_i(t)}{dt} = f_i\ [x_1(t), x_2(t), \ldots, x_n(t), r_1(t), r_2(t), \ldots, r_p(t)]\ ; \qquad (i = 1, 2, \ldots, n)$$

where $r_1(t), r_2(t), \ldots, r_p(t)$ are the inputs, and f_i denotes the ith linear functional relationship.

The outputs are expressed as

$$c_k(t) = g_k[x_1(t), x_2(t), \ldots, x_n(t), r_1(t), r_2(t), \ldots, r_p(t)]\ ;(k = 1, 2, \ldots, q)$$

where g_k denotes the kth linear functional relationship.

The matrix form of the state equations and the output equations for a system with p inputs and q outputs is

$$\dot{\mathbf{x}}(t) = \mathbf{A}\mathbf{x}(t) + \mathbf{B}\mathbf{r}(t) = \mathbf{f}[\mathbf{x}(t), \mathbf{r}(t)]$$

$$\mathbf{c}(t) = \mathbf{D}\mathbf{x}(t) + \mathbf{E}\mathbf{r}(t) = \mathbf{g}[\mathbf{x}(t), \mathbf{r}(t)]$$

$\mathbf{r}(t)$ is a $p \times 1$ matrix of input variables, usually called the input vector; $\mathbf{B}$ is an $n \times p$ matrix; $\mathbf{c}(t)$ is a $q \times 1$ matrix of output variables, usually called the output vector; $\mathbf{D}$ is a $q \times n$ matrix; and $\mathbf{E}$ is a $q \times p$ matrix.

The solution of the state equations is called the ***state transition equation***.

$$\mathbf{X}(s) = (s\mathbf{I} - \mathbf{A})^{-1}\mathbf{x}(0^+) + (s\mathbf{I} - \mathbf{A})^{-1}\mathbf{B}\mathbf{R}(s)$$

22.4 STATE TRANSITION MATRIX

$$(s\mathbf{I} - \mathbf{A})^{-1} = \frac{\mathbf{I}}{s} + \frac{\mathbf{A}}{s^2} + \frac{\mathbf{A}^2}{s^3} + \ldots$$

$$\pounds^{-1}[(s\mathbf{I} - \mathbf{A})^{-1}] = \mathbf{I} + \mathbf{A}\,t + \frac{\mathbf{A}^2t^2}{2!} + \ldots + \frac{\mathbf{A}^n t^n}{n!} + \ldots \text{ for } t \geq 0$$

Thus, $\quad \pounds^{-1}[(s\mathbf{I} - \mathbf{A})^{-1}] = e^{\mathbf{A}t} = \mathbf{\Phi}(t), \qquad t \geq 0$

$\mathbf{\Phi}(t)$ is called the state transition matrix of $\mathbf{A}$.

$\mathbf{\Phi}(t)$ has the properties that

$$\mathbf{\Phi}(0) = \mathbf{I}$$

$$\mathbf{\Phi}(t_2 - t_1)\,\mathbf{\Phi}(t_1 - t_o) = \mathbf{\Phi}(t_2 - t_o) \quad \text{for any } t_o, t_1, t_2$$

$$\mathbf{\Phi}^{-1}(t) = \mathbf{\Phi}(-t)$$

$$\mathbf{x}(t) = \mathbf{\Phi}(t - t_o)\,\mathbf{x}(t_o) + \int_{t_o}^{t} \mathbf{\Phi}(t - \tau)\mathbf{B}\,\mathbf{r}(\tau)\,d\tau \qquad (t \geq t_o)$$

22.5 RELATIONSHIP BETWEEN THE STATE EQUATIONS AND THE TRANSFER FUNCTIONS

We have $\quad \dot{\mathbf{x}}(t) = \mathbf{A}\mathbf{x}(t) + \mathbf{B}\mathbf{r}(t) = \mathbf{f}[\mathbf{x}(t), \mathbf{r}(t)]$

$$\mathbf{c}(t) = \mathbf{Dx}(t) + \mathbf{Er}(t) = \mathbf{g}[\mathbf{x}(t), \mathbf{r}(t)] \text{ and}$$

$$\mathbf{X}(s) = (s\mathbf{I} - \mathbf{A})^{-1}\mathbf{x}(0^+) + (s\mathbf{I} - \mathbf{A})^{-1}\mathbf{BR}(s)$$

If zero initial conditions are assumed as required to define transfer functions,

$$\mathbf{X}(s) = (s\mathbf{I} - \mathbf{A})^{-1}\mathbf{BR}(s)$$

Also,
$$\mathbf{C}(s) = \mathbf{DX}(s) + \mathbf{ER}(s)$$

$$= \mathbf{D}(s\mathbf{I} - \mathbf{A})^{-1}\mathbf{BR}(s) + \mathbf{ER}(s)$$

$$= [\mathbf{D}(s\mathbf{I} - \mathbf{A})^{-1}\mathbf{B} + \mathbf{E}]\mathbf{R}(s)$$

giving the transfer function matrix

$$\mathbf{G}(s) = \mathbf{D}(s\mathbf{I} - \mathbf{A})^{-1}\mathbf{B} + \mathbf{E}$$

From this expression of transfer function, one can see that the ***characteristic equation*** of the system is also given by

$$| s\mathbf{I} - \mathbf{A} | = 0$$

The roots of the characteristic equation are called the ***eigenvalues*** of the matrix **A**.

Chapter 23

TRANSFER FUNCTIONS AND TIME DOMAIN ANALYSIS

23.1 GENERAL

A control system may include control elements of several types, *viz*.—mechanical, electrical, electromechanical, hydraulic, pneumatic, and various other types. To find the overall transfer function, or to develop the state equations of the control system, one should first obtain the governing equations relating the variables of interest of individual elements.

Considering ***translational motion***; the equations relating force $f(t)$ and the displacement $y(t)$ for a mass M, for a spring of stiffness constant K, and for a friction element (dashpot) of viscous friction constant B are respectively

$$f(t) = M\frac{d^2y(t)}{dt^2}$$

$$f(t) = K\,y(t)$$

$$f(t) = B\frac{dy(t)}{dt}$$

If we take $f(t)$ as analogous to electric voltage $e(t)$, and the displacement $y(t)$ as analogous to electric charge $q(t)$, we find that the mass and the inductance, the spring and the capacitance, and the dashpot and the resistor are analogous quantities as

$$e(t) = L\frac{d^2q(t)}{dt^2}$$

$$e(t) = \frac{1}{C}q(t)$$

$$e(t) = R\frac{dq(t)}{dt}$$

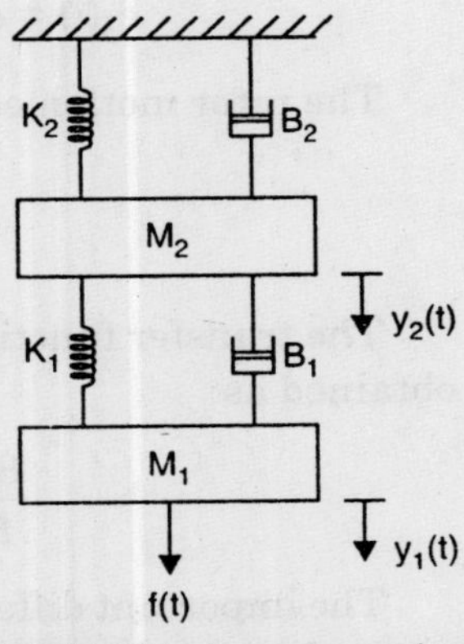

FIGURE 23.1. A mass, spring and dashpot system.

For the mechanical system shown, comprising two mass elements, two springs, and two dashpots, one can obtain the force balance equations as

$$f(t) = M_1 + \frac{d^2y_1(t)}{dt^2} + B_1\left[\frac{dy_1}{dt} - \frac{dy_2}{dt}\right] + K_1\,[y_1(t) - y_2(t)]$$

$$0 = -B_1\left[\frac{dy_1}{dt} - \frac{dy_2}{dt}\right] - K_1[y_1(t) - y_2(t)] + M_2\frac{d^2y_2(t)}{dt^2} + B_2\frac{dy_2(t)}{dt} + K_2y_2(t)$$

These can be converted into four state equations by defining the four state variables

$$x_1 = y_1, \quad x_2 = y_2, \quad x_3 = \frac{dx_1}{dt}, \text{ and } x_4 = \frac{dx_2}{dt}$$

23.2 TRANSFER FUNCTION OF A DC MOTOR

(*a*) ***Field Controlled :*** For such a control, the armature current I_a is constant, the gap flux $\phi(t) = K_f\,i_f(t)$, the torque developed $T_m(t) = K_m'I_a\,\phi(t) = K_m'I_a\,K_f\,i_f(t) = K_m\,K_f\,i_f(t)$, where $K_m = K_m'I_a$.

The field circuit equation is $\quad e_f(t) = R_f\,i_f(t) + L_f\dfrac{di_f(t)}{dt}$

The rotor motion equation is $\quad T_m(t) = B_m\dfrac{d\theta_m(t)}{dt} + J_m\dfrac{d^2\theta_m(t)}{dt^2}$

The transfer function between the output variable $\theta_m(t)$ and the input variable $e_f(t)$ can be obtained as

$$\frac{\Theta_m(s)}{E_f(s)} = \frac{K_mK_f}{L_fJ_ms^3 + (R_fJ_m + L_fB_m)s^2 + R_fB_ms} = \frac{K_mK_f}{R_fB_ms(1+T_ms)(1+T_fs)};$$

where $\quad T_m = J_m/B_m$, and $T_f = L_f/R_f$

(*b*) ***Armature controlled :*** In this case, the field current and hence the air gap flux φ is constant. The torque $T_m(t) = K_i\,i_a(t)$.

The armature circuit equation is

$$e_a(t) = e_b(t) + R_a\,i_a(t) + L_a\frac{di_a(t)}{dt} = K_b\frac{d\theta_m(t)}{dt} + R_a\,i_a(t) + L_a\frac{di_a(t)}{dt}$$

The rotor motion equation remains the same, *i.e.*,

$$T_m(t) = B_m\frac{d\theta_m(t)}{dt} + J_m\frac{d^2\theta_m(t)}{dt^2}.$$

The transfer function between the output variable $\theta_m(t)$ and the input variable $e_a(t)$ can be obtained as

$$\frac{\Theta_m(s)}{E_a(s)} = \frac{K_i}{L_aJ_ms^3 + (R_aJ_m + L_aB_m)s^2 + (R_aB_m + K_bK_i)s}$$

The important difference between the field controlled motor and the armature controlled motor is that the former is an open loop system whereas the latter is a feedback system via back emf $e_b(t)$. Also, the armature controlled motor has relatively better stability and better damping characteristics.

23.3 TRANSFER FUNCTION OF AN AC SERVOMOTOR

This is an ac two-phase squirrel cage induction motor. As in a single-phase *IM*, there are two windings with their axes 90° electrical apart in space. The fixed phase is supplied from a constant ac voltage source. The control phase gets supply from the output of a servo amplifier. This voltage has a phase difference of 90° with the fixed phase voltage. The magnitude of this voltage varies in proportion to the error signal. With no error signal, the torque developed is zero.

The rotor is designed with high resistance so that the torque-speed characteristics are linear right from standstill.

$$k = \frac{\text{blocked rotor torque at control voltage } E_2 = E_1}{\text{rated control voltage } E_1} = \frac{T_0}{E_1}$$

$$m = -\frac{\text{blocked rotor torque at a particular control voltage}}{\text{no load speed at the same control voltage}} = -\frac{T_0}{n_0}$$

The torque $\quad T_m(t) = m\,\dfrac{d(\theta_m)}{dt} + k\,e_2(t)$

Also, $\quad T_m(t) = J_m\,\dfrac{d^2(\theta_m)}{dt^2} + B_m\,\dfrac{d(\theta_m)}{dt}$

It leads to $\quad \dfrac{\Theta_m(s)}{E_2(s)} = \dfrac{K_m}{s(1+\tau_m s)}$, where $K_m = \dfrac{k}{B_m - m}$, $\tau_m = \dfrac{J_m}{B_m - m}$

If m is positive and $> B_m$, negative damping occurs, and the motor becomes unstable.

AC servomotors are used for low power control applications.

AC tachometer is similar to the two phase servomotor except that the control winding now instead of being supplied from the error signal, develops an emf across it proportional to the rotor speed and it serves as the output signal.

$$e_t(t) = K_t\,\frac{d(\theta_m)}{dt}$$

$$\frac{E_t(s)}{\Theta_m(s)} = K_t\,s$$

23.4 EFFECTS OF FEEDBACK

As already stated, feedback exists in a system if a closed sequence of cause and effect relationship is found to exist among variables of the system. Feedback causes improvement in some performance parameters of the control system, while a few are affected adversely.

Feedback reduces the gain of the system over one range of frequency, while it increases it over another range. The gain of the system is given by $|M(j\omega)|$.

Feedback reduces sensitivity with respect to a parameter located in the forward path of a loop. The larger the loop gain, the more effective feedback is in reducing the sensitivity. The sensitivity is defined as the ratio of the fractional change in some specific system quantity to the fractional change in one of the system parameters. For example,

$$S^M{}_k = \frac{d\,Ln\,M}{d\,Ln\,k} = \frac{dM/M}{dk/k}$$

Unwanted signals from external or internal sources corrupt the true signal and thus cause distortion in the output. The effect of such extraneous signals can be mitigated with feedback. The extent of mitigation will depend on the point at which the extraneous signal enters the system. The effect of feedback on distortion is brought out by change in signal to noise ratio due to feedback.

The bandwidth is the characteristic of the frequency response of a system. It is a measure of the ability of the system to reproduce the input signal faithfully, and of the noise rejection characteristic. The feedback has the effect of reducing the gain and increasing the bandwidth.

With reference to a pair of terminals, if the feedback signal is in series with the terminals, the feedback is said to be series feedback; and if the feedback signal is in parallel with the pair, shunt feedback exists. It is found that shunt feedback decreases the impedance, while series feedback increases the impedance.

The negative feedback generally gives improved characteristics for sensitivity, distortion and others, but it may have an adverse effect on system stability if the loop gain exceeds unity.

23.5 TIME DOMAIN ANALYSIS

The time domain response $c(t)$ of a control system is split into transient response $c_t(t)$ and the steady state response $c_{ss}(t)$, *i.e.*, $c(t) = c_t(t) + c_{ss}(t)$. $c_t(t)$ approaches zero as t approaches infinity. Thus, $c_{ss}(t)$ is the response when t approaches infinity. If the output steady state response does not agree with the input exactly, the system is said to have steady state error.

In the time domain analysis, three standard input signals are used; *viz*.- step function $Ru(t)$, ramp function $Rtu(t)$, and parabolic function $Rt^2u(t)$.

The various parameters of transient performance of a control system are defined for step function input

(i) ***Overshoot*** is an indication of the largest error between input and output during the transient state.

Percent overshoot = ((maximum overshoot)/(final desired value)) × 100

(ii) ***Time delay*** is the time required for the response to reach 50% of the final value.

(iii) ***Rise time*** is the time required for the response to rise from 10% to 90% of the final value.

(iv) ***Settling time*** is the time for the response to reach a value which is not more than a certain %age of the final value; usually 105%. Thus, the % overshoot would have fallen to 5%.

Characteristic equation of a closed loop transfer function is the denominator expression of M(s), equated to zero. As $M(s) = \dfrac{G(s)}{1+G(s)\,H(s)}$, the Characteristic equation is

$1 + G(s)\,H(s) = 0.$

Let $M(s)$ be expressed as

$$\frac{K(s+z_1)(s+z_2)\ldots(s+z_m)}{\prod_{i=1}^{i=k}(s+\sigma_i)\prod_{j=1}^{j=q}(s^2+2\alpha_j s+\omega^2_{nj})}$$

This function has m zeros, k simple poles, and q pairs of complex conjugate poles.

The step function response can be expresses as

$$c(t) = \pounds^{-1}\left[\frac{A}{s}+\sum_{i=1}^{i=k}\frac{B_i}{(s+\sigma_i)}+\sum_{j=1}^{j=q}\frac{C_j s+D_j}{s^2+2\alpha_j s+\omega^2_{nj}}\right]$$

This output response is seen to be

$$c(t) = A\,u(t)+\sum_{i=1}^{i=k}B_i e^{-\sigma_i t}+\sum_{j=1}^{j=q}D_j E_j\, e^{-\alpha_j t}\sin\left(\sqrt{\omega^2_{nj}-\alpha^2_j}\,t+\varphi_j\right)$$

where $E_j = \sqrt{\dfrac{1-2a_j\alpha_j+a_j^2\omega^2_{nj}}{\omega^2_{nj}-\alpha^2_j}}$; and $\phi_j = \tan^{-1}\dfrac{a_j\sqrt{\omega^2_{nj}-\alpha^2_j}}{1-a_j\alpha_j}$; a_j being C_j/D_j

One can see the relation between the roots of the characteristic equation, *i.e.*, the poles of the system; and the transient response of the system. The real parts of the roots determine how quickly the transient decays. Farther from the *jw* axis (imaginary axis) are the roots located in the left half of the complex *s*-plane, faster is the rate of decay of the transient. The roots which are closest to the *jw* axis are called dominant roots as all other transient components would decay more rapidly.

Also, farther from the σ axis (real axis) are the roots located, higher is the frequency of the corresponding sinusoid in the time domain.

The roots in the right half of the complex *s*-plane imply transients rising with time. This further means that the system is unstable.

The zeros of the system function do not affect the time behavior as such, but these play role in determining the residues, *i.e.*, B_i, C_i, and D_i.

23.6 SECOND-ORDER SYSTEM

The closed loop transfer function of a second-order system can be expressed as

$$\frac{C(s)}{R(s)} = \frac{\omega_n^{\;2}}{s^2+2\delta\omega_n s+\omega_n^{\;2}} = \frac{\omega_n^{\;2}}{s^2+2\alpha s+\alpha^2+\omega_c^{\;2}}$$

The roots of the characteristic equation are :

$$-\delta\omega_n \pm j\omega_n\sqrt{1-\delta^2} = -\alpha \pm j\omega_c$$

δ is the damping ratio or the damping factor, and equals $\cos\theta = \alpha/\omega_n$

α is the damping constant, equal to $\delta\omega_n$, and is the negative of the real part of the complex roots

ω_n is the undamped natural frequency, and is the radial distance of the complex roots from the origin in the complex s-plane

ω_c is the conditional or the damped frequency, equal to $\omega_n\sqrt{1-\delta^2}$, and is the imaginary part of the complex roots

$\delta = 0$ means no damping, and the response is completely oscillatory (undamped oscillations)

$0 < \delta < 1$ means under damped case, and the response is damped (damped oscillations)

$\delta = 1$ means critically damped case, and the response is damped and non oscillatory

$\delta > 1$ means over damped case, and the response is damped and non oscillatory

We may recall that

$$\text{£}^{-1}\frac{1}{(s^2+2\delta\omega_n s+\omega^2{}_n)} = \frac{e^{-\delta\omega_n t}}{\omega_n\sqrt{1-\delta^2}}\sin\left[\omega_n\sqrt{1-\delta^2}\,t\right]$$

or

$$\text{£}^{-1}\frac{1}{(s^2+2\alpha s+\alpha^2+\omega^2{}_c)} = \frac{e^{-\alpha t}}{\omega_c}\sin\omega_c t$$

$$\text{£}^{-1}\frac{\omega^2{}_n}{(s^2+2\delta\omega_n s+\omega^2{}_n)} = \frac{\omega_n\, e^{-\delta\omega_n t}}{\sqrt{1-\delta^2}}\sin\left[\omega_n\sqrt{1-\delta^2}\,t\right]$$

$$\text{£}^{-1}\frac{\omega^2{}_n}{s(s^2+2\delta\omega_n s+\omega^2{}_n)} = 1+\frac{e^{-\delta\omega_n t}}{\sqrt{1-\delta^2}}\sin\left[\omega_n\sqrt{1-\delta^2}\,t-\varphi\right],$$

where
$$\varphi = \tan^{-1}\frac{\sqrt{1-\delta^2}}{-\delta}, (t \geq 0)$$

The step response of the second-order system is given by

$$c(t)\,\text{£}^{-1}\frac{\omega^2{}_n}{s(s^2+2\delta\omega_n s+\omega^2{}_n)}$$

$$= 1+\frac{e^{-\delta\omega_n t}}{\sqrt{1-\delta^2}}\sin\left[\omega_n\sqrt{1-\delta^2}\,t-\varphi\right],$$

where
$$\varphi = \tan^{-1}\frac{\sqrt{1-\delta^2}}{-\delta}, (t \geq 0)$$

we see that when $\delta = 0$, $\varphi = -90°$, $\omega_c = \omega_n$

and when $0 < \delta < 1$, $-90° < \varphi < 0°$, $\omega_n < \omega_c < 0$; *i.e.*, as δ increases from 0 to 1, φ increases from $-90°$ to $0°$, and ω_c decreases from ω_n to 0.

In the under damped case, the occurrence of the overshoots, or the undershoots is found by equating $\frac{dc(t)}{dt}$ to zero. This occurs at time given by

$$t = \frac{n\pi}{\omega_n\sqrt{1-\delta^2}} = \frac{n\pi}{\omega_c}$$

The overshoots occur for $n = 1, 3, 5, \ldots,$

The value of the maximum overshoot (first overshoot) is $e^{-\frac{\pi\delta}{\sqrt{1-\delta^2}}}$ and occurs at $t = \dfrac{\pi}{\omega_n\sqrt{1-\delta^2}}$.

The overshoot is seen to be a function of the damping ratio δ alone.

The settling time is found to be

$$T_s = \frac{-Ln\,k}{\delta\omega_n},$$

where k is the fraction of the ss value within which from the ss value the response is to remain.

The ramp response of a second order system is

$$c(t) = \pounds^{-1}\frac{\omega^2{}_n}{s^2(s^2 + 2\delta\omega_n s + \omega^2{}_n)}$$

$$= t - \frac{2\delta}{\omega_n} + \frac{e^{-\delta\omega_n t}}{\omega_n\sqrt{1-\delta^2}}\sin\left[\omega_n\sqrt{1-\delta^2}\,t - \varphi\right],$$

where $$\varphi = 2\tan^{-1}\frac{\sqrt{1-\delta^2}}{-\delta}, \qquad (t \geq 0)$$

The ss response of the system is given by $t - \dfrac{2\delta}{\omega_n}$. A ss error of $\dfrac{2\delta}{\omega_n}$ is present in the output.

Increasing the gain reduces δ. ss error is thus reduced by increasing the gain. Small values of δ may make the system oscillatory, and the system may even go unstable for further increase in the gain. Thus, a compromise has to be struck.

23.7 IMPROVING TRANSIENT AND STEADY-STATE PERFORMANCE

A proportional type of feedback is one which develops a correcting signal proportional to the magnitude of the error signal. During the transient period, the error signal is time dependent. We may add a signal proportional to the time derivative of the error signal to produce an actuating signal $K\{r(t) - c(t)\} + \tau_d \dfrac{d}{dt}\{r(t) - c(t)\}$. With the inclusion of the derivative signal, it is found that the transient performance of the system is improved, in the sense that the value of effective δ increases and thus the overshoot is reduced.

Also, the response is faster, *i.e.*, the rise time becomes shorter.

The damping is also improved if rate feedback (Tachometer) control instead of the derivative control is provided. In this arrangement we add the derivative of the output. Thus, the actuating signal becomes $K\{r(t) - c(t)\} - \tau_d \dfrac{d}{dt} c(t)$

Derivative control does not affect the *ss* error if it does not vary with time. However, if the *ss* error increases with time, the derivative control reduces *ss* error also.

If a signal proportional to the time integral of the error signal is added to the correcting signal to produce the actuating signal $K\{r(t) - c(t)\} + K_I \int \{r(t) - c(t)\}dt$, it is found that the *ss* performance of the system is improved. The transient performance is affected a little for a suitable choice of the proportionality constant K_I of the integrator controller. For a few types of systems and inputs, the *ss* error even vanishes.

23.8 STEADY-STATE ERROR

As the term implies, the steady-state error ε_{ss} is the error $\varepsilon(t)$ found in $c(t)$ as time t approaches infinity. If $b(t)$ is the feedback signal, the error signal is $\varepsilon(t) = r(t) - b(t)$. $\varepsilon_{ss} = \lim_{t \to \infty} \varepsilon(t)$.

Assuming that $sE(s)$ has no multiple poles on the imaginary axis, and no poles in the RHP,

$\varepsilon_{ss} = \lim sE(s)$ as s tends to zero. $E(s) = R(s) - B(s) = R(s) - H(s)\, C(s)$. The ss error is usually evaluated with $H(s) = 1$, *i.e.*, assuming unity feedback. Thus, $E(s) = R(s) - C(s) = R(s) - R(s)G(s)/\{1+G(s)\} = R(s)/\{1+G(s)\}$.

$$\varepsilon_{ss} = \lim_{s \to 0} \frac{sR(s)}{1+G(s)}.$$

Further, the ε_{ss} is generally evaluated for the three standard input signals, *viz.*, $Ru(t)$, $Rtu(t)$, and $Rt^2u(t)/2$. For these three signals, the respective transforms are : $\dfrac{R}{s}$, $\dfrac{R}{s^2}$, and $\dfrac{R}{s^3}$.

Thus, for $r(t) = Ru(t)$, *i.e.*, step displacement input,

$$\varepsilon_{ss} = \frac{R}{1 + \lim_{s \to 0} G(s)} = \frac{R}{1+K_p},$$

where $K_p = \lim_{s \to 0} G(s)$. For $r(t) = Rtu(t)$, *i.e.*, step velocity input,

$$\varepsilon_{ss} = \frac{R}{\lim_{s \to 0} sG(s)} = \frac{R}{K_v}, \text{ where } K_v = \lim_{s \to 0} sG(s),$$

and for $r(t) = Rt^2\, u(t)/2$, *i.e.*, step acceleration input,

$$\varepsilon_{ss} = \frac{R}{\lim\limits_{s \to 0} s^2 G(s)} = \frac{R}{K_a},$$

where $K_a = \lim\limits_{s \to 0} s^2 G(s)$

K_p, K_v, and K_a are respectively called positional error constant, velocity error constant, and the acceleration error constant. These depend upon the 'type' of $G(s)$. $G(s)$ may be expressed as

$$G(s) = \frac{K(1+T_1 s)(1+T_2 s)\ldots(1+T_m s)}{s^j\,(1+T_1{}' s)(1+T_2{}' s)\ldots(1+T_n{}' s)}.$$

The value of j in s^j determines the 'type'of $G(s)$. The $G(s)$ is type 0, 1, 2,...for j = 0, 1, 2,... respectively.

We see that

$K_p = K$ for type 0, and ∞ for all higher order types of $G(s)$.

$K_v = 0$ for type 0, $= K$ for type 1, and ∞ for all higher order types of $G(s)$.

$K_a = 0$ for type 0 and for type 1, $= K$ for type 2, and ∞ for all higher order types of $G(s)$.

All the pertinent information can be given in the tabular form.

Error Constants and Steady-state Error Relations

Type of system	*Error constants*			*Various inputs and the corresponding steady-state errors*		
	$K_p =$	$K_v =$	$K_a =$	$r(t) = Ru(t)$	$r(t) = Rtu(t)$	$r(t) = Rt^2\, u(t)/2$
j	$\lim\limits_{s\to 0} G(s)$	$\lim\limits_{s\to 0} sG(s)$	$\lim\limits_{s\to 0} s^2G(s)$	$\varepsilon_{ss} = \frac{R}{1+K_p}$ *Position error*	$\varepsilon_{ss} = \frac{R}{K_v}$ *Velocity error*	$\varepsilon_{ss} = \frac{R}{K_a}$ *Acceleration error*
0	K	0	0	$\frac{R}{1+K}$	$A + Bt = \infty$	$C + Dt + Et^2 = \infty$
1	∞	K	0	0	R/K	$D + Et = \infty$
2	∞	∞	K	0	0	R/K
3	∞	∞	∞	0	0	0

Chapter 24 STABILITY ANALYSIS OF CONTROL SYSTEMS

24.1 GENERAL

Stability of a system means that its response does not increase indefinitely for any finite input to it. For any bounded input, its response remains bounded. Mathematically, it can be stated as

$$| c(t) | \leq M \leq \infty \text{ for } t \geq t_0, \text{ if } | r(t) | \leq N \leq \infty \text{ for } t \geq t_0.$$

It can also be proved that a stable system has finite value of the integral of the impulse response, *i.e.*, $\int_0^\infty |g(\tau)| d\tau \leq P < \infty$; where $g(\tau)$ is the impulse response of the system whose transfer function is $G(s)$.

For the above relation to hold, the impulse response of the system should remain bounded; which further implies that all the roots of the characteristic equation of the system function must lie only in the left half of the complex s-plane. Stated in other words, the roots of the characteristic equation must all have negative real parts.

Instead of actually solving the characteristic equation to see whether its all roots are in LHP or not, one can apply any one of the various criteria which have been developed to test the location of the roots from stability point of view. One of these, given next, is the Routh-Hurwitz criterion.

24.2 ROUTH-HURWITZ CRITERION

Let the characteristic equation be

$$F(s) = a_0 s^n + a_1 s^{n-1} + a_2 s^{n-2} + \ldots + a_{n-1} s + a_n = 0$$

The necessary conditions for the roots of the characteristic equation not to have positive real parts are that (*i*) all the coefficients of the polynomial have the same sign, and (*ii*) none of the coefficients vanishes. These checks can be readily applied merely by inspecting the equation.

The sufficiency condition is that the polynomial's Hurwitz determinants D_k must be all positive. A Hurwitz determinant D_k is defined as

$$\begin{bmatrix} a_1 & a_3 & a_5 & \cdot & \cdot & \cdot & a_{2k-1} \\ a_0 & a_2 & a_4 & \cdot & \cdot & \cdot & a_{2k-2} \\ 0 & a_1 & a_3 & \cdot & \cdot & \cdot & a_{2k-3} \\ 0 & a_0 & a_2 & \cdot & \cdot & \cdot & a_{2k-4} \\ 0 & 0 & a_1 & \cdot & \cdot & \cdot & a_{2k-5} \\ \cdot & \cdot & \cdot & \cdot & \cdot & \cdot & \cdot \\ \cdot & \cdot & \cdot & \cdot & \cdot & \cdot & \cdot \\ 0 & 0 & 0 & \cdot & \cdot & \cdot & a_{2k-k} \end{bmatrix} ; \; k = 1, 2, \ldots, n.$$

There is a short-cut procedure to test that all the Hurwitz determinants are positive. A Hurwitz array H is required to be constructed.

The polynomial coefficients are arranged in two rows; the first row consists of first, third, fifth, ..., coefficients and the second row consists of the second, fourth, sixth, ..., coefficients of the polynomial. Some of these elements may be zeros even.

$$\begin{matrix} a_o & a_2 & a_4 & a_6 & a_8 & \cdot & \cdot & \cdot \\ a_1 & a_3 & a_5 & a_7 & a_9 & \cdot & \cdot & \cdot \end{matrix}$$

The Hurwitz array H will consist of a total of $(n + 1)$ rows, and $\left(\frac{n}{2}+1\right)$ columns for n even, and $\left(\frac{n}{2}+\frac{1}{2}\right)$ columns for n odd. The additional $(n - 1)$ rows are constructed successively. Say, an element h_{jk} (of jth row and k^{th} column of H) is to be found out. $(j-1)$ rows are available. Let $h_{(j-2)\,1}$, $h_{(j-2)\,(k+1)}$, $h_{(j-1)\,1}$, and $h_{(j-1)\,(k+1)}$ be denoted by A, B, C, and D respectively. Then h_{jk} is given by $\dfrac{CB-AD}{C}$. In the process, one or more of the elements may be found to be zeros.

If, at a particular stage, C becomes zero, then this zero is replaced by a small positive number denoted by ε for the subsequent operations and one proceeds to construct the rows in the normal way.

If, at a particular stage, all the elements in a constructed row turn out to be zeros, then this row is replaced by the coefficients of the polynomial obtained by differentiating the auxiliary polynomial wrt s. The auxiliary polynomial is built from the coefficients in the row just above the row of zeros. For example, if $n = 7$, the array would have 8 rows and 4 columns. If the 5th row (from the top) happens to contain all zeros, and the elements in the 4th row are denoted by h_{41}, h_{42}, h_{43}, and h_{44}; the auxiliary polynomial is $h_{41}s^4 + h_{42}\,s^2 + h_{43}\,s^0$. The zeros row is replaced by the elements $h_{51} = 4 \times h_{41}$, $h_{52} = 2 \times h_{42}$, $h_{53} = 0$, and $h_{54} = 0$.

Such a situation indicates the presence of pairs of conjugate roots (it includes equal and opposite roots also). The number of pairs of roots is given by the degree of the auxiliary polynomial divided by 2.

After having constructed the complete array, the signs of the elements in the first column of this array are inspected. If all these elements have the same sign, there are no roots of the characteristic equation in the RHP of the complex s-plane and the corresponding control system is stable. If there are r sign changes of the elements in the first column, there are r roots in the RHP and the corresponding control system is unstable.

The Routh-Hurwitz criterion determines whether the system is stable or unstable. It gives no information about the relative stability of the system, or the degree of stability of the system.

24.3 THE NYQUIST CRITERION

This criterion finds the number of roots of the characteristic equation located in the RHP of the complex s-plane. As already noted, the closed loop transfer function is

$$M(s) = \frac{G(s)}{1 + G(s)H(s)}$$

and the characteristic equation of the system is

$$F(s) = 1 + G(s)H(s) = 0$$

Let $$F(s) = \frac{(s + z_1)(s + z_2)\ldots(s + z_m)}{s^j (s + p_1)(s + p_2)\ldots(s + p_n)}$$

This function has m zeros located at $s = -z_1, -z_2, \ldots, -z_m$; and $j + n$ poles, *viz*., a pole of multiplicity j at the origin, and n poles located at $s = -p_1, -p_2, \ldots, -p_n$. The zeros of $F(s)$ must not be in the RHP if the closed loop transfer function $M(s)$ is to be stable.

If we write $G(s)H(s)$ as $\dfrac{N(s)}{D(s)}$, then $F(s) = \dfrac{D(s) + N(s)}{D(s)}$. We see that the poles of the loop transfer function $G(s)H(s)$ are the same as those of the function $F(s)$. The zeros of the two functions, in general, would differ.

Now, if there is a closed path Γ marked in clockwise or counterclockwise sense in the complex s-plane, then any function $F(s)$ will also trace a closed path in the complex $F(s)$-plane corresponding to the points on the closed path Γ in the s-plane. If Γ encloses Z zeros and P poles of $F(s)$ in the s-plane, then $F(s)$ plot encircles the origin of $F(s)$-plane $(Z - P)$ times. If $(Z - P)$ is positive, then the encirclement of the origin in the $F(s)$-plane by the $F(s)$ plot corresponding to Γ is in the same sense as the sense of Γ, and if $(Z - P)$ is negative, the encirclements of the origin are in a sense opposite to that of Γ.

Since, our aim is to check that $F(s)$ does not have zeros in the right half of the complex s-plane, we choose Γ in the s-plane such that it covers whole of the RHP. It is thus infinite in extent. It is a semicircle centered at the origin, diameter being whole of the imaginary axis, and the radius being infinite, covering the whole of the RHP of the s-plane. In imagining this closed path Γ, the poles of $F(s)$ on the imaginary axis in the s-plane if any, are excluded from this closed path by drawing tiny semicircles around the poles. The infinite closed path is assigned counterclockwise sense. It can be regarded to be consisting of four sections-one straight segment along the imaginary axis from $+j\infty$ to $+j0$, second semicircle of 0 radius from $+j0$ to $-j0$, third straight segment from $-j0$ to $-j\infty$, and the fourth, a semicircle from $-j\infty$ to $+j\infty$ and of infinite radius. This closed path has been called the Nyquist path.

If there are Z zeros and P poles of $F(s)$ in the RHP, *i.e.*, if the Nyquist path encloses Z zeros and P poles of $F(s)$, then the plot of $F(s)$ in the $F(s)$-plane corresponding to the Nyquist path will encircle the origin $(Z-P)$ times. We can also say that the plot of $(F(s)-1)$ will encircle the point $(-1,\ j0)$, $(Z-P)$ times. $(F(s)-1)$ is $G(s)H(s)$, and the poles of $F(s)$ and those of $G(s)H(s)$ are identical. If the closed loop transfer function $M(s)$ is to be stable, Z should be 0. We, therefore, conclude that for stability of the closed loop transfer function $M(s)$, the plot of the loop transfer function $G(s)H(s)$, called the Nyquist plot, corresponding to the Nyquist path, should encircle the critical point $(-1, j0)$ $-P$ times ($-$ sign means encirclement in the clockwise sense) in the complex $G(s)H(s)$-plane, where P is the number of poles of $G(s)H(s)$ (or of $F(s)$) in the RH of the complex s-plane. Usually, the loop transfer function $G(s)H(s)$ is stable, *i.e.*, $P = 0$. Under this condition, for $M(s)$ to be stable, the encirclement of the critical point $(-1, j0)$ by the $G(s)H(s)$ plot corresponding to the Nyquist path must be zero. Thus, if $P = 0$, the GH plot must not encircle the critical point $(-1, j0)$ for the closed loop transfer function to be stable. However, if $P \neq 0$, and the GH Nyquist plot does not encircle the critical point, the corresponding closed loop system is unstable.

The Nyquist plot is a polar plot of $G(s)H(s)$. Drawing the Nyquist plot itself requires considerable practice. Many a times drawing the plot for section 1 only, *i.e.*, from $+j\infty$ to $+j0$, is adequate, and one can conclude on stability. Usually, one need to find limit of GH as s approaches 0, or ∞.

For a function like

$$G(s)H(s) = \frac{K}{(1+sT_1)(1+sT_2)\ldots(1+sT_n)}$$

the Nyquist locus goes through a total of n quadrants, or as many quadrants as the number of time constants in the $D(s)$, as the frequency is varied from ∞ to 0; the value of the function GH being 0 at $\omega = \infty$, and K at $\omega = 0$. The locus for $\dfrac{K}{(1+sT_1)}$ lies in the 4th quadrant, that for $\dfrac{K}{(1+sT_1)(1+sT_2)}$ extends into 3rd quadrant, and for $\dfrac{K}{(1+sT_1)(1+sT_2)(1+sT_3)}$ extends into 2nd quadrant and so on. The locus is thus extending in the clockwise direction. Adding finite poles like this makes the closed loop system less stable.

If $\dfrac{K}{(1+sT_1)}$ is multiplied by $1/s^n$, the Nyquist plot at infinite and zero frequencies is rotated by an angle $n\pi/2$ in the clockwise direction. Adding poles at the origin also makes the closed loop system less stable.

If additional zeros are introduced by multiplying the given GH function by factors like $(1 + sT_d)$, the corresponding closed loop function's stability improves. Adding zeros is equivalent to applying derivative control.

The frequency at which the GH plot crosses the negative real axis is designated critical frequency of the system, denoted by ω_c. At this frequency, the imaginary part of $G(j\omega)\,H(j\omega)$ is zero. For the closed loop system to be stable, $|\,G(j\omega_c)\,H(j\omega_c)\,|$ should be less than 1.

24.4 ROOT LOCUS TECHNIQUE

A root locus diagram is the locus or the plot of the roots of the characteristic equation, (or the poles) of the closed loop transfer function $M(s)$ if one or more of the parameters of the open loop functions $G(s)$ or $H(s)$ is/are varied from $-\infty$ to $+\infty$. From such a diagram, one can quickly see the range of the parameter over which the system goes unstable.

Root locus is usually confined to varying one parameter only, say the gain K. When K varies from 0 to ∞, the plot is called the root locus diagram, and when it varies from 0 to $-\infty$, it is called inverse root locus diagram.

The root locus is the plot of the roots of the characteristic equation of the closed loop transfer function $M(s)$, *i.e.*, $1 + G(s)H(s) = 0$ as K is varied. Thus, every point s on the locus will satisfy the above equation, or shall satisfy the equation $G(s)H(s) = -1$. This in turn means

$$|\,G(s)H(s)\,| = 1, \quad \text{and} \quad \angle(G(s)H(s)) = (2k+1)\pi\,;\ k = 0, \pm 1, \pm 2, \ldots$$

Let $$G(s)H(s) = \frac{K(s+z_1)(s+z_2)\ldots(s+z_m)}{(s+p_1)(s+p_2)\ldots(s+p_{m+n})}$$

The above stated two conditions then become

$$|\,G(s)H(s)\,| = \frac{|K|\prod_{i=1}^{m}|s+z_i|}{\prod_{j=1}^{m+n}|s+p_j|} = 1\ (-\infty < K < \infty)$$

and $$\angle G(s)H(s) = \angle K + \sum_{i=1}^{m}\angle(s+z_i) - \sum_{j=1}^{m+n}\angle(s+p_j) = (2k+1)\pi, \quad k = 0, \pm 1, \pm 2, \ldots$$

A point on the root locus will thus satisfy above two equations.

By analyzing these two equations, one can discover several properties of the root locus diagram.

Stated below are some of the properties and rules to construct the root locus diagram.

- $K = 0$ points

The K = 0 points on the root loci are at the poles of $G(S)H(s)$. The loci start at the poles.

- $K = \pm\infty$ points

The $K = \pm\infty$ points on the root loci are at the zeros of $G(S)H(s)$. The loci end at the zeros.

- Number of separate loci

If P and Z are respectively the number of finite poles and the number of finite zeros of G(S)H(s), then if $P > Z$, the number of separate loci is P; else it is Z.

- Symmetry property

The root locus plots of systems with rational transfer functions with constant coefficients are symmetrical with respect to the real axis.

- Asymptotes of root loci as $s \to \infty$

For large values of s, the root loci (positive K) are asymptotic to straight lines with angles given by $\theta_k = \dfrac{(2k+1)\pi}{P-Z}$, and for the inverse loci (negative K), $\theta_k = \dfrac{2k\pi}{P-Z}$; where $k = 0$, 1, 2, . . . , $(P - Z - 1)$

- Intersection of the asymptotes

The point of intersection of the asymptotes is on the real axis, and is given by

$$\sigma_1 = \frac{\sum \text{Poles of } G(s)H(s) - \sum \text{zeros of } G(s)H(s)}{P-Z}$$

- Root loci on the real axis

On a given section of the real axis in the s-plane, root locus is found only if the total number of real poles and real zeros of $G(s)H(s)$ to the right of the section is odd. If this number is even, inverse loci would be found.

- Intersection of the root loci with the imaginary axis

By applying the Routh-Hurwitz criterion, or from the Bode plot, one can find the value of K and ω at which the loci crosses the imaginary axis.

- Breakaway points

The points on the root loci where multiple roots of the characteristic equation are found are called the breakaway points or the saddle points of the root loci. The root loci must approach and leave a breakaway point on the real axis at an angle of $180°/n$ apart, where n is the number of root loci approaching and leaving the point. The breakaway points are found by finding the roots of $dK/ds = 0$. These points can be found from $d((G(s)H(s))/ds = 0$ also; this equation however, is only the necessary condition for the breakaway points.

- Calculation of K at a point on the root loci

The absolute value of K for any point s_1 on the root loci is found as

$$|K| = \frac{1}{|G(s_1)H(s_1)|} = \frac{\prod \text{All vector length drawn from the poles of } G(s)H(s) \text{ to } s_1}{\prod \text{All vector lengths drawn from the zeros of } G(s)H(s) \text{ to } s_1}$$

Chapter 25 Frequency Response Characteristics

25.1 GENERAL

Because of the greater computational effort involved in the time domain analysis, it is usually difficult to obtain time response of the feedback control systems. Frequency response is often obtained by means of graphical methods and the time domain behavior is found from the relationships between the frequency response and the time response. As such it is pertinent to study frequency response of the control systems.

In the frequency response study, s in the expression for the transfer function $M(s)$ is substituted by $j\omega$ and we get the expression of $M(j\omega)$; the resulting expression is simplified to obtain the magnitude response function $M(\omega)$, and the phase response function $\phi(\omega)$.

Unity feedback path gain is assumed.

$$M(s) = \frac{C(s)}{R(s)} = \frac{G(s)}{1+G(s)} \quad \text{(Unity feedback path gain assumed)}$$

$$M(j\omega) = \frac{G(j\omega)}{1+G(j\omega)} = M(\omega)\ \angle\phi(\omega)$$

$$M(\omega) = \left|\frac{G(j\omega)}{1+G(j\omega)}\right| \quad \text{and} \quad \phi(\omega) = \angle G(j\omega) - \angle(1 + G(j\omega))$$

If the output is to follow the input exactly, $M(\omega)$ should be unity at all frequencies and phase shift must be a linear function of frequency. For $M(\omega)$ to be unity at all frequencies, the loop gain of the system must be infinite at all frequencies. This is neither physically possible, nor desirable from stability, and signal-to-noise ratio points of view. Typically, $M(\omega) - \omega$ curve may look as shown in Fig. 25.1 (*a*). The gain, after an initial increase with frequency, decreases as frequency increases. This is due to inertia and inductive elements present in the system.

A typical $\phi(\omega) - \omega$ curve may look as shown in Fig. 25.1 (*b*). At high frequencies, $\phi(\omega)$ is almost constant. It is so because at high frequencies the real and imaginary parts of $G(j\omega)$ become nearly constant and hence $\angle G\,(j\omega)$ and $\angle(1 + G\,(j\omega))$ are constant.

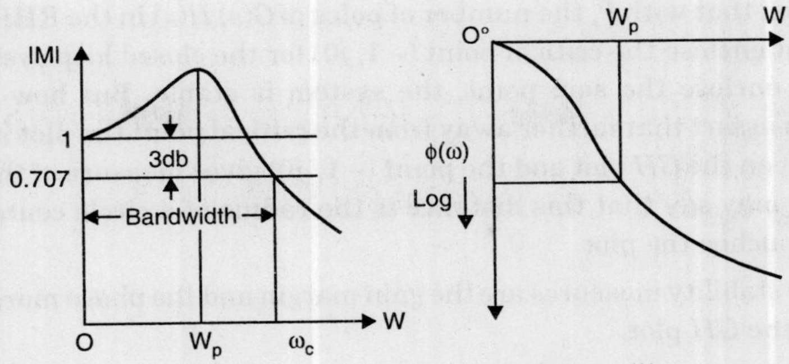

FIGURE 25.1. Typical frequency response characteristics of a control system. (*a*) Gain characteristic, (*b*) Phase characteristic.

25.2 FREQUENCY DOMAIN SPECIFICATIONS

1. ***Bandwidth*** is the frequency ω_c at which $M(\omega)$ drops to 70.7 % of the value of $M(\omega)$ at zero frequency. This also means that the gain drops by 3 db.

If the bandwidth is large, the response is faster and the overshoot is high.

2. ***Peak resonance*** M_p is the maximum value of $M(\omega)$. High M_p corresponds to a large overshoot in the time response. M_p is usually kept between 1.1 and 1.5.

3. ***Resonant frequency*** ω_p is the frequency at which the peak resonance M_p occurs.

4. ***Cut-off rate*** $dM/d\omega$ is defined for frequencies beyond ω_c. High cut-off rate implies high degree of rejection of noise signal.

For a ***second-order system*** for which we express $M(s)$ as $\dfrac{\omega_n^2}{s^2 + 2\delta\omega_n s + \omega_n^2}$, we find

$$M(u) = \frac{1}{\left[(1-u^2)^2 + (2\delta u)^2\right]^{1/2}}, \text{ and } \phi(u) = -\tan^{-1}\frac{2\delta u}{1-u^2} \quad \text{where } u = \frac{\omega}{\omega_n}.$$

$$\omega_p = \omega_n\sqrt{1-2\delta^2} \quad \text{and} \quad M_p = \frac{1}{2\delta\sqrt{1-\delta^2}}$$

It is noted that there is no resonance peak for $\delta > 0.707$. With increase in δ; M_p and ω_p both decrease, though with different rates of decrease.

25.3 RELATIVE STABILITY

If it is found that the system is stable, one would like to know how stable is it? How can the stability condition of the system be improved?

We have seen that with P, the number of poles of $G(s)\,H(s)$ in the RHP, equal to zero, the GH plot should not enclose the critical point $(-1, j0)$ for the closed loop system to be stable. If the plot does not enclose the said point, the system is stable. But how stable is it? Well, intuitively, we can assert that farther away from the critical point the plot is, more stable it is. The distance between the GH plot and the point $(-1, j0)$ gives measure of the relative stability of the system. We may say that this distance is the radius of a circle centered at the critical point which just touches the plot.

The relative stability measures are the gain margin and the phase margin. These margins are defined w.r.t. the GH plot.

The gain margin is the measure of the closeness of the phase crossover point to the critical point. And the phase crossover point is the point where the phase of the GH plot becomes $-180°$. The frequency at which the phase crossover occurs is the phase crossover frequency, ω_c.

Then the GM in db is $0 - 20 \log |G(j\omega_c)\,H(j\omega_c)| = 20 \log 1/(|G(j\omega_c)\,H(j\omega_c)|)$. Thus, the gain margin is the amount of gain in decibels that can be allowed to increase in the loop before the system reaches instability.

The phase margin is the measure of the closeness of the gain crossover point to the critical point. And the gain crossover point is the point where the gain of the GH plot becomes 1. The frequency at which the gain crossover occurs is the gain crossover frequency, ω_h.

Then the $\Phi M = \angle(G(j\omega_h)\,H(j\omega_h)) - 180°$. Thus, the phase margin is the angle through which the Nyquist locus of $G(s)\,H(s)$ must be rotated in order that the unity magnitude point on the locus passes through the critical point.

These margins can be shown on the Bode plot of the GH function. Larger the values of the GM and the ΦM, more stable the system is.

25.4 CONSTANT M LOCI IN THE G-PLANE

M versus ω curve can be obtained from the $G(j\omega)$ plot.

If we set $G(j\omega) = x + jy$, then $M(j\omega) = (x + jy)/(1+(x + jy))$. It is then seen that

$$|M(j\omega)| = M(\omega) = \frac{\sqrt{x^2 + y^2}}{\sqrt{(1+x)^2 + y^2}}.$$

If M is treated as a parameter, we can show that

$$\left(x - \frac{M^2}{1-M^2}\right)^2 + y^2 = \left(\frac{M}{1-M^2}\right)^2;$$

which represents a family of circles in the G-plane with x = real of $G(j\omega)$, and y = imaginary of $G(j\omega)$. The centre is at $x = \frac{M^2}{1-M^2}$, and $y = 0$. The radius of the circle is $\frac{M}{1-M^2}$.

The constant M loci are symmetrical w.r.t. the real axis and the $M = -1$ circle, which is a straight line. The M circles to the left of this line correspond to values of $M > 1$, and those to the right of this line are for $M < 1$.

25.5 CONSTANT PHASE SHIFT LOCI IN THE G-PLANE

We have
$$\phi(\omega) = \angle G(j\omega) - \angle(1 + G(j\omega)) = \tan^{-1}\left(\frac{y}{x}\right) - \tan^{-1}\left(\frac{y}{1+x}\right)$$

Let
$$N = \tan \phi$$

We then have $N = \dfrac{y}{x^2 + x + y^2}$. This can also be reduced to a family of circles with N as a parameter.

$$\left(x + \frac{1}{2}\right)^2 + \left(y - \frac{1}{2N}\right)^2 = \left(\frac{1}{4} + \frac{1}{4N^2}\right)$$

The centre is at $x = -\dfrac{1}{2}$, and $y = \dfrac{1}{2N}$. The radius of the circle is $\sqrt{\dfrac{1}{4} + \dfrac{1}{4N^2}} = \sqrt{\dfrac{N^2 + 1}{4N^2}}$

The constant N loci are symmetrical w.r.t. the real axis and the $x = -1/2$ straight line. The N circles above the real axis correspond to values of $\phi > 0$, the angle decreasing in value upward; and those below the real axis are for $\phi < 0$, the absolute angle decreasing downward.

25.6 NICHOLS CHART

The Nichols charts are the relationships between the gain of $G(j\omega)$ in decibels and the phase shift of $G(j\omega)$ in degrees; and are plotted in the plane called the gain-phase plane. These are obtained from the constant M and constant N circles. A particular M circle in the polar plot of G is transformed into a gain-phase plot of G in the gain-phase plane. This plot is also a closed curve but not a circle.

Nichols charts are more convenient to use in control systems design problems.

Chapter 26 ADDITIONAL ASPECTS OF CONTROL SYSTEMS

26.1 GENERAL

A control system with only its minimum required components seldom gives satisfactory performance. Compensating networks are often needed in the forward/ feedback path to improve the performance of the system.

For *dc* or unmodulated signals, the frequency range usually lies between 0 and 20 cycles per second; while for *ac* or modulated signals, the range is from 60 to 400 cycles per second.

On *dc* signals, the compensation networks are usually $R-C$ networks which can introduce phase lead or phase lag.

26.2 PHASE-LEAD NETWORK

This network is simply two branches in series; one branch being a parallel circuit of a resistor R_1 and a capacitor C, and the other being only a resistor R_2. The input signal is E_1, and the output is E_2, across R_2.

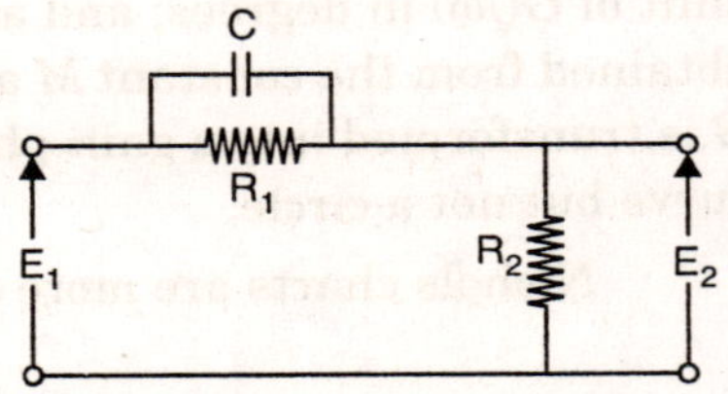

FIGURE 26.1. A phase-lead network.

It can be found that

$$\frac{E_2(s)}{E_1(s)} = \frac{1}{a}\,\frac{1+aTs}{1+Ts},$$

where $$a = \frac{R_1+R_2}{R_2}\ (>1), \quad \text{and} \quad T = \frac{R_1R_2}{R_1+R_2}C$$

The general effects of the phase lead compensation are :

- The phase shift in the vicinity of the resonant frequency is increased in the positive direction.
- For a given relative stability, the velocity constant is usually increased.

- For a given gain constant K, the relative stability of the system is usually improved. It implies that the phase margin is increased and the resonance peak M_p is reduced.
- The bandwidth is generally increased.
- The overshoot is reduced and the response rise is faster.

26.3 PHASE-LAG NETWORK

This network is simply two branches in series; one branch being a resistor R_1, and the other a series circuit of a resistor R_2 and a capacitor C. The input signal is E_1, and the output is E_2, across the series combination of R_2 and C.

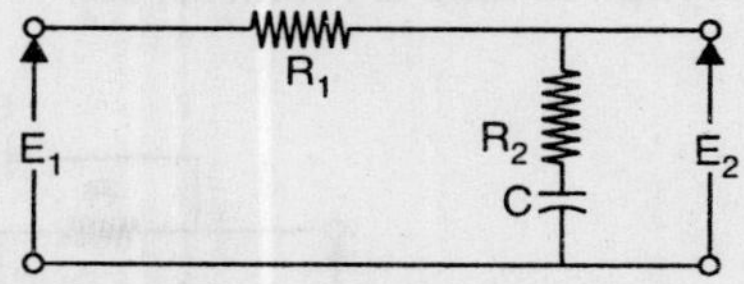

FIGURE 26.2. A phase-lag network.

It can be found that

$$\frac{E_2(s)}{E_1(s)} = \frac{1 + aTs}{1 + Ts},$$

where $$a = \frac{R_2}{R_1 + R_2} \;(< 1), \quad \text{and} \quad T = (R_1 + R_2)\,C$$

The general effects of the phase-lag compensation are :

- For a given relative stability, the velocity constant is increased.
- The gain crossover frequency is decreased; thus, the bandwidth of the system is reduced.
- For a given gain constant K, the magnitude curve of $G_1(s)$ is attenuated at the low frequencies, thus, causing an improvement of the phase margin and the resonance peak of the system.
- The time response usually is slower.

26.4 PHASE LAG-LEAD NETWORK

A phase lag-lead network is shown in Fig. 26.3. It consists of two branches in series; one branch being a resistor R_1 in parallel with a capacitor C_1, and the other, a resistor R_2 in series with a capacitor C_2. The input signal is E_1, and the output is E_2, across the series combination of R_2 and C_2.

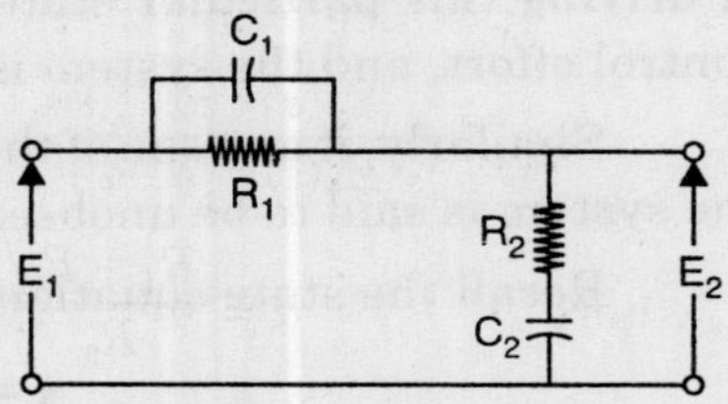

FIGURE 26.3. A phase lag-lead network.

For this network, the transfer function is

$$\frac{E_2(s)}{E_1(s)} = \frac{(1 + R_1C_1s)(1 + R_2C_2s)}{1 + (R_1C_1 + R_1C_2 + R_2C_2)s + R_1R_2C_1C_2s^2}$$

This network has the advantages of the both, and some disadvantages of the former networks are eliminated also.

26.5 BRIDGED-T NETWORK

When the signals are *ac* or modulated, a twin-*T* network, or a bridged-*T* network can be applied for compensation. By properly adjusting its parameters, it is possible to cancel the unwanted poles in the original system responsible for excessive oscillatory response. The two basic types of bridged-*T* networks, type 1, and type 2 are shown in Fig. 26.4. In the type 1 network, each of the two arms is a resistor R, the pillar is a capacitor C_1, and the bridge is also a capacitor, C_2. In the type 2 network, each of the two arms is a capacitor C, the pillar is a resistor R_1, and the bridge is also a resistor, R_2.

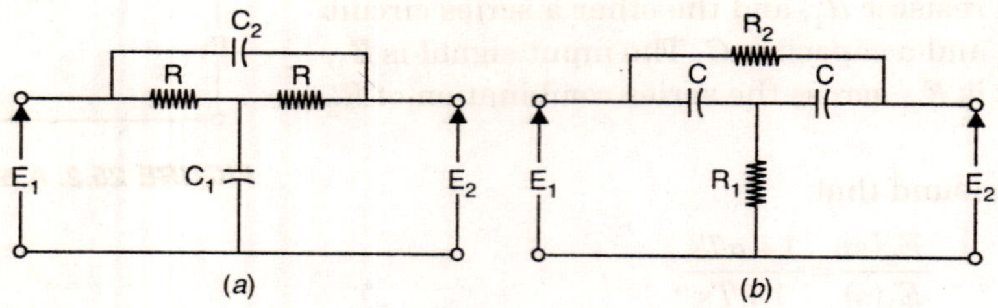

FIGURE 26.4. Bridged-*T* networks. (*a*) Type 1 network, (*b*) Type 2 network.

We can find that

$$\frac{E_2(s)}{E_1(s)} = \frac{1+2RC_2s+R^2C_1C_2s^2}{1+R(C_1+2C_2)s+R^2C_1C_2s^2} \quad \text{for type 1 network, and}$$

$$\frac{E_2(s)}{E_1(s)} = \frac{1+2R_1Cs+C^2R_1R_2s^2}{1+C(R_1+2R_2)s+C^2R_1R_2s^2} \quad \text{for type 2 network.}$$

26.6 CONTROLLABILITY AND OBSERVABILITY

These terms are generally applied in state variables analysis of linear multivariable systems.

If any one of the state variables is independent of the input signal, then there is no way of driving this particular state variable to a desired state in finite time by means of some control effort, and the system is said to be uncontrollable.

Similarly, if any one of the states cannot be found from the measurements of the outputs, the system is said to be unobservable.

Recall the state equations, reproduced below

$$\dot{\mathbf{x}} = \mathbf{A}\mathbf{x}(t) + \mathbf{B}\mathbf{m}(t)$$

$$\mathbf{c}(t) = \mathbf{D}\,\mathbf{x}(t) + \mathbf{E}\,\mathbf{m}(t) \qquad \text{where}$$

$\mathbf{x}(t)$ is $n \times 1$ state vector

$\mathbf{m}(t)$ is $r \times 1$ input vector

$\mathbf{c}(t)$ is $p \times 1$ output vector

A, B, D, and **E** are constant matrices of order $n \times n$, $n \times r$, $p \times n$, and $p \times r$ respectively.

We form an array **S** as

$$\mathbf{S} = [\mathbf{B} \quad \mathbf{AB} \quad \mathbf{A}^2\mathbf{B} \quad . \; . \; . \quad \mathbf{A}^{n-1}\mathbf{B}]$$

which is a matrix of order $n \times nr$

The system is controllable if **S** contains n linearly independent column vectors, *i.e.*, **S** is of rank n. If r equals 1, **S** is an $n \times n$ square matrix, and the condition becomes that the determinant of **S** is nonzero.

Alternatively, if we find that

$$\mathbf{P}^{-1}\mathbf{AP} = \begin{bmatrix} \lambda_1 & 0 & 0 & \dots & 0 \\ 0 & \lambda_2 & 0 & \dots & 0 \\ 0 & 0 & \lambda_3 & \dots & 0 \\ . & . & . & \dots & . \\ . & . & . & \dots & . \\ 0 & 0 & 0 & \dots & \lambda_n \end{bmatrix},$$ then the system is controllable. Here

λ_i, $i = 1, 2, 3, \ldots, n$; are distinct eigenvalues of **A**, and

$$\mathbf{P} = [\, \mathbf{X}_1 \quad \mathbf{X}_2 \quad \dots \quad \mathbf{X}_i \quad \dots \quad \mathbf{X}_n]$$

where $\mathbf{X}_i$ is the eigenvector of the eigen value λ_i.

The eigenvector is obtained from $(\lambda_i \mathbf{I} - \mathbf{A})\, \mathbf{X}_i = \mathbf{0}$.

We can always select the first element of any $\mathbf{X}_i$ to be unity.

The condition for observability is that the rank of **T** is n, *i.e.*, there is a set of n linearly independent column vectors in **T**, where

$$\mathbf{T} = [\, \mathbf{D}^t \quad \mathbf{A}^t\mathbf{D}^t \quad (\mathbf{A}^t)^2\, \mathbf{D}^t \ldots \quad (\mathbf{A}^t)^{n-1}\, \mathbf{D}^t]$$ is an $n \times np$ matrix.

If p equals 1, **T** is an $n \times n$ square matrix, and the condition becomes that the determinant of **T** is nonzero, *i.e.*, **T** is nonsingular.

A system is completely output controllable if the $p \times (n + 1)r$ matrix

$$[\mathbf{DB} \quad \mathbf{DAB} \quad \mathbf{DA}^2\, \mathbf{B} \quad \ldots \quad \mathbf{DA}^{n-1}\, \mathbf{B} \quad \mathbf{E}]$$ has a set of p linearly independent columns, where p = number of outputs.

Section 7 ELECTRICAL MACHINES

Chapter 27

Magnetic Circuits

27.1 FLUX LINES AND MAGNETIC CIRCUITS

A magnetic field is that in which a moving charge, a current-carrying element, and a magnetic needle, all experience a force. The source of magnetic field is basically electric current. The magnetic effect of a given current element will have different values in different media/materials. The effect is represented graphically by lines, called the field- or the ***flux lines***; the tangent to which at any point gives the direction along which the magnetic needle will orient when placed at that point, and the relative strength of the effect is depicted by the relative closeness of the lines. ***A magnetic flux line closes on itself***.

The closed path followed by a group of flux lines which constitutes the flux φ, is the ***magnetic circuit.***

The magnetic flux in practical magnetic circuits is produced by a winding, or a coil, of several turns (N), and carrying current I.

27.2 MAGNETIC CIRCUITS PARAMETERS AND VARIABLES

If these ampere-turns, called the ***magnetomotive force*** (mmf), denoted by F, act over a magnetic path of length l, of uniform area of cross-section a, and of ***permeability*** μ, then the following relations hold :

Magnetic ***reluctance***, $R = \dfrac{l}{\mu a}$

Magnetic reluctance is analogous to electric resistance.

Magnetic **flux**, $\varphi = \dfrac{F}{R} = \dfrac{NI}{R}$

The magneto motive force, and the magnetic flux are analogous respectively to the electromotive force and the electric current in electric circuits.

Magnetic ***flux density***, $B = \frac{\varphi}{a} = \mu \frac{NI}{l} = \mu H$

H is called the magnetic field strength, or the magnetizing force; which is the mmf per unit length.

It has been assumed that the flux lines pass through the area a in axial direction, and the flux density over area a is uniform.

The turns N are so placed that the flux φ passes through the hollow of each of the turns. μ, relating B and H, is expressed as $\mu_r \mu_o$; μ_r is the ***relative permeability*** of the material, and μ_o is the permeability of air, or vacuum and has a value of $4\pi \times 10^{-7}$ H/m.

μ_r is the ratio of the flux density produced in a material to the flux density produced in air by the same magnetic field strength. μ_r may have values from a few hundreds to several thousands, depending on the material properties.

27.3 B-H CURVE

The magnetic characteristic of a magnetic material is given by its magnetization curve, which is a plot of B against H. For ferromagnetic materials, the curve is non-linear. However, it is quite linear in the lower range of B, and gradually becomes non-linear as B increases. For a given value of H, the permeability μ of the material is simply the ratio of B to H, as found from the ***B-H curve***.

For most of the materials commonly employed in magnetic circuits, μ_r (or μ) varies considerably with H (or B). μ_r increases with H, shows a maximum, and then decreases at a slow rate.

Generally, the flux density would be different in different parts of a magnetic circuit. The values of μ for different parts, therefore, have to be found accordingly.

Two or more parts of a magnetic circuit are in series if the same flux enters the next part as left the preceding part; and similarly, two or more parts of a magnetic circuit are in parallel if the same mmf acts across each of them.

27.4 MAGNETIC CIRCUITS CALCULATIONS AND LOSSES

For ***analyzing magnetic circuits***, reluctances in series, or in parallel, can be reduced to a single equivalent reluctance. Further, the electric circuits' laws like Ohm's law, and Kirchhoff's laws can be applied to the analogous magnetic quantities.

When the current in the magnetizing coil is alternating, and hence the resulting H and B too; the effective value of μ is taken to be the ratio of peak value of B to the peak value of H.

Under alternating current conditions, the *B-H* plot takes the shape of a loop, called the ***hysteresis loop***. The area of the hysteresis loop is a measure of the energy loss (due to hysteresis) incurred in the material every cycle of the current (or *H*).

The ***hysteresis loss*** is given by $P_h = k_h f B_m^{1.6}$

Under ac conditions, emf is induced in the infinitely large number of closed paths of the material by the component alternating flux acting along the axes of these closed paths; and hence causes currents in these paths. These currents, called the eddy currents, cause additional power loss known as ***eddy-current loss***.

To reduce eddy-current loss, the area a through which the flux passes is split into a large number of smaller areas by employing a stack of what is called ***stampings*** of sufficiently small thickness. These stampings are insulated from each other by a thin coating of varnish or oxide. Further, the material should be a high resistivity material. The hysteresis loss is not affected by the use of stampings.

The eddy-current loss is given by $P_e = k_e t^2 f^2 B_m^{\ 2}$

f is the frequency, and *t* is the thickness of the material (stamping), supposed to be quite small. It is assumed that the form factor of the flux wave remains unchanged.

In many magnetic circuits, used to serve as reactors, or as lifting electromagnets, an air gap is introduced in the circuit. In these air gaps, there is magnetic energy stored.

27.5 ENERGY DENSITY AND MAGNETIC PULL

The ***energy density*** in the gap is given by $\frac{1}{2}BH = \frac{1}{2}\mu_0 H^2 = \frac{1}{2}\frac{B^2}{\mu_0}$ J/m^3.

On the opposite faces of the gap, there is a pull acting on them tending to shorten the gap.

The ***magnetic pull*** is given by $\frac{1}{2}BH = \frac{1}{2}\mu_0 H^2 = \frac{1}{2}\frac{B^2}{\mu_0}$ N/m^2.

Chapter 28 TRANSFORMERS

28.1 PRINCIPLE OF WORKING

A transformer works on the principle of mutual induction between two circuits linked by common magnetic flux.

28.2 EMF PER TURN

The rms value of the induced **emf per turn** in an inductor, or a transformer winding is,

$$E_t = 4.44\, f\, B_m\, a_i$$

f is the frequency of pulsation of flux in the core

B_m is the peak value of the flux density in the core

a_i is the net cross-sectional area of the core

It is assumed that the flux in the core varies sinusoidally.

28.3 STACKING FACTOR

Stacking factor is the ratio of net iron area a_i to the gross area a (physical or geometrical area). a_i is slightly less than a, because of the varnish, or the oxide coating on the stampings. As laminations thickness decreases, the stacking factor also decreases.

In larger transformers, steps are taken to make the area of cross-section of the core close to circular with a view to lower the mean length of the turn, which in turn leads to less copper requirement, less copper loss, and better voltage regulation.

28.4 TURNS RATIO

The ***turns ratio***, *t*, or the voltage ratio of a transformer is $\frac{N_1}{N_2} = \frac{E_1}{E_2}$; where N_1 and N_2 are respectively the number of turns on the primary and the secondary windings.

28.5 IDEAL TRANSFORMER

In an ***ideal transformer***, the core material is assumed to be of infinite permeability, the hysteresis loop vanishes, the resistivity of the core material is infinite, and the windings do not have any resistance. It follows that no magnetizing current is required to set up the flux in the core; there is no hysteresis and eddy-current loss. Further, there is no leakage flux, no voltage drop in the windings and hence the voltage regulation is zero. The efficiency of the transformer is 100%. It can be represented simply by two windings.

For such a transformer, $N_1\mathbf{I}_1 \pm N_2\mathbf{I}_2 = 0$, *i.e.*, the net mmf acting on the core is zero. If the two mmf's act in the same direction then + sign is applicable; else the negative sign.

Applied voltage $\mathbf{V}_1$ = induced voltage $\mathbf{E}_1$, secondary voltage $\mathbf{V}_2 = \mathbf{E}_2$.

A large well designed and well constructed transformer is very close to the ideal transformer in performance.

28.6 CORE TYPES

A ***shell type*** core construction, in which the magnetic circuit surrounds the winding, has relatively lower leakage flux as compared to a ***core type*** core construction, in which the winding surrounds the core.

28.7 INDUCED EMF

It is generally said that the ***induced emf lags*** the flux inducing the emf by 90°.

Before we make such a statement three items have to specified :

(*i*) the reference direction of the flux Φ in the core,

(*ii*) the sense of the winding wound on the core, and

(*iii*) the reference polarity of the induced emf.

In the figure shown, $\mathbf{E}_1$ and $\mathbf{E}_2$ both lead the flux Φ by 90°. If any of the three items is reversed, the ***lead/lag relation*** will also reverse.

We can, however, say that the emf tending to set up flux in the same direction as the direction of the flux inducing the emf, lags the flux by 90°.

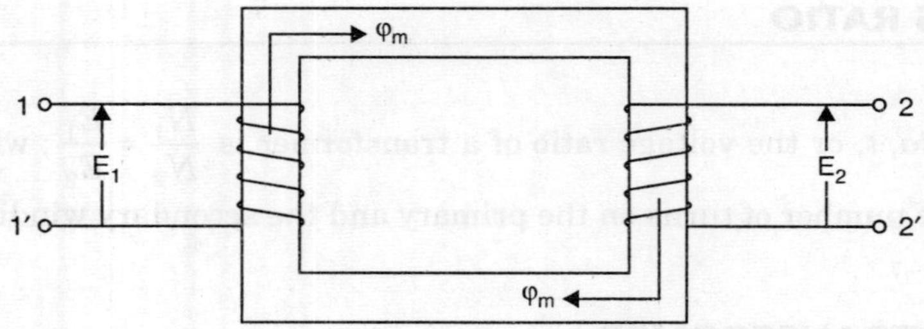

FIGURE 28.1. Mutual flux and emf's in a transformer (E_1 and E_2 both lead the flux ϕ).

To find the induced emf in a coil, or in a winding :

(*i*) Designate the two ends of the coil as end 1, and end 2.

(*ii*) Let the reference direction of the current be from end 1 through the coil to end 2. Let this current be called i_{12}.

(*iii*) Obtain the expression for the flux linkages λ of the coil for that direction of the flux in which the coil reference current i_{12} produces the flux. This would need the knowledge of the sense of the winding. One convention is that the coil produces flux in the axial direction of the current itself.

(*iv*) Let e_{12} denote the emf induced in the coil with end 1 positive and end 2 negative, *i.e.*, the potential of end 1 wrt the potential of end 2. Then $e_{12} = \frac{d\lambda}{dt}$.

If the coil is acting as a power absorber, we can write the terminal voltage

$$v_{12} = e_{12} + i_{12}\, r = \frac{d\lambda}{dt} + i_{12}\, r$$

Keeping the reference direction of the current the same, the coil acts as a power generator with reversed voltage polarity, *i.e.*, end 2 positive, and end 1 negative.

$$v_{21} = e_{21} - i_{12}\, r = -\,e_{12} - i_{12}\, r = -\frac{d\lambda}{dt} - i_{12}\, r$$

Alternatively, keeping the reference voltage polarity to be the same, the coil acts as a power generator with reversed reference current direction, *i.e.*, i_{21}. The expression for λ for the reversed reference direction should now be used.

$$v_{12} = e_{12} - i_{21}\, r = -\frac{d\lambda}{dt} - i_{21}\, r$$

Thus for a power absorber we may write $v = \frac{d\lambda}{dt} + i\,r$, and

for a power generator we write $v = -\frac{d\lambda}{dt} - i\,r$

with appropriate voltage polarity, the current and the flux directions.

To do away with the showing of the core and the sense of the winding explicitly, ***dot convention*** is adopted. A dot each is placed at one of the two ends of the various windings to indicate that the emf's of the dotted end wrt the undotted end in all the windings shall be in phase.

28.8 EQUIVALENT CIRCUIT

The performance evaluation of an actual transformer is facilitated by using its ***equivalent circuit***, which incorporates in it the various deviations from the ideal transformer. The core imperfections are represented on the primary side by a parallel branch comprising an inductive reactance X_o which takes into account the finite permeability of the core; and a resistance R_o in parallel with X_o, which takes into account the core losses. Closer the core is to that of an ideal transformer, higher are the numerical values of the reactance, and the resistance. The windings resistances and the leakage reactances $r_1, x_1; r_2, x_2$ are shown as series elements on the respective sides. Closer the windings are to those of an ideal transformer; smaller are the numerical values of the resistances and the reactances.

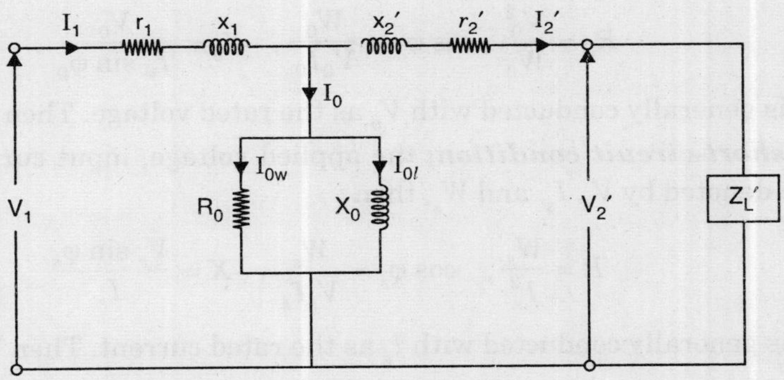

FIGURE 28.2. The equivalent circuit of a transformer referred to the primary circuit.

The secondary quantities are referred to the primary side using the turns ratio t as given below. The referred quantities are denoted by dashed symbols.

$$\mathbf{V}_2' = t\,\mathbf{V}_2, \quad \mathbf{E}_2' = t\mathbf{E}_2, \quad \mathbf{I}_2' = \mathbf{I}_2/t, \quad r_2' = t^2\, r_2, \quad x_2' = t^2 x_2, \quad \mathbf{Z}_2' = t^2\mathbf{Z}_2$$

Generally, we find that $r_2' \approx r_1, x_2' \approx x_1, V_2' \approx V_1$, and $I_2' \approx I_1$.

With quantities referred to the primary side, the equivalent circuit looks a T-circuit. Sacrificing the accuracy a little by shifting the shunt circuit to the extreme left, r_2' and x_2' can be clubbed with r_1, and x_1 respectively. Then

$$R_1 = r_1 + r_2', \quad X_1 = x_1 + x_2'$$

If it is understood that the equivalent circuit is always referred to the primary side, the suffix 1 may even be dropped from R_1 and X_1.

Also, $R << R_o$, and $X << X_o$.

For voltage regulation studies, the high impedance shunt circuit is altogether neglected. The equivalent circuit then is simply a series impedance $\mathbf{Z} = R + jX$.

28.9 OPEN-CIRCUIT AND SHORT-CIRCUIT TESTS

In transformers, the ***no load*** (*i.e.*, with the secondary circuit ***open***) input current with full applied voltage, may be around 5% of the rated current; and the power factor about 0.2 lagging. The wave shape of the exciting current (no-load current) is non-sinusoidal and looks peaky as the *B-H* curve of the core material is not linear. It predominantly contains 3rd harmonic. Since, the current is very small, the loss in the winding which is proportional to the square of the current is truly insignificant; however, as the voltage applied is normal, the flux density in the core is normal and as such the input power is the normal voltage ***core*** or the ***iron loss***, W_i.

Under ***short-circuit*** (*i.e.*, with the secondary circuit shorted) ***test*** condition with rated current flowing, the applied voltage need be around 5%, and the input power factor may be about 0.4 lagging. Since, the applied voltage and hence the flux density is very small, the loss in the core, which is proportional to the square of the flux density is really insignificant; however, as the rated current is circulating in both the windings, the input power is the rated current or the full load power loss in the windings, called the ***copper loss***, W_c.

If in a single-phase transformer, under ***open-circuit condition***; the applied voltage, input current, and the power, respectively are denoted by V_o, I_o, and W_o, then

$$R_o = \frac{V_0^2}{W_0}, \quad \cos\varphi_o = \frac{W_0}{V_0 I_0}, \quad X_o = \frac{V_0}{I_0 \sin\varphi_0}.$$

The test is generally conducted with V_o as the rated voltage. Then $W_o = W_i$.

If under ***short-circuit condition***; the applied voltage, input current, and the power respectively are denoted by V_s, I_s, and W_s, then

$$R = \frac{W_s}{I_s^2}, \quad \cos\varphi_s = \frac{W_s}{V_s I_s}, \quad X = \frac{V_s \sin\varphi_s}{I_s}$$

The test is generally conducted with I_s as the rated current. Then $W_s = W_c$.

28.10 CONSTANCY OF FLUX

For an actual transformer in operation, as the applied voltage remains substantially constant, and as the load varies the voltage drop in the windings (though it varies directly as load) remains quite small; the induced voltage and hence the ***flux*** also ***remains nearly constant***. It implies that as the secondary current varies, the primary current adjusts itself to cancel the ampere-turns of the secondary in accordance with $N_1\mathbf{I}_1 \pm N_2\mathbf{I}_2 = 0$.

28.11 PER UNIT VALUES

Quite often, the numerical values of various quantities are expressed in ***per unit***. The per unit value of a quantity is simply the actual value divided by the base value of that quantity; and the base values of various quantities are generally taken the rated values (as per the equipment) of those quantities.

28.12 VOLTAGE REGULATION

The ***voltage regulation*** of a transformer refers to the effect of load current on the terminal voltage. It is defined for full load condition, and at a specified load power factor cos φ.

% *VR* at load power factor cos φ =

rise in the terminal voltage, expressed as % of the rated voltage, when full load at the rated voltage and at the specified power factor is thrown off; the primary applied voltage kept constant.

With all the quantities referred to the primary circuit,

$$\mathbf{V}_1 = \mathbf{V}_2 + \mathbf{I}(R + jX)$$
$$V_1^2 = (V_2 \cos\varphi + IR)^2 + (V_2 \sin\varphi + IX)^2$$

The various forms of this equation can be easily derived.

% *VR* is approximately given by

$$\frac{IR\cos\varphi + IX\sin\varphi}{V} \times 100 = (R_{pu}\cos\varphi + X_{pu}\sin\varphi) \times 100.$$

The voltage regulation may be as good as 3%.

The voltage ***regulation is zero*** at a leading load power factor angle given by $\tan^{-1}(R/X)$; and is ***the poorest*** at the lagging power factor angle given by $\tan^{-1}(X/R)$.

28.13 EFFICIENCY

The efficiency (the ***power efficiency***), is the ratio of the output power to the input power. At a fractional loading *k*, and power factor cos φ, the efficiency is given by

$$\eta = \frac{kS\cos\varphi}{kS\cos\varphi + k^2W_c + W_i},$$ where *S* is the *VA* rating of the transformer.

The efficiency at a given power factor is maximum when the variable loss, *i.e.*, the loss that varies with load, $k^2 W_c$, equals the constant loss, which in this case is W_i.

Thus, the ***maximum efficiency*** loading condition is $k_m = \sqrt{(W_i/W_c)}$.

The ***all-day efficiency*** is the energy efficiency, *i.e.*, the ratio of the energy output to the energy input during the period from 00.00 hr to the next 00.00 hr. The term is used in context with the distribution transformers which remain energized all the time and the load varies within wide limits.

$$\text{All-day efficiency} = \frac{\sum_{i=1}^{n}(k_iS\cos\phi_i)T_i}{\sum_{i=1}^{n}(k_iS\cos\phi_i + k_i^2W_c)T_i + 24 \times W_i}$$

k_i is the fractional, or pu loading during the ith interval, of duration T_i hours; *n* is the total number of intervals during the 24-hr period.

28.14 PARALLEL OPERATION

Any two or more power circuits to be connected in parallel should have identical emf's, *i.e.*, the frequency should be the same, the voltages must be equal in magnitude and phase and only like polarity terminals should be connected together. For three-phase transformers, the phase sequence also must be the same.

Two transformers A and B, of emf's $\mathbf{E}_A$ and $\mathbf{E}_B$, and impedances $\mathbf{Z}_A$ and $\mathbf{Z}_B$, ***operating in parallel***, share the load current **I** flowing in load impedance $\mathbf{Z}_L$, at voltage **V** (all quantities on per phase basis) as

$$\mathbf{I}_A = \frac{\mathbf{Z}_B}{\mathbf{Z}_A + \mathbf{Z}_B}\mathbf{I} + \frac{(\mathbf{E}_A - \mathbf{E}_B)}{(\mathbf{Z}_A + \mathbf{Z}_B)}, \quad \text{and} \quad \mathbf{I}_B = \frac{\mathbf{Z}_A}{\mathbf{Z}_A + \mathbf{Z}_B}\mathbf{I} + \frac{(\mathbf{E}_B - \mathbf{E}_A)}{(\mathbf{Z}_A + \mathbf{Z}_B)}$$

The alternative relations are :

$$\mathbf{I}_A = \frac{(\mathbf{E}_A - \mathbf{V})}{\mathbf{Z}_A}, \quad \text{and} \quad \mathbf{I}_B = \frac{(\mathbf{E}_B - \mathbf{V})}{\mathbf{Z}_B}$$

Or

$$\mathbf{I}_A = \frac{\mathbf{E}_A\mathbf{Z}_B + (\mathbf{E}_A - \mathbf{E}_B)\mathbf{Z}_L}{\mathbf{Z}_A\mathbf{Z}_B + \mathbf{Z}(\mathbf{Z}_A + \mathbf{Z}_B)}, \quad \text{and} \quad \mathbf{I}_B = \frac{\mathbf{E}_B\mathbf{Z}_A + (\mathbf{E}_B - \mathbf{E}_A)\mathbf{Z}_L}{\mathbf{Z}_A\mathbf{Z}_B + \mathbf{Z}(\mathbf{Z}_A + \mathbf{Z}_B)}$$

$$\mathbf{V}\left(\frac{1}{\mathbf{Z}_L} + \frac{1}{\mathbf{Z}_A} + \frac{1}{\mathbf{Z}_B}\right) = \frac{\mathbf{E}_A}{\mathbf{Z}_A} + \frac{\mathbf{E}_B}{\mathbf{Z}_B}$$

For convenience, all the above quantities are referred to the load circuit.

The above relations apply equally well to two synchronous generators operating in parallel.

28.15 AUTOTRANSFORMER

An ***autotransformer*** is a single- winding transformer, with tapping (*s*), or continuously variable contact provided on it. For a step down transformer, the tapped part of the winding is the secondary winding; and for a step up transformer the tapped part of the winding is the primary winding. If the voltage ratio is close to unity, it requires considerably less conductor material. There is a saving in insulation too.

For a given *kVA* rating of the transformer,

$$\text{Conductor material } \frac{\text{in autotransformer}}{\text{in normal two winding transformer}} = \frac{t-1}{t}$$

Where t is V_h/V_l and, therefore, greater than or equal to 1

In an auto-transformer, of the total power transferred to the secondary, the fraction $1/t$ is transferred conductively, and the remaining, *i.e.*, $(t-1)/t$, is transferred inductively.

Neglecting the magnetizing branch, the winding resistance R, and the leakage reactance X as seen on the high voltage side of the auto-transformer are

$$R = r_1 + (t-1)^2 r_2, \quad X = x_1 + (t-1)^2 x_2$$

r_1, and x_1 are respectively the resistance, and the leakage reactance of the non-common section of the winding; and r_2, and x_2 are respectively the resistance, and the leakage reactance of the common section of the winding.

If $t = 1$, then r_1 is zero, and $R = 0$; similar statement applies to x also.

With $t = 1$, in fact, no transformer is needed, and hence the results as above.

If the winding is uniform and r, and x are respectively the resistance, and the leakage reactance per turn of the winding, then

$$R = r(N_h - N_l)\,[1 + (N_h - N_l)] \quad X = x(N_h - N_l)\,[1 + (N_h - N_l)]$$

If a normal two-winding transformer of voltage ratio V_h/V_l and volt-ampere rating S is connected as an autotransformer, then the possible voltage, and the volt-ampere ratings are

$(V_h + V_l)/V_h$	$S(1 + t)$
$(V_h + V_l)/V_l$	$S(1 + 1/t)$
$(V_h - V_l)/V_h$	$S[abs(1 - t)]$
$(V_h - V_l)/V_l$	$S(1 - 1/t)$

An auto-transformer may be a single-phase, or a three-phase (one winding for each phase) transformer.

If due to any reason the common point connection of the autotransformer gets opened, the low voltage side is subjected to full high voltage - a very serious condition.

28.16 THREE-PHASE TRANSFORMERS

Three single-phase transformers can be connected to operate as a ***three-phase transformer***, or a single three-phase transformer unit can be there for the purpose.

In a core-type core (three-limbed core) of a three-phase unit, the flux in any limb is 1.5 times the flux produced by the winding on that limb. As such, for a given voltage, the magnetizing current in the unit is 67% of that required in the bank of three single-phase transformers. Also, a unit has less weight, requires less space, and is less costly.

Star-star connection-adopted in small high voltage transformers.

Delta-delta connection-adopted in large low voltage transformers.

Star-delta connection-adopted in large transformers used in transmission systems, star connection being in the *hv* circuit to effect saving in insulation cost.

Delta-star connection-adopted in distribution transformers used in distribution systems, star connection being in the *lv* circuit to serve single-phase, as well as three-phase loads.

In a bank of three single-phase transformers, connected in delta-delta, even if one of the transformers is removed, the balanced operation continues. However, the capacity now reduces to 58% of the original capacity. This condition is called the open delta, or ***V-connection***.

The ***connections of three-phase transformers*** indicating the phase relation of the line to neutral emf's of the corresponding phases in the *hv* winding and in the *lv* winding are grouped as

$Yy0, Dd0, Dz0$
$Yy6, Dd6, Dz6$
$Dy1, Yd1, Yz1$
$Dy11, Yd11, Yz11$

The first letter signifies the hv side connection, the second letter signifies the *lv* side connection, and the numbers 0, 6, 1, 11 show the phase relation between the emf's of the two sides. These respectively mean that the *lv* emf lags the corresponding *hv* emf by 0º, 180º, 30º, and – 30º.

The three-phase transformers within the same phase group only can be operated in parallel.

28.17 SCOTT-CONNECTION

Scott-connected transformers are used to transform power from three-phase to two-phase, or vice versa. The set consists of two transformers, one is called the main transformer, and the other, the teaser transformer. If the primary winding of the main transformer has N_1 turns, the primary of the teaser should have $\frac{\sqrt{3}}{2}$ (= 87%) N_1 turns. The secondary windings of both the transformers have N_2 turns each. One end of the primary of the teaser transformer is connected to one of the three phases of the three-phase circuit, and the other to the mid point of the primary of the main transformer. The remaining two phases of the three-phase circuit are connected to the two terminals of the primary of the main transformer. With this arrangement of connection, it can be found that the equal emf is induced in each turn of both the transformers, the secondary voltages constitute two-phase voltages wherein the voltages have a phase difference of 90º, and that if one side of the circuit is balanced, the other side too will be balanced. The neutral on the three- phase side is located on the primary of the teaser transformer at $\frac{1}{2\sqrt{3}}$ (= 29%) from the junction.

28.18 HARMONICS

As already stated, because of the non-linear *B-H* curve of the core material, the ***exciting current wave shape*** in a single-phase transformer departs from sinusoidal, and predominantly contains third harmonic component. In a three-phase system, the third harmonic currents in all the three phases are in phase with each other. As such, these currents can flow only if a return path is available. Absence of a connection between the transformer neutral, and the source neutral prohibits the flow of these currents. As a consequence, exciting current wave is constrained to remain more or less sinusoidal, instead the flux wave becomes flat topped and the induced emf wave peaky. Thus, an isolated star connection on the primary side causes third harmonic emf in the phases(in addition to the fundamental of course) which leads to what is called '***oscillation of the neutral***', *i.e.*, the transformer neutral acquires a potential equal to the phase third harmonic emf and its phase is continuously changing. The transformer neutral otherwise should have been at ground potential. There cannot be any third harmonic emf between the lines as these would cancel out.

However, if the third harmonic currents can flow in the transformer windings by virtue of the neutral connections or the delta connection, then the flux wave and the emf wave are sinusoidal.

28.19 THREE-WINDING TRANSFORMERS

Three-winding transformers are generally in the three-phase transformers category only. The windings are designated as primary, secondary, and tertiary windings. The tertiary winding is invariably connected in delta. The major use of the delta-connected tertiary winding is to suppress harmonic voltages, to limit voltage unbalance, and to allow sufficient earth fault current to flow for operation of the protective relaying.

The *kVA* rating of the primary winding is somewhat less than or equal to the sum of the ratings of the secondary, and the tertiary windings. In fact, the loading of the primary circuit is the phasor sum of the secondary, and the tertiary circuits loadings.

As the simplified equivalent circuit on per phase basis of a normal two-winding three-phase transformer is simply the primary winding impedance, and the secondary winding impedance in series; the same for a three-winding transformer is obtained simply by connecting the tertiary winding impedance at the junction point, giving the appearance of a star. The impedances are all referred to the primary circuit.

The impedances z_1, z_2, and z_3 of the three star branches can be determined experimentally by conducting three short-circuit tests, in each of which one winding is excited, one is shorted, and the third one is left open.

S No	*winding excited*	*winding shorted*	*winding open*	*impedance determined*
1	1	2	3	$\mathbf{z}_{12} = \mathbf{z}_1 + \mathbf{z}_2$
2	1	3	2	$\mathbf{z}_{13} = \mathbf{z}_1 + \mathbf{z}_3$
3	2	3	1	$\mathbf{z}_{23} = \mathbf{z}_2 + \mathbf{z}_3$

$\mathbf{z}_{23}$ is the impedance referred to the primary circuit, though this measurement has been made in the secondary circuit.

$$\mathbf{z}_1 = (\mathbf{z}_{12} + \mathbf{z}_{13} - \mathbf{z}_{23})/2$$
$$\mathbf{z}_2 = (\mathbf{z}_{12} - \mathbf{z}_{13} + \mathbf{z}_{23})/2$$
$$\mathbf{z}_3 = (-\mathbf{z}_{12} + \mathbf{z}_{13} + \mathbf{z}_{23})/2$$

28.20 INRUSH OF CURRENT

When a transformer is switched on, a sudden ***inrush of current*** occurs. The inrush is maximum if the switch happens to be closed at an instant when the applied voltage has a value equal to zero. The effect is accentuated if there is some residual magnetism present in the core. The degree of the effect will depend on the magnitude and the direction of the residual flux.

The inrush occurs because at every instant, the Faraday's law, *i.e.*, $e = -N(d\Phi/dt)$ has to be satisfied, and this may entail very high value of the flux (far in the saturation region) and consequently very high exciting current. The current even becomes unidirectional for quite a few cycles.

Chapter 29

ELECTROMECHANICAL ENERGY CONVERSION-GENERAL

29.1 ENERGY CONVERSION PRINCIPLE

An electromechanical energy conversion process involves a coupling field between the electrical system and the mechanical system; and the tendency of the energy in the coupling field to release itself and do work is the basis of the conversion process. The interchange of energy between an electrical system and a mechanical system is possible if the energy in the coupling field is influenced by the configuration of the mechanical elements. The electromechanical energy conversion occurs if the change in flux in the coupling field is associated with mechanical motion.

29.2 ENERGY AND COENERGY

In a singly excited magnetic system, the ***energy stored*** in the magnetic field when a current i in the exciting winding, or the flux φ in the magnetic field, or the flux linkage λ is established is given by

$$W_{fld} = \int_0^{\varphi} F(\varphi)d\varphi = \int_0^{\lambda} i(\lambda)d\lambda,$$

and the ***coenergy*** is given by

$$W'_{fld} = \int_0^{F} \varphi(F)dF = \int_0^{i} \lambda(i)di$$

When the flux φ and the and the mmf F are directly proportional, then

$$W_{fld} = W'_{fld} = \tfrac{1}{2} F\varphi = \tfrac{1}{2} i\lambda$$

Bringing in the reluctance R, the permeance $P(= 1/R)$ of the magnetic flux path, and the inductance $L(= N^2 P)$ of the exciting winding of N turns,

$$W_{fld} = W'_{fld} = ½\,\varphi^2 R = ½\,F^2\,P = ½\,i^2 L$$

In a doubly excited magnetic system,

$$W_{fld} = ½\,L_{11}\,i_1^{\,2} + ½\,L_{22}\,i_2^{\,2} + L_{12}\,i_1\,i_2$$

And in a triply excited system, it is

$$W_{fld} = ½\,L_{11}\,i_1^{\,2} + ½\,L_{22}\,i_2^{\,2} + ½\,L_{33}\,i_3^{\,2} + L_{12}\,i_1\,i_2 + L_{13}\,i_1\,i_3 + L_{23}\,i_2\,i_3$$

In matrix form, for n coils we can write

$$W_{fld} = \frac{1}{2}\,[\mathbf{i}]^t[\mathbf{L}][\mathbf{i}] \quad \text{or in expanded form}$$

$$W_{fld} = \frac{1}{2}\begin{bmatrix} i_1 & i_2 & \cdots & i_n \end{bmatrix}\begin{bmatrix} L_{11} & L_{12} & \cdots & L_{1n} \\ L_{21} & L_{22} & \cdots & L_{2n} \\ \vdots & \vdots & & \vdots \\ \vdots & \vdots & & \vdots \\ L_{n1} & L_{n2} & \cdots & L_{nn} \end{bmatrix}\begin{bmatrix} i_1 \\ i_2 \\ \vdots \\ \vdots \\ i_n \end{bmatrix}$$

29.3 FORCE AND THE TORQUE

The value of the ***component force f*** acting in any direction on any part of the energy conversion system can be determined by imagining the part in question to be given small displacement, say, dx, in that direction and evaluating

$$-\frac{\partial W_{fld}}{\partial x}(\varphi, x) = -\frac{\partial W_{fld}}{\partial x}(\lambda, x) \quad \text{or} \quad \frac{\partial W'_{fld}}{\partial x}(F, x) = \frac{\partial W'_{fld}}{\partial x}(i, x)$$

When ***flux and mmf*** are ***linearly proportional***,

$$f = \frac{\partial W_{fld}}{\partial x}(F, x) = \frac{\partial W_{fld}}{\partial x}(i, x)$$

For a ***singly excited system***, with flux and mmf assumed linearly proportional,

$$f = -\,½\,\varphi^2\,\frac{dR}{dx} = ½\,F_2\,\frac{dP}{dx} = ½\,i^2\,\frac{dL}{dx}$$

The force acts in a direction to tend to decrease the stored energy at constant flux, to increase the stored energy and coenergy at constant mmf, to decrease the reluctance, or to increase the permeance, and the inductance.

The ***torque*** can be determined by giving the part in question a small angular displacement $d\theta$ and finding the corresponding derivatives wrt θ. In all the above expressions, x gets replaced by θ. For instance,

$$T = -\,½\,\varphi^2\,\frac{dR}{d\theta}\,; \quad \text{the angle } \theta \text{ is the mechanical angle}$$

29.4 RELUCTANCE MOTOR

For a ***single-phase reluctance motor***, with the assumption of sinusoidal variations of flux and the reluctance, and the rotor angular speed equal to the angular frequency of the flux, the instantaneous torque acting on the rotor is given by

$$T = -\tfrac{1}{4}\,\varphi^2_{max}\,(R_q - R_d)\,[\sin(2\,\omega t - 2\delta) + \tfrac{1}{2}\sin(4\,\omega t - 2\delta) + \tfrac{1}{2}\sin(-2\delta)]$$

Where $-\delta$ is the angle which the rotor axis makes with the stator pole axis in the direction of rotation at $t = 0$; R_d, and R_q are respectively the direct-axis, and the quadrature-axis reluctances of the magnetic circuit ($R_q > R_d$).

The ***average value of the torque*** is $T_{av} = (1/8)\ \varphi^2_{max}\,(R_q - R_d)\sin 2\,\delta$

In ***multiply excited systems***, the tendency of the magnetic field energy to change when one group of windings moves wrt the other gives rise to mechanical forces/torques.

For a ***doubly excited*** system,

$$T = \tfrac{1}{2}\,i_1^2\,\frac{dL_{11}}{d\theta} + \tfrac{1}{2}\,i_2^2\,\frac{dL_{22}}{d\theta} + i_1\,i_2\,\frac{dL_{12}}{d\theta}$$

And for a triply excited system, it is

$$T = \tfrac{1}{2}\,i_1^2\,\frac{dL_{11}}{d\theta} + \tfrac{1}{2}\,i_2^2\,\frac{dL_{22}}{d\theta} + \tfrac{1}{2}\,i_3^2\,\frac{dL_{33}}{d\theta} + i_1\,i_2\,\frac{dL_{12}}{d\theta} + i_1\,i_3\,\frac{dL_{13}}{d\theta} + i_2\,i_3\,\frac{dL_{23}}{d\theta}$$

For an n-tuply excited system,

$$T = \frac{1}{2}\,[\mathbf{i}]^t\,\frac{dL(\theta)}{d\theta}\,[\mathbf{i}]$$

29.5 ROTATING MACHINES AND THE TORQUE DEVELOPMENT

Every ***rotating machine*** has a stationary member, and a rotating member; both being basically of cylindrical shape, and coaxial. A small uniform air gap between the rotor surface, and the stator surface has to be provided to permit rotation of the rotor. Generally, the rotor and the stator constitute two flux producing systems (excepting some special machines like reluctance motor, hysteresis motor). Also, except the permanent magnet machines, the rotor as well as the stator have suitable arrangement of interconnected conductors on them, which when carrying currents produce magnetic flux in the air gap. The member that carries the electric power converted from mechanical to electrical, or to be converted into mechanical is called the armature. The armature may be the stator, or the rotor; generally it's the stator, as it is easier to handle large currents and high voltages if the circuits are on the stationary member rather than on a rotating member. In dc machines, of necessity, the armature is on the rotor.

In all the machines, the ***average electromagnetic torque*** is developed only if

(*i*) the magnetic fields of the stator and the rotor are constant in amplitude,

(*ii*) the relative speed between the two is zero, and

(*iii*) they are not aligned with each other.

In fact, we can say that the tendency of the two fields to align themselves gives rise to the electromagnetic torque.

The electromagnetic torque is developed in a motor as well as in a generator; in the former it causes rotation of the rotor, and in the latter it opposes the motion of the rotor. The torque acts on the rotor, as well as on the stator; that on the rotor is associated with the motion, and that on the stator is transmitted to the foundation of the machine.

In ***dc machines***, the stator has a concentrated winding forming the field or the excitation system; and the rotor has distributed winding, forming the armature. The field may be separately excited from a dc source, or it may be self-excited, *i.e.*, by the armature itself. When operating, the flux of the armature also is stationary, and the angle between the field flux and the armature flux is 90°.

In ***induction machines***, the stator has a distributed winding forming the armature; and the rotor may have a distributed winding, or an assembly of bars, shorted at the ends with end rings. In the closed rotor circuit, the currents are induced by the stator by induction. When operating, the fields of both rotate at a constant speed, called the synchronous speed.

In ***synchronous machines***, the stator has a distributed winding forming the armature; and the rotor may have a concentrated winding with poles projecting, or more appropriately termed-salient poles; else a cylindrical shape having winding accommodated in slots, all the turns of which produce flux in the same direction. The rotor thus has the field system, or the excitation system. The field is excited from an independent dc source. When operating, the fields of both rotate at a constant speed, called the synchronous speed.

Generally, the armatures in induction machines, as well as in synchronous machines, are wound for three-phase operation. Synchronous machines are made of much higher ratings than the induction motors.

29.6 WINDINGS AND THE MMF WAVES

As stated above, the armatures have ***distributed windings***, *i.e.*, the winding is spread uniformly over a number of slots around the periphery of the armature. The slots, which are parallel to the shaft, house insulated conductors/coil sides interconnected in such a way that all make a positive contribution to the desired result. In an S-slot, P-pole armature, the pole pitch in terms of number of slots, is S/P. One pole pitch is equivalent to an electrical angle of π radians (= 180°), or a mechanical angle of $2\pi/P$ radians (= 360°/P). The slot pitch is $P\pi/S$ radians electrical, or (P/S) × 180° electrical. Usually, we resort to what is called a double layer winding which facilitates end connections. In this type of winding, coil sides are laid in two layers in the slot. The two sides of a coil occupy slots which are generally one pole pitch apart. For every pole pair, a certain number of coils in successive slots are connected in series, forming a group. All the groups of a phase may be connected in series, or in parallel, or in series parallel combination, depending on the voltage and current need. The armature winding is thus a series of belts of conductors wrapped longitudinally on the armature iron, which carry oppositely directed currents, and the successive displacement of the belts equals a slot pitch. If the number of slots is very large, it tends to a wrapped ***uniform current sheet***.

An ***mmf wave*** shows the armature ampere turns acting at various points in the air gap around the periphery. The mmf due to a group of coils in series is maximum at the two points which are on the axis of the group, and is zero at the two points 90° electrical away from those points. The maximum value equals $N_g i_c/2$, N_g being the total number of turns in series in the group, and i_c is the coil or the conductor current. As we cross a slot, the mmf changes by $2N_c i_c$ (double layer winding) amperes, N_c being the number of turns in a coil. It results in a ***stepped triangular mmf wave***. The steps almost disappear, as the number of slots becomes very large. It means that a uniform current sheet gives rise to a **triangular mmf wave**. Incidentally, a ***sinusoidal current sheet*** shall cause a ***sinusoidal mmf wave.***

A triangular mmf wave can be resolved into its Fourier series, the fundamental of which has the peak value equal to $8/\pi^2$ times the peak value of the triangular wave.

The above principle can be applied to draw the mmf wave of any distributed winding.

29.7 TORQUE AND OTHER RELATED EXPRESSIONS

We assume that the rotor and stator surfaces are smooth surfaces; in other words, the air gap length g is same throughout. At all points magnetic reluctance is the same. The torque is given by

$$T = -\frac{P}{2} L_{sr}\, i_s\, i_r \sin \frac{P}{2}\theta_m = -\frac{P}{2} L_{sr}\, i_s\, i_r \sin \theta$$

P number of poles

L_{sr} mutual inductance between the stator, and the rotor windings when these are aligned with each other

i_s stator winding current

i_r rotor winding current

θ_m angle, in mechanical radians, by which rotor axis is ahead of the stator axis in the direction of rotation

θ the corresponding angle in electrical radians, $\theta = (P/2)\,\theta_m$

Inductances

$$L_{ss} = \frac{\mu_0}{2}\frac{\pi d\, l}{g}\left(\frac{4}{\pi}k_s \frac{N_s}{P}\right)^2 + L_{si}$$

$$L_{rr} = \frac{\mu_0}{2}\frac{\pi d\, l}{g}\left(\frac{4}{\pi}k_r \frac{N_r}{P}\right)^2 + L_{ri}$$

$$L_{sr} = \frac{\mu_0}{2}\frac{\pi d\, l}{g}\left(\frac{4}{\pi}k_\delta \frac{N_\delta}{P}\right)^2 + \left(\frac{4}{\pi}k_r \frac{N_r}{P}\right)$$

d average diameter at the air gap

l the axial length of the gap

k_s the winding factor for the space fundamental of the stator winding

N_s the number of series turns in the stator winding

k_r the winding factor for the space fundamental of the rotor winding

N_r the number of series turns in the rotor winding

The other torque expressions are :

$$T = \frac{P}{2}\frac{\pi}{2}\frac{\mu_0 d\, l}{g} F_s F_r \sin\theta$$

$$T = -\frac{P}{2}\frac{\pi}{2}\frac{\mu_0 d\, l}{g} F_{sr} F_r \sin\delta_s$$

$$T = -\frac{P}{2}\frac{\pi}{2}\frac{\mu_0 d\, l}{g} F_{sr} F_r \sin\delta_r$$

$$T = -\frac{P}{2}\frac{\pi d\, l}{2} B_{sr} F_r \sin\delta_r$$

$$T = -\left(\frac{P}{2}\right)^2 \frac{\pi}{2} \Phi_{sr} F_r \sin\delta_r$$

F_s peak value of the fundamental component of the stator spatial mmf wave

$$= \frac{4}{\pi} k_s \frac{N_s}{P} i_s$$

F_r peak value of the fundamental component of the rotor spatial mmf wave

$$= \frac{4}{\pi} k_r \frac{N_r}{P} i_r$$

F_{sr} peak value of the resultant of the fundamental components of the stator and the rotor spatial mmf waves, $F_{sr}^2 = F_s^2 + F_r^2 + 2F_s F_r \cos\theta$

δ_s angle between the stator mmf and the resultant mmf

δ_r angle between the rotor mmf and the resultant mmf $\delta_s + \delta_r = \theta$

B_{sr} peak value of the spatial resultant flux density wave, $= \mu_o F_{sr}/g$

Φ_{sr} resultant flux per pole

As the stator current i_s and the rotor current i_r are functions of time, all the quantities dependent on i_s and i_r, like F_s, B_{sr}, Φ_{sr} etc., are also functions of time. As such, above torque expressions are for instantaneous value of the torque.

The negative sign means that the electromagnetic torque developed tends to reduce the angle between the two axes.

If the sign of any of the three variables, *i.e.*, i_s, i_r and θ (or δ_r) is reversed, the torque direction also reverses.

29.8 INDUCED VOLTAGES AND THE WINDING FACTORS

In rotating machinery, the coil side or the armature conductor; the magnetic field; and the conductor velocity; constitute three mutually perpendicular directions; as such the emf developed in the coil side is given by $N_c Blv$. The same shall be the magnitude of the emf in the other side of the coil, it being one pole pitch away. The coil emf is thus $2N_c Blv$. Alternatively, we may write the expression for the flux linkages of the coil as a function of time, and find the coil emf applying Faraday's law. Since N_c, l, v are constant, the instantaneous value of the emf will be exact reproduction of the spatial distribution of B. If the spatial distribution of flux density B is flat-topped, as in *dc* machines, the emf wave shape will also be flat-topped; if it is sinusoidal, emf wave shape too is sinusoidal.

In one complete revolution of the rotor, $P/2$ cycles of the spatial B wave are met by the conductor, therefore, in one revolution, $P/2$ cycles of the emf are generated in the armature conductors. Thus, ***frequency of emf*** generated is

$$f = (P/2)(N/60) = NP/120 \text{ cycles/second,}$$

N being the speed of rotation in rpm.

Assuming ***sinusoidal spatial flux density wave***, the peak value of the emf developed in the coil,

$$E_m = 2\,N_c\,B_m\,l\,v = 2\,N_c\,B_m\,l\,\omega_{\text{mech}}\,r = 2\,N_c\,B_m\,l\,\omega\,r\,2/P = 2\pi\,f\,N_c\,\Phi,$$

where Φ is the flux per pole.

The rms value of the emf developed in the coil,

$$E = 4.44\,f\,N_c\,\Phi$$

This compares with the expression for the induced voltage in a transformer.

The result can be applied to a distributed winding having N_s series turns,

$$E = 4.44\,f\,k_w\,N_s\,\Phi, \quad \text{where } k_w \text{ is the winding factor.}$$

Assuming ***flat-topped (rectangular) spatial flux density wave***, as in *dc* machines, the coil emf is

$$E_c = 2\,N_c\,B\,l\,v = 2\,N_c\,B\,l\,\omega_{\text{mech}}\,r = 2\,N_c\,B\,l\,r\,2\pi\,N/60 = 2\,N_c P\,\Phi\,N/60$$

This emf is of rectangular wave shape (replica of the B spatial wave).

Because of the commutation action, the emf appearing at the brushes is unidirectional.

For a distributed winding of N_s series turns,

$$E = 2\,N_s P\,\Phi\,N/60$$

If there are A parallel paths in the armature winding of Z conductors, then $2\,N_s = Z/A$,

$$E = (Z/A)\,P\,\Phi\,N/60 = \Phi Z\,NP/(60A) \text{ [rearranged only]}$$

Assuming sinusoidal spatial flux density wave, it has been seen that the rms value of the induced emf in any phase is

$$E_{ph} = 4.44\,f\,k_w\,N_{ph}\,\Phi$$

If there are ***harmonics*** present in the space flux density wave, those harmonics will also induce their own emf's in the armature winding. We may modify the notation slightly.

$$E_{ph1} = 4.44\, f\, k_{w1}\, N_{ph}\, \Phi_1, \quad \text{for the fundamental}$$

For the nth harmonic, there are $n \times P$ poles

$$E_{phn} = 4.44\, n\, f\, k_{wn}\, N_{ph}\, \Phi_n, \quad \Phi_n \text{ is the flux per } n^{th} \text{ harmonic pole}$$

$$E_{phn} / E_{ph1} = (k_{wn}\, B_n)/(k_{w1}\, B_1)$$

Only ***triplen*** harmonics are present. As such

$$E_{ph} = \sqrt{(E_{ph1}^{\ 2} + E_{ph3}^{\ 2} + E_{ph5}^{\ 2} + \ldots + E_{phn}^{\ 2})}$$

The line voltages do not contain any triplen harmonic voltages, these being in phase, cancel out in the line voltage.

The winding factor $k_{wn} = k_{dn}\ k_{pn}$, where k_{dn} is the distribution-, or also called the breadth factor, and k_{pn} is the pitch, or the coil span factor; both for the emf term of frequency nf. These factors are reduction factors; each has a value less than or equal to 1.

The distribution factor takes into account the fact that the coils in the successive slots have a phase difference between the induced emf's in them; consequently, the phasor sum of the emf's of the coils is less than their algebraic sum.

$$\mathbf{k_{dn}} = \frac{\sin(mn\gamma/2)}{m \sin(n\gamma/2)},$$

where m, assumed integer, is the number of slots per pole per phase, and γ is the slot pitch in electrical radians. $\sigma = m\, \gamma$, is the phase spread in radians.

$$\mathbf{k_{pn}} = \cos(n\, \varepsilon/2),$$

where ε is the angle by which the coil pitch falls short of π radians.

A coil which is not a full pitch coil is called a short-pitched coil, or a chorded coil. Clearly, for full pitch coils, $k_{pn} = 1$.

29.9 THE ROTATING MMF

An armature is said to have balanced n-phase winding, if all the phases are identical and each phase is spaced from the next one in sequence by an electrical angle of $2\pi/n$ radians. If such a winding is excited from a balanced n-phase supply of frequency f, a ***rotating mmf*** of constant amplitude equal to $n/2$ times the peak value of mmf of any one phase, rotating at $\omega = 2\pi f$ electrical radians per second, will be produced. For a three-phase machine, the rotating mmf can be expressed as

$$F(\theta, t) = 3/2\ F_{max,ph} \cos(\theta - \omega t)$$

The speed and the orientation of the rotating mmf are such that when a particular phase current reaches its peak value, the rotating mmf reaches the axis of that phase and is oriented along it; and if the phase sequence is $a - b - c$, it rotates from phase a-axis to phase b-axis and to phase c-axis.

29.10 LOSSES IN ROTATING MACHINES

In all the windings of a machine, the $I^2 R$ power loss, usually called ***copper loss***, occurs. For the armature windings, this loss varies with the armature current, and for the field windings (except the series, interpole, and commutating windings, in *dc* machines), this loss is independent of the armature current and is thus constant. Let the total variable copper loss be denoted by k^2P_c, where P_c is the full load copper loss, and k, the pu value of the loading. A small power loss at the carbon brushes = $k\ P_b$ also occurs, where P_b is the loss at both the brushes under full load condition, and is taken as 2 × full load armature current.

Let the constant copper loss be denoted by P_{cc}.

In computing the copper losses, the dc resistance at 75° C should be the used.

The loss in the regulating resistances is not considered a part of the losses of the machine.

Whenever any part of the machine has pulsating flux in it, or the part has a relative speed wrt the flux, then in that part, eddy current loss and hysteresis loss occur. Both these losses taken together are called the core, or the ***iron loss***. This loss occurs even under unloaded condition, and it changes slightly when the machine is loaded. Let the iron loss under unloaded condition be denoted by P_{io}. In *dc* machines, this loss occurs in the armature iron and the pole faces; in induction machines, it occurs mainly in the armature iron and a little in the rotor iron; and in synchronous machines, this loss occurs in the armature iron. Since, the machines are generally operated at constant (rated) voltage and constant speed (rated), iron loss P_{io} is treated as a constant loss. This loss occurs due to rotation; as such it forms a part of the rotational losses.

The rotation, in addition, causes ***mechanical losses***, P_m. The mechanical losses comprise (*i*) the power loss at the rubbing surfaces like that between the brushes and the commutator surface, or the slip ring surface; and at the bearings; (*ii*) the power loss due to air friction which occurs at the rotor surface and in ventilating ducts; and (*iii*) the power consumed by the cooling fans provided, if any. P_m, which depends on the speed, is also treated as a constant loss.

The ***stray load loss***, P_{sl}, is the additional loss in the copper due to non-uniform distribution of the current, and in the iron due to the changes in the flux which occur due to load current. This is taken as 1% of the output for a dc machine, and 0.5% of output for induction, and synchronous machines; and is treated as a variable loss.

The no load rotational losses, therefore, are $P_{rot} = P_{io} + P_m$

The total constant losses are $P_k = P_{cc} + P_{rot}$

These constant losses are also called the no load losses, as the no load input power equals approximately the constant losses. Measuring the no load power at rated speed and various winding resistances forms the ***Swinburne's test*** to determine the efficiency of a *dc* machine.

The total variable losses are $P_v = k^2P_c + kP_b + P_{sl}$

Knowing all the losses at any loading condition, and the input, or the output; the efficiency of the machine acting as a generator, or as a motor, can be found.

$$\eta = \frac{kS}{kS + k^2 P_c + kP_b + P_{sl} + P_k}$$, where S is the machine rating.

For a motor it is more convenient to measure input electrical power (= VI), rather than the output mechanical power; as such we may write

$$\eta = \frac{VI - (k^2 P_c + kP_b + P_{sl} + P_k)}{VI},$$

When the variable losses equal the constant losses, the efficiency is maximum. The corresponding loading k_{max} can be found.

Chapter 30 DC MACHINES

30.1 ARMATURE WINDING PITCHES

In a double layer winding, the number of coils is an integer multiple of the number of slots. As the junction of the finish of one coil and the start of the next coil is connected to a commutator segment, the number of commutator segments equals the number of coils. As such, a dc machine armature winding is a ***closed winding***.

The commutator end is the front end, and the other end—the back end.

All the slots are numbered consecutively. If one side of a coil is placed in slot number i, and the other side of that coil is in slot number x, then the coil span y_{cs}, in terms of number of slots, is $x - i$. y_{cs} is kept equal to the nearest lower integer of S/P.

All the coil sides are numbered in a way that the upper layer successive coil sides bear consecutive odd numbers, and the lower layer coil sides bear even numbers.

The relevant pitches are-the ***back pitch*** y_b, the ***front pitch*** y_f, the ***resultant pitch*** y_r, and the ***commutator pitch*** y_c.

y_f, y_b, and y_r are found in terms of number of coil sides.

If there are n_{cs} coil sides per slot, then the back pitch in terms of number of coil sides is $y_b = n_{cs} \times y_{cs} + 1$, and is an odd number. It means that the coil side lying in the upper layer and numbered i is connected at the back end to the coil side lying in the lower layer and numbered $i + y_b$.

There are two types of armature windings; the ***lap winding***, and the ***wave winding***.

For ***progressive*** lap winding,

$$y_f = y_b - 2, y_r = y_b - y_f = 2, \text{ and } y_c = 1$$

For ***retrogressive*** lap winding

$$y_f = y_b + 2, \ y_r = y_b - y_f = -2, \text{ and } y_c = -1$$

In lap winding, when looking into the winding at the output brushes, we find P number of identical parallel paths.

For all the parallel paths to have equal number of coils, it is necessary that the number of coils per pole pair is an integer, *i.e.*, $2C/P$ is an integer, C being the total number of coils.

Due to slight asymmetry which may exist between the pole pairs, the emf's of all the parallel paths may not be exactly equal. To obviate the ill effects of this, the supposed to be equipotential points, are connected together with ***equalizer rings*** at the back end. The equipotential points are 2π radians apart, and will be found only if $2S/P$ is an integer.

In ***wave winding***

$$y_c = \frac{2(C \pm 1)}{P},$$

C being the total number of coils; + sign is for progressive winding, and – sign for the retrogressive winding. y_c must be an integer.

$$y_r = y_f + y_b = 2\,y_c$$

In wave winding, when looking into the winding at the output brushes, we find 2 identical parallel paths.

In this winding, equalizer rings are not required as each of the two parallel paths spans all the pole pairs.

Dummy coils (not connected to any other coil) may sometimes be needed for securing mechanical balance of the armature.

30.2 DEVELOPED emf AND TORQUE

Developed emf in the armature winding, *i.e.*, the emf available across the output brushes (the brushes assumed to be located on the commutator so that the field axis passes through them)

$$\mathbf{E_a} = \frac{\mathbf{\Phi ZNP}}{\mathbf{60A}}$$

Φ flux per pole

Z total number of armature conductors

N speed of rotation of the armature in revolutions per minute

P number of poles

A number of parallel paths in the armature winding

= 2 for a wave wound armature

= P for a lap wound armature

Thus, for a given machine,

$$E_a \propto \Phi N$$

$$E_a = K_e \Phi N, \quad K_e = \frac{ZP}{60A}$$

If the flux per pole and the field winding current producing the flux, bear a linear relation

$$\Phi = K_f i_f, \text{ then}$$

$$E_a = K_e' i_f N, \quad K_e' = K_e K_f = \frac{ZP}{60A} K_f$$

Further, if i_f is constant,

$$E_a = K_e'' N, \quad K_e'' = K_e' i_f = K_e K_f i_f = \frac{ZP}{60A} K_f i_f$$

The internal, or the electromagnetic power associated with the armature is $P_{\text{arm}} = E_a I_a$. If the speed of rotation of the armature is N rpm, then the electromagnetic torque in N m is

$$\mathbf{T} = \frac{ZP}{2\pi A} \Phi I_a = K_t \Phi I_a, \; K_t = \frac{ZP}{2\pi A}$$

If the polarity of the armature terminals, and that of the field winding terminals is kept unchanged, then the direction of rotation of the rotor remains the same, irrespective of the fact whether the machine runs as a motor, or as a generator.

30.3 COMMUTATION

As the emf developed in a coil is alternating emf, some way to obtain dc voltage at the output terminals is required. In dc machines, this is achieved through commutation. It may, however, be stated that the nature of the emf in a coil is not altered, *i.e.*, it remains alternating.

The commutation is the process of reversal of the current in the coil undergoing commutation. The widths of the commutator segments, mica insulation, and the carbon brushes are such that two or more commutator segments get short-circuited by the brush. During the course of rotation, every coil gets shorted in turn. The duration of short circuit period will depend on the various widths, and the speed of rotation; typically it is 2 ms. Full current in the coil has to reverse in such a short time.

If the change of current with time is linear, then it is linear commutation. Linear commutation is quite satisfactory.

The process of current reversal is, however, hindered by the emf induced in the coil undergoing commutation by the cross component of the armature reaction. This emf is known as the reactance voltage. Because of this, when the short circuit is over, the current would not have completely reversed, and it causes sparking at the brushes. This is overcome by providing ***interpoles***, also, called ***compoles***; located between the main poles. The idea is to induce an emf in the coil undergoing commutation which cancels the reactance voltage, and provides a little extra too, to hasten the current reversal. The compole windings are connected in series with the armature. The compoles *AT* per pole is kept about 1.2 to 1.3 times that of the armature. For a generator, the polarity of the compole must be the same as the polarity of the main pole immediately ahead in the direction of rotation.

30.4 GENERATORS

A dc generator may be a self-excited one, or it may be separately excited.

In the self-excited category, we have shunt generator, series generator, and compound generator.

In a shunt generator, the field winding (shunt field winding) is connected across the armature; while in a series generator, the field winding (series field winding) is connected in series with the armature. The compound generator has the shunt field winding, as well as the series field winding. In long shunt connection, the shunt field winding is connected across the branch comprising the armature and the series field winding in series; while in the short shunt connection, the shunt field winding is connected across the armature.

The field winding of a separately excited generator is excited from a suitable independent dc source.

The ***magnetization*** or the ***open circuit characteristic*** of a dc machine is the curve showing the variation of the field flux with field current i_f, obtained at the rated speed. As the armature emf E is directly proportional to the flux Φ, the shape of this curve is the same as that of E vs i_f. The initial portion of this curve is quite linear, but at higher field currents as saturation sets in the iron of the magnetic flux path, the curve becomes non-linear. This characteristic is almost the same in all dc machines.

For voltage build up in self excited generators, it is necessary that there is some residual flux of the correct direction present in the air gap, and that the field circuit resistance is less than the so-called critical field resistance. The critical field resistance of the machine for a particular speed is the slope of the tangent drawn from the origin to the magnetization curve obtained at that speed.

The curve showing the variation of the terminal voltage with load current is termed the ***load*** or the ***external characteristic*** of the generator. Again, the speed is assumed to be maintained constant.

This characteristic is different for different types of dc machines.

The characteristic is dependent on the following.

(*i*) The increase in the voltage drop in the armature circuit with load current, which is due to the armature winding resistance, the series field winding resistance, if any, and the constant brush contact drop, V_b. This causes the curve to droop somewhat.

(*ii*) Variation of the air gap flux with load current.

The air gap flux would vary with load current, (*a*) if the generator series field winding is in the circuit, (*b*) due to the reduction in the shunt field current and hence the flux on account of fall in the terminal voltage with load current, and (*c*) due to armature reaction.

The voltage change due to the series field will depend upon the relative ampere turns of the series field and the shunt field, and whether the series field aids, or opposes the shunt field. The ***armature reaction*** is the effect of armature ampere turns upon the value and the distribution of the magnetic flux in the air gap. The effect is ***demagnetizing*** and ***distorting***. The armature reaction always causes a fall in the terminal voltage.

The armature terminal voltage

$$V_t = E_a - I_a R_a - V_b,$$

E_a is the internal voltage due to the air gap flux, and R_a is the total armature circuit resistance (excluding the brushes).

V_b may be negligible, or may be merged with $I_a R_a$. Then

$$V_t = E_a - I_a R_a$$

If there is no series field winding, and if the effect of the reduction in the shunt field current, and that of the armature reaction is neglected, then E_a has a fixed value equal to the no load developed voltage at the given speed.

$$E_a = K_e \Phi N, \text{ and } V_t = E_a - I_a R_a$$

are the vital equations for analyzing the performance of dc generators.

For a shunt generator, $I_a = I_l + I_{sh}$; I_l, I_{sh} are respectively the load current, and the shunt field current, and $I_{sh} = V_t/R_{sh}$, where R_{sh} is the shunt field circuit resistance.

For a series generator, $I_a = I_l = I_{se}$, and in the relation $V_t = E_a - I_a R_a$, R_a includes the series field winding resistance.

For a long shunt generator, $I_a = I_l + I_{sh}$; $I_{se} = I_a$, and $I_{sh} = (V_t + I_a R_{se})/R_{sh}$

For a short shunt generator, $I_a = I_l + I_{sh}$; $I_{se} = I_l$, and $I_{sh} = V_t/R_{sh}$

If the series field winding AT are additive to the shunt field winding *AT*, then the generator is cumulatively compounded; and if subtractive, it's differentially compounded. If the degree of cumulative compounding is such that the load characteristic is only slightly better than that of the shunt generator, then it is under compounded. If the no load and the full load voltages are equal, then it is level compounded; and if the full load voltage is greater than the no load voltage, then the generator is over compounded. A series generator's load characteristic is a curve rising from a very small voltage to the full load voltage. A differentially compounded generator has a steeply falling characteristic. A series generator and a differentially compounded generator are not suited for normal applications. A series generator is used as a voltage booster over long dc feeders, and a differentially compounded generator is used in dc welding applications.

30.5 MOTORS

For a motor, $V_t = E_a + I_a R_a$.

$E_a = K_e \Phi N$, $V_t = E_a + I_a R_a$, and $T = K_t \Phi I_a$ ***are the vital equations for analyzing the performance of dc motors.***

$$N = \frac{V_t - I_a R_a}{K_e \Phi}$$

For a ***shunt motor***, $I_a = I_l - I_{sh}$, I_l, I_{sh} are respectively the line current, and the shunt field current; and $I_{sh} = V_t/R_{sh}$, and is very nearly constant.

Since, Φ is very nearly constant, and $I_a R_a$ is very small as compared to V_t, the speed N of a shunt motor falls only a little with I_a. It is considered a constant speed motor. Torque becomes proportional to I_a. Therefore, the speed vs torque characteristic is similar to the speed vs I_a characteristic.

For a ***series motor***, $I_a = I_l = I_{se}$, and in the relation $V_t = E_a + I_a R_a$, R_a includes the series field winding resistance.

If we consider Φ to be proportional to I_{se} $(=I_a)$, then the torque T is proportional to I_a^2.

Because of this relation, in comparison to a shunt motor designed to have the same full load torque as the series motor, a series motor exerts less torque for currents less than the full load current, and a larger torque for currents more than the full load current. Since, the starting current is much larger than the full load current, a series motor is able to exert much larger torque at starting than that a corresponding shunt motor can exert.

Setting $\quad \Phi = K_{se} I_a$, $\quad N = \dfrac{V_t}{K_e K_{se} I_a} - \dfrac{R_a}{K_e K_{se}}$

Since, R_a is quite small, the second term is relatively quite small; as such N is approximately inversely proportional to I_a, or T. For this reason, the no load speed tends to be dangerously high, and as a precaution, a series motor should not be started without some load on its shaft. For higher values of I_a, Φ increase little with I_a due to saturation; and the characteristics tend to those of the shunt motor.

30.6 SPEED CONTROL

The principles of speed control are contained in the equation $N = \dfrac{V_t - I_a R_a}{K_e \Phi}$

The speed can be made higher than the normal speed, by making the flux less than the normal flux. Of course, we are not to make the speed quite far above the rated speed. The flux can be reduced by inserting a rheostat in the field circuit, adopted in shunt motors and the method called ***shunt field rheostat control***; or by cutting out the turns of the field winding, adopted in series motors. With field reduction, the torque developed reduces proportionately, but the power remains constant. A weak field, however, leads to poor commutation.

Inserting an external resistance in the armature circuit lowers the value of the numerator of the speed equation. Therefore, we are able to lower the speed by inserting a suitable rheostat in the armature circuit, and the method is called the ***armature circuit resistance control*** method. In this method, the torque remains constant and the power is reduced. As the armature currents are much larger than the shunt field currents, the method is wasteful of power by way of $I_a^2 R_{a\,\text{ext}}$ loss.

In both the above methods, the range of speed control is not wide.

The value of the numerator of the speed equation is lowered by lowering the value of V_t also. This idea leads to the ***armature terminal voltage control*** method. That means we should have a dc power source of variable voltage. Such an elaborate scheme is used in the

Ward-Leonard method of speed control. A separate motor dc generator set is used in which the polarity and the magnitude of the generator voltage can be controlled by controlling its field circuit current. The speed range thus, available is very wide, with the possibility of reversal of the direction of rotation. If the main motor's field also is controlled, we can have supernormal speeds too.

30.7 COMPENSATING WINDING

When a dc machine is subjected to rapidly changing loads, momentary high emf's are induced in coils located near the middle of poles due to armature reaction. This leads to high potential difference between these coil's commutator segments, which may cause breakdown of the air and eventually a flashover around the commutator. To prevent it, compensating winding embedded in slots on the pole faces, and connected in series with the armature is provided. It has the magnetic effect opposite to that of the adjoining armature winding and the AT per pole of this winding is about 70% of that of the armature.

30.8 CROSS-FIELD MACHINES

A dc armature winding is a closed winding. In a normal dc machine one pair of brushes per pole pair is provided. These brushes physically fall on the centre lines of the poles. The armature reaction produce field perpendicular to the main field and in the diagrams we show the brushes along this line. The main field axis is the direct axis (d-axis), and the armature reaction field axis is the quadrature axis (q-axis), and as such, the brushes are q-axis brushes. If we place additional pair of brushes and close its circuit, it will produce its own armature reaction. In cross-field machines, this additional brush axis is perpendicular to the q-axis and brushes are called the d-axis brushes. The load is connected across this new pair; and the q-axis brushes are short circuited. The armature reaction produced is in a direction opposite to that of the d-axis and is completely demagnetizing, because of which the load terminal voltage falls steeply with load. As such, it is necessary to provide compensating winding.

The cross-field machines fall in the general category of ***metadynes***, and the one provided with compensating winding is called the ***amplidyne***.

The two vital equations for analysing the cross-field machines are the two emfs equations.

$$E_{aq} = K_e \Phi_d N$$

$$E_{ad} = K_e \Phi_q N$$

These machines are two-stage rotating amplifiers and have applications where high power amplification, or high speed of response is required.

Chapter **31** INDUCTION MACHINES

31.1 GENERAL

Induction machines are more often used as motors than as generators and the induction motors are the most widely used type of motors. These work on the principle of mutual induction like transformers, but here the induction is by rotating flux rather than by pulsating flux. As already stated, the primary , or the armature is on the stator, and the secondary on the rotor. The armature has a balanced three-phase winding. When the armature is connected to a balanced three-phase supply, a rotating mmf of constant amplitude rotating at synchronous speed is produced in the air gap. The rotor conductors are cut by this rotating mmf, hence, an emf at slip frequency is induced in the rotor circuit. The rotor circuit being closed, rotor currents flow. The field produced by the armature currents and that produced by the rotor currents interact to develop the torque.

At standstill, the rotor currents frequency is the same as that of the armature currents, and as the rotor picks up speed, the magnitude and the frequency of the induced emf and the current fall; reaching zero at the synchronous speed (if the rotor at all could acquire that speed). Two types of rotors are used, the cage rotor and the wound rotor; the former being more common. Also, the cage rotor adapts to any number of pole pairs. The wound rotor is relatively costlier, but it enables us to control the starting torque and the speed.

In induction machines, magnetizing current drawn, as compared with that in transformers, is relatively greater because of the air gap in the magnetic flux path.

31.2 THREE-PHASE MOTORS

Notation :

p number of pole pairs

f	supply frequency
n_s	synchronous speed of the rotating mmf in rps = f/p
Φ	rotating air gap flux
n_r	rotor speed in rps
N_r	rotor speed in rpm = 60 n_r
s	slip
f_2	frequency of rotor emf and current
N_1	number of stator turns per phase
N_2	number of rotor turns per phase
V	applied voltage on per phase basis
E_1	primary (stator) induced voltage on per phase basis
E_2	secondary (rotor) induced voltage on per phase basis
k_{w1}	winding factor of the stator winding
k_{w2}	winding factor of the rotor winding

All the resistances, reactances, torques, and powers are expressed on per phase basis.

Rotor slip, $$s = \frac{n_s - n_r}{n_s} = 1 - \frac{n_r}{n_s}$$

The following are the ***speed ranges*** and the corresponding ***operating modes*** :

Slip, s	***speed, n_r/n_s***	***speed range***	***operating mode***
$s < 0$	$n_r/n_s > 1$	super synchronous	generating
$s = 0$	$n_r/n_s = 1$	synchronous	bordering generating and motoring
$0 < s < 1$	$1 > n_r/n_s > 0$	sub synchronous	motoring
$s = 1$	$n_r/n_s = 0$	zero	standstill
$s > 1$	$n_r/n_s < 0$	opposite rotation	braking

$$n_r = n_s (1 - s) = f(1 - s)/p$$

$$N_r = 60\, f(1 - s)/p$$

$$f_2 = s\, f$$

$$E_1 = 4.44\, k_{w1} f N_1 \Phi$$

$$E_2 = 4.44\, k_{w2}\, s\, f N_2 \Phi$$

$$E_2{}' = sE_1 = \frac{K_{w1} N_1}{K_{w2} N_2} E_2$$

The rotor current $$I_2{}' = \frac{sE_1}{\sqrt{r_2{}'^2 + s^2 x_2{}'^2}}$$

The rotor power $$P_2 = (1/s) E_2{}' I_2{}' \cos \varphi_2$$

$$P_2 = (I_2{}'^2 r_2{}')/s \text{ also}, \quad \text{or} \quad I_2{}'^2 r_2{}' = s P_2$$

The mechanical power, $P_m = (1 - s) P_2$

$$\mathbf{P_2 : P_m : I_2^2 r_2 = 1 : (1 - s) : s}$$

Torque, $T = P_2/2\pi\, n_s\ Nm = P_2$ synchronous watts (SW)

$$\text{Torque, } \mathbf{T} = \frac{\mathbf{E_1}^2}{\mathbf{x_2}'}\frac{\mathbf{s}\alpha}{\mathbf{s}^2 + \alpha^2}\ \mathbf{SW}, \qquad \text{where } \alpha = r_2/x_2$$

$$\text{Torque,} \qquad T = K_t \frac{s\alpha}{s^2 + \alpha^2}\ SW, \text{ where } K_t = \frac{E_1^2}{x_2'}$$

Maximum torque occurs at a slip

$$s_{\max T} = \alpha$$

$$T_{\max} = \tfrac{1}{2} K_t = \tfrac{1}{2}\frac{E_1^2}{x_2'}$$

The ***stable range*** of operation is from $s = 0$ to $s = s_{\max T}$, and the operation is unstable or transitory from $s = s_{\max T}$ to $s = 1$.

Operating induction motors at as small a slip as possible nets all the advantages.

At small slips (*i.e.*, the rotor speeds close to synchronous speed),

$$\mathrm{T} \approx s\,K_t/\alpha = \frac{s}{\alpha}\frac{E_1^2}{x_2'} = \frac{s}{r_2}\mathrm{E}_1^2$$

At large slips (*i.e.*, the rotor speeds small),

$$\mathrm{T} \approx \alpha\,K_t/s = \frac{\alpha}{s}\frac{E_1^2}{x_2'} = \frac{1}{s}\frac{r_2'}{s_2'^2}E_1^2$$

Increasing the rotor resistance, shifts the slip $s_{\max T}$ towards 1, *i.e.*, speed$_{\max T}$ towards zero; and it does not affect the value of $T_{\max}$.

The mechanical power,

$$P_m = K_t \frac{s\alpha(1-s)}{s^2 + \alpha^2}$$

Maximum mechanical power occurs at a slip

$$s_{\max Pm} = -\alpha^2 \pm \alpha\sqrt{(1+\alpha^2)};$$

+ sign for motoring mode, and – sign for the generating mode.

The maximum mechanical power

$$P_{m\max} = \tfrac{1}{2} K_t\,(z_2 - r_2)/x_2, \text{ where } z_2 = \sqrt{r_2^2 + x_2^2}$$

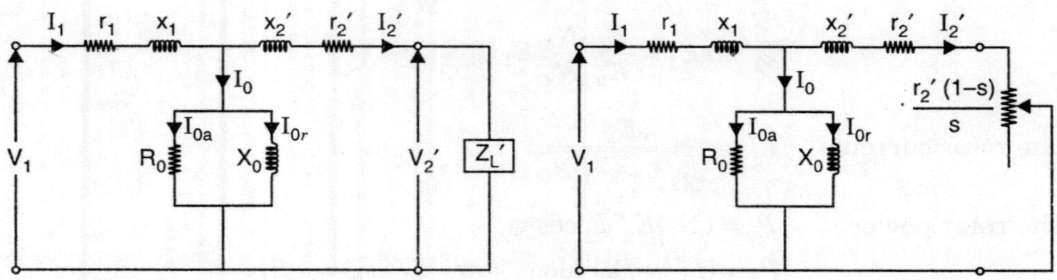

(*a*) (*b*)

FIGURE 31.1. Equivalent circuit of (*a*) a transformer, and (*b*) an induction motor.

The equivalent circuit of an induction machine drawn referred to the primary differs from that of a transformer to the extent that the circuit is now closed through a resistance r_2' $(1 - s)/s$, which accounts for the mechanical power, P_m. In the case of a transformer, the output terminals are closed through the load impedance $\mathbf{Z}_L'$. Though the circuits are similar, the parameters values can be significantly different, particularly those of shunt branch.

31.3 POWER COMPONENTS

Input power P_1 to the armature windings, supplies:

(*i*) $I^2 R$ loss in the armature winding

(*ii*) Iron loss in the armature core

(*iii*) Power transferred to the rotor via magnetic field in the air gap, P_2.

The rotor power P_2 supplies:

(*i*) $I^2 R$ loss in rotor winding

(*ii*) Iron loss in the rotor core (very small at small slips)

(*iii*) mechanical power developed, P_m.

Developed mechanical power, P_m furnishes:

(*i*) friction loss at bearings and slip rings (if any)

(*ii*) windage loss

(*iii*) useful mechanical power output at the rotor shaft.

31.4 STARTING CURRENT, STARTING TORQUE AND STARTERS

I_n armature current at normal voltage under running condition, or the rated armature current

s_n normal running slip

I_{sc} short-circuit (standstill) armature current at normal voltage

x fraction of the normal voltage applied to the armature

I_s short-circuit armature current at a voltage x times the normal voltage $= x\, I_{sc}$

T_s starting torque at a voltage x times the normal voltage

T_n torque at normal voltage under running condition, or the rated torque

$$\frac{T_s}{T_n} = \frac{I_s^2}{I_n^2} S_n = x^2 \left(\frac{I_{sc}}{I_n}\right)^2 S_n$$

In case of a ***stator resistor or inductor starter***, which are simply in series with the stator, the starting line current and the starting torque become respectively x times and x^2 times of those when started with normal voltage (*i.e.*, with $x = 1$)

In case of a ***star-delta starter***, the starting line current and the starting torque both become 1/3 times of those when started directly with delta connection (the motor should be of design of delta-connection normal running).

In case of an ***auto-transformer starter***, the starting line current and the starting torque both become x^2 times of those when started with normal voltage (*i.e.*, with $x = 1$).

31.5 HIGH TORQUE CAGE MOTORS

At starting, the frequency of rotor emf and current is high; as such the currents tend to flow more in the upper layer of the rotor conductors. And under normal running condition, the current is quite uniformly distributed over the cross-section of the conductors. The rotor thus, offers high resistance at starting and relatively lower resistance under normal running condition. This is advantageous as it provides good starting torque. To enhance this effect, one way is to use deep and narrow slots and accordingly the rotor bars also of such shape; and the other way is to have double cage rotor. In double cage rotor, the outer cage has high resistance and low leakage reactance; while the inner one has relatively low resistance and high leakage reactance. This arrangement provides high starting torque in comparison to that with normal type of rotor.

The ratio of the torque due to the outer cage to that due to the inner cage is

$$\frac{T_o}{T_i} = \frac{s^2 + \alpha_i^2}{s^2 + \alpha_o^2} \frac{r_{2o}}{r_{2i}} \left(\frac{x_{2i}}{x_{2o}} \right)^2$$

At starting, $s = 1$,

$$\frac{T_{os}}{T_{is}} = \frac{z_{2i}^2}{z_{2o}^2} \frac{r_{2o}}{r_{2i}}$$

While when running at slip s

$$\frac{T_{or}}{T_{ir}} = \frac{s^2 x_{2i}^2 + r_{2i}^2}{s^2 x_{2o}^2 + r_{2o}^2} \frac{r_{2o}}{r_{2i}} = \frac{x_{2i}^2 + r_{2i}^2 / s^2}{x_{2o}^2 + r_{2o}^2 / s^2} \frac{r_{2o}}{r_{2i}}$$

31.6 SPEED CONTROL

The principles of speed control of an induction motor are contained in the expression for torque at small slips :

$$\mathrm{T} \approx \frac{s}{r_2'} E_1^2, \text{ which leads to}$$

$$n_r = \frac{f}{p}\left(1 - \frac{T\, r_2'}{E_1^2}\right)$$

For a given torque T, we can control any one of, or any combination of f, p, r_2', and E_1. If the frequency and the voltage are to be used as controlling parameters, we have to maintain E/f ratio constant so that the machine operates under constant flux condition.

Then $$n_r = \frac{f}{p} - \frac{T\, r_2'\, f_0^{\,2}}{f\, E_0^{\,2}} = \frac{f}{p} - K\frac{r_2'}{f},$$

where f_o, E_o are the normal frequency and voltage respectively; and $K = (T\, f_o^{\,2}/E_o^{\,2})$

We see that with the increase in frequency, the rotor speed increases.

By changing the number of poles also we can change the rotor speed, but discrete control only can be exercised. Increasing the number of poles, lowers the speed. Suitable tapings from the windings are brought out so that on reconnecting the winding appropriately, the number of poles can be changed.

Speed control can also be exercised with the control of the rotor circuit resistance, which is possible only in slip-ring induction motors. Increasing r_2', lowers the speed; but then this method is wasteful of power.

31.7 INDUCTION REGULATOR

An induction regulator has a rotor which does not rotate but can be turned upto an angle of 360°. The magnitude of the induced voltage in the rotor winding is independent of the rotor angular position w.r.t the armature, but its phase depends on it. Unlike the induction motor, the rotor winding is not short circuited. On connecting the armature winding with the rotor winding appropriately, the output voltage is the phasor sum of the stator applied voltage and the rotor induced voltage. Depending on the rotor position, the output voltage has a value between $V_1 - V_2$ and $V_1 + V_2$. We thus are able to achieve voltage regulation action.

31.8 INDUCTION GENERATOR

If the rotor speed is super synchronous, the slip s becomes negative. Examining the torque and mechanical power expressions, we see that both become negative, thereby meaning that the rotor now has to be rotated externally. Also, the phase of the rotor current I_2 reverses, and the rotor electrical power P_2 becomes negative. The rotor, therefore, is furnishing through the air gap electrical power to the armature circuit. In the armature winding, the phase of the referred current I_2' reverses and therefore, power flows from the armature to the mains rather than from the mains to the armature. The magnetizing current, however, continues to flow as before.

This is the induction generator action.

An induction generator may be mains excited, wherein it draws the magnetizing current from the mains; or it may be a self-excited one, wherein a capacitor connected across it furnishes the exciting current. In both the cases, the electrical power flows from the armature terminals into the mains, or to the electrical circuit connected to it (the armature). For a mains connected IG, the voltage and frequency are independent of the speed and are the same as those of the mains; but in a self-excited generator, the voltage and the frequency are dependent on the speed of rotation of the rotor. Further, it is necessary that there is some residual magnetism in the core, and that the capacitor value be above a certain value, called the critical value. The value of the critical capacitive reactance X_C can be determined experimentally from the magnetizing characteristic of the machine in the same way as we find critical field circuit resistance for a dc shunt generator. We can also find from the magnetizing characteristic the value of the capacitance required to generate a particular voltage.

31.9 SINGLE-PHASE MOTORS

The armature of a pure single-phase induction motor has one winding, and the rotor is similar to that in a three-phase induction motor. When connected to a single-phase supply, a pulsating rather than rotating, field is produced. As such, no inherent starting torque is developed. The pulsating field can however, be regarded as two equal and oppositely revolving fields. Both these fields develop their own torques in opposite directions; their sum being zero at standstill. If the rotor is rotating, the sum of the torques however, is not zero. Therefore, if the rotor is turned in either direction, it would continue to rotate in that direction. Even at synchronous speed some torque is available, but in opposite direction; in other words, the torque becomes zero at a speed slightly less than the synchronous speed.

The equivalent circuit of such a single-phase induction motor would contain the rotor equivalent in two parts; one due to the forward rotating field and the other due to the backward rotating field. If the slip for the forward field is s, then that for the backward field, it is 2-s. For a three-phase induction motor, the rotor part equivalent circuit contains the leakage reactance x_2' in series with the resistance r_2'/s; for the single-phase induction motor, the part due to the forward field has $0.5x_2'$ in series with $0.5r_2'/s$, and that due to the backward field has $0.5x_2'$ in series with $0.5r_2'/(2-s)$. Each of these circuits is shunted by $0.5R_0$ and $0.5X_0$, where R_0 and X_0 respectively account for the iron losses and the magnetizing reactance of the motor. The circuit can be simplified by neglecting R_0, or both R_0 and X_0.

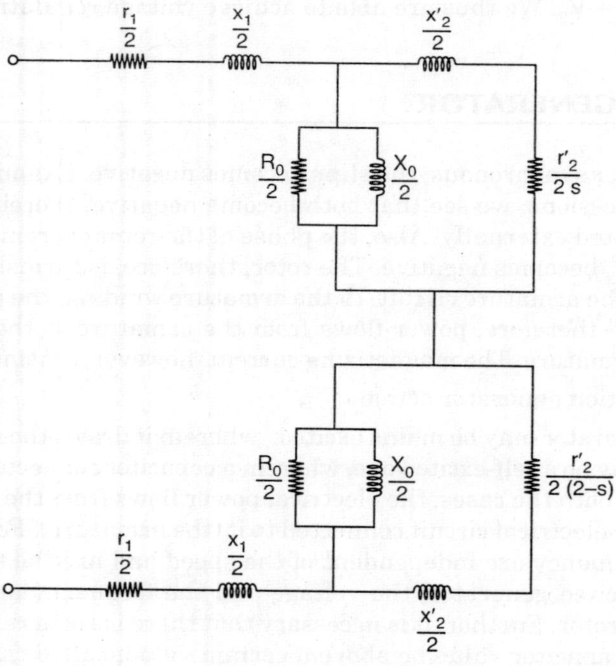

FIGURE 31.2. The equivalent circuit of a pure single-phase induction motor.

Because of the absence of starting torque, the armature winding has to be modified.

Two windings, instead of one, displaced from each other by 90° electrical in space, are provided on the armature. The two windings are made different from each other to the extent that when both are connected to the source, the currents flowing in them have a phase difference nearing 90°; and the magnitudes nearly equal. We are thus, in fact, having a two-phase motor. The windings being 90° apart in space, and their fluxes having phase difference; a rotating field component, apart from pulsating, is produced. This enables the motor to self-start, and enhances the running torque too. The rotor rotates from the leading flux towards the lagging flux. To reverse the direction of rotation, the polarity of one of the two windings should be reversed.

The windings characteristic can be made different, by having their resistance to reactance ratio different. This can be achieved by providing windings with wires of different gauges; or by introducing a capacitor in one of the windings. The winding having higher resistance to reactance ratio, or having the capacitor is called the auxiliary winding, and the other the main winding. Once the motor has picked up speed, the auxiliary winding may be cut out of the circuit, or it may remain permanently connected.

Different arrangements of the auxiliary winding are available for meeting different starting—and the running torque requirements.

Shaded pole motors

These motors are fractional hp motors; a typical value being 1/50 hp. The armature has a concentrated winding on two poles opposite to each other; it being similar to the *dc* machine poles. The rotor is similar to that in an ordinary single-phase induction motor. When connected to a single-phase supply, a pulsating field is produced. This field is split in two components so that these components have a phase difference between them, and therefore, can develop starting torque. This phase splitting of the flux is achieved by using a shading coil, which is nothing but a single turn coil around each of the two narrow and diagonally opposite small parts of the two poles. The flux under the shaded part of the pole lags the flux under the unshaded part, and the rotor rotates from the leading flux towards the lagging flux, *i.e.,* from the unshaded part towards the shaded one. We cannot reverse the direction of rotation by changing the polarity of the winding.

Chapter **32** SYNCHRONOUS MACHINES

32.1 GENERAL

The construction of a synchronous machine has already been described briefly. Here we consider various performance features of the machine. We shall first consider a three-phase *cylindrical rotor* machine, operating as generator under steady state condition. Saturation of iron, the losses in it, and the harmonics are neglected. It implies that the proportionality law can be applied to $B - H$ relation, and that the flux density space wave is a sine wave.

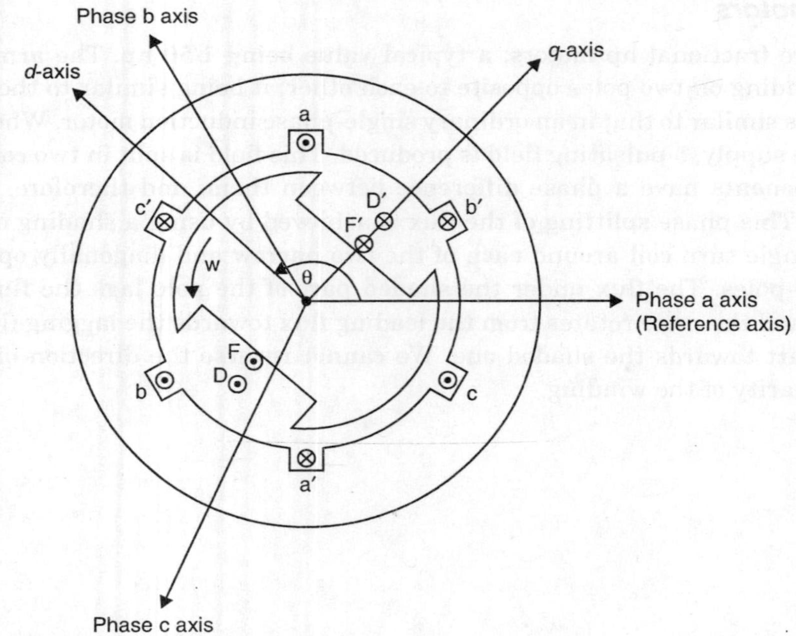

FIGURE 32.1. A Synchronous machine. The winding axes, and the *d*, *q* axes.

During rotation, the phases seen by the rotor in turn are designated as phase a, phase b, and phase c. The orientation of the winding axes, and the reference directions of currents in the windings are in accordance with the right hand screw rule. All the angles are in electrical units. The field axis is also called the direct-axis (***d-axis***); and an axis 90° behind (relative to the direction of rotation) in space is designated as the quadrature axis (***q-axis***). In some literature, q-axis is taken 90° ahead of the d-axis. We shall, however, follow the former convention. Clearly, the d-, and q-axes rotate along with the rotor. The rotor position at any time t as measured in the direction of rotation from phase a-axis (the reference axis) is denoted by $\theta = \omega t + \theta_o$; ω being the rotor speed in radians per second, and θ_o being the rotor angle at $t = 0$. θ_o is often expressed as $\pi/2 + \delta$, δ being the q-axis angle at $t = 0$. The rotor angles with phase b and phase c axes are respectively $(\theta - 2\pi/3)$ and $(\theta - 4\pi/3)$.

32.2 PHASE RELATIONS

For bringing out the phase relations between the voltages, fluxes, and the currents, we may focus on any of the three phases, say phase a. When the field axis and the phase axis are aligned, the field flux linkages with that phase are maximum; and when the two axes come at 90° to each other, the linkages are reduced to null. Thus, the flux linking phase a can be expressed as $\Phi_f \cos\theta = \Phi_f \cos(\omega t + \theta_o)$. The field flux linking phase a, can therefore, be treated as a phasor $\mathbf{\Phi}_f$, its phase angle from the real axis being θ_o. The field mmf causing $\mathbf{\Phi}_f$ is the phasor $\mathbf{F}_f$, the two being in phase with each other.

Remember θ_o is the space angle between the rotor axis and the reference phase (phase a) axis at time $t = 0$. On the phasor diagram also, we show the d-axis along the phasor $\mathbf{\Phi}_f$, and the q-axis behind it by 90°. The emf $\mathbf{E}_f$ induced by $\mathbf{\Phi}_f$ in phase a, lags it by 90°, *i.e.*, when the rotor axis is along the axis of phase a and therefore its flux linkages maximum, the instantaneous value of the emf is zero; and when the rotor is at an angle of 90° and therefore the flux linkages zero, the value of emf is positive maximum. The phasor $\mathbf{E}_f$ is thus behind phasor $\mathbf{\Phi}_f$ by 90°, *i.e.*, it is along the q-axis.

When balanced three-phase currents flow in the armature, these produce a uniformly rotating mmf of constant amplitude in the air gap. During rotation, the mmf is found oriented along the axis of that phase which has peak value of the current flowing in it. When phase a current i_a is peaked, the rotating mmf is along its axis, and when phase b is peaked, it reaches phase b axis, and so on. Thus, in space, under unity power factor condition, *i.e.*, $\mathbf{E}_f$ and $\mathbf{I}_a$ in phase, the rotating armature mmf is behind the field axis by 90° (space angle), *i.e.*, it is along the q-axis. If the armature current phase angle is α lagging, the rotating armature mmf falls further behind the q-axis by α.

As the field mmf could be treated as a phasor $\mathbf{F}_f$, the rotating armature mmf can also be treated as a phasor $\mathbf{F}_a$. As already noted that when current in phase a has the maximum value, the rotating armature mmf is oriented along its axis and therefore has maximum effect on it, and since we are focusing on phase a; the rotating armature mmf $\mathbf{F}_a$ is considered in phase with $\mathbf{I}_a$.

The peak value of the induced emf E_f is $\omega M_f i_f$, where M_f is the maximum value of the mutual inductance between a stator phase and the field winding, which occurs when the two axes are aligned.

We, therefore, can summarize that if the armature current $\mathbf{I}_a$ lags the corresponding emf $\mathbf{E}_f$ by angle α, the phase difference between the field flux $\mathbf{\Phi}_f$ (or $\mathbf{F}_f$) and the current $\mathbf{I}_a$ is $(90° + \alpha)$; and the space difference between the field axis and the rotating armature mmf is also (90° + α); which is reflected as phase difference between the field mmf phasor $\mathbf{F}_f$ and the armature mmf phasor $\mathbf{F}_a$.

32.3 AIR GAP EMF

As $\mathbf{F}_f$ causes $\mathbf{\Phi}_f$ and $\mathbf{E}_f$, $\mathbf{F}_a$ shall similarly cause $\mathbf{\Phi}_a$ and $\mathbf{E}_a$. The resultant armature phase emf $\mathbf{E}_r$ is the phasor sum of $\mathbf{E}_f$ and $\mathbf{E}_a$, *i.e.*, $\mathbf{E}_r = \mathbf{E}_f + \mathbf{E}_a$. Alternatively, we can first find the resultant flux $\mathbf{\Phi}_r = \mathbf{\Phi}_f + \mathbf{\Phi}_a$, and then the resultant emf $\mathbf{E}_r$. It may be mentioned that these statements are valid only under the assumption of linear $B - H$ relation. $\mathbf{E}_f$ is called the excitation emf as it is due to the exciting (field) winding flux (or mmf), $\mathbf{E}_a$ is called the armature reaction emf as it is due to the reaction of the armature mmf, and $\mathbf{E}_r$ is called the **air gap emf**, as it is due to the resultant air gap flux (or mmf). It may again be stated that every induced emf lags the corresponding inducing flux (or the mmf) by 90°.

32.4 SYNCHRONOUS IMPEDANCE

Considering that the armature winding has per phase resistance r_a and leakage reactance x_l, the terminal voltage (generator action) is

$$\mathbf{V}_t = \mathbf{E}_r - \mathbf{I}_a (r_a + j\,x_l) = \mathbf{E}_f + \mathbf{E}_a - \mathbf{I}_a (r_a + j\,x_l)$$

$$\mathbf{E}_a = -j\,4.44\,f\,k_w\,N_s\,\mathbf{\Phi}_a = -j\,4.44\,f\,k_w\,N_s\,K_{\Phi a}\,\mathbf{F}_a$$

$$= -j\,4.44\,f\,k_w\,N_s\,K_{\Phi a}\,\frac{3}{2}\,\frac{4}{\pi}\,k_w\,\frac{N_s}{P}\,2\,\mathbf{I}_a = -j\,x_{ar}\,\mathbf{I}_a \text{ (say)}$$

We see that the effect of armature currents, *i.e.*, the armature reaction can be taken into account by an equivalent fictitious reactance x_{ar}, which can be called the ***armature reaction reactance***.

$$\mathbf{V}_t = \mathbf{E}_f - j\,x_{ar}\,\mathbf{I}_a - \mathbf{I}_a (r_a + j\,x_l) = \mathbf{E}_f - \mathbf{I}_a r_a - j\mathbf{I}_a (x_{ar} + x_l) = \mathbf{E}_f - \mathbf{I}_a (r_a + j\,x_s),$$

where $x_s = x_{ar} + x_l$, and is called the ***synchronous reactance***. The word 'synchronous' signifies that the armature reaction reactance, as explained above, is valid only when the field and the armature mmf rotate in synchronism with each other.

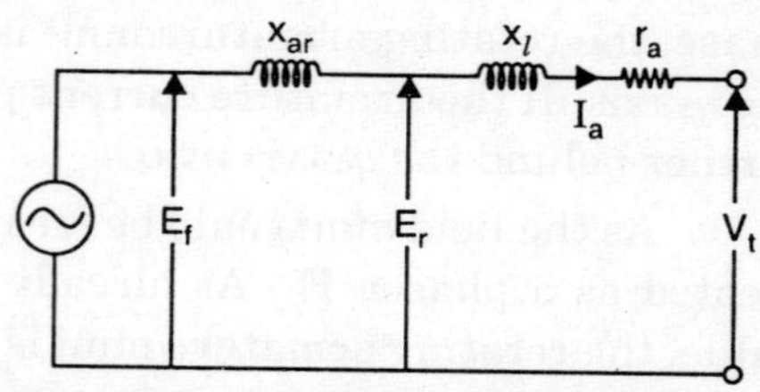

FIGURE 32.2. Circuit model (on per phase basis) of a cylindrical rotor synchronous generator.

x_s actually equals $\omega(L_s + M_s)$, where L_s is the self inductance of a stator phase and M_s is the mutual inductance between any two stator phases, both being positive numbers. $x_s \approx 1.5\,\omega L_s$. The synchronous reactance of an isolated machine is then directly proportional to the machine speed.

The ***synchronous impedance***, $\quad \mathbf{z}_s = r_a + j\,x_s$.

Under rated current conditions, $I_a r_a$, $I_a x_l$, and $I_a x_s$ are respectively of the order of 1%, 10%, and 100% to 200% of the rated voltage.

Thus, whole of the voltage drop is practically caused by the armature reaction reactance alone, *i.e.*, mainly due to armature reaction.

Therefore, $\mathbf{E}_r$ ***very nearly equals*** $\mathbf{V}_t$, which further means that if the terminal voltage is constant, the air gap flux (resultant flux) in the machine remains constant.

It may be reminded that in arriving at above relations, linearity in $B - H$ curve was assumed. It means that if $\mathbf{F}_f$ acting alone produces $\mathbf{\Phi}_f$ and $\mathbf{E}_f$; and $\mathbf{F}_a$ acting alone produces $\mathbf{\Phi}_a$ and $\mathbf{E}_a$; then $\mathbf{F}_f$ and $\mathbf{F}_a$ acting together produce $(\mathbf{\Phi}_f + \mathbf{\Phi}_a)$ and $(\mathbf{E}_f + \mathbf{E}_a)$.

32.5 SATURATION

If non-linearity in the *B-H* curve, *i.e.*, the saturation is to be accounted for, above statement needs to be modified. If $\mathbf{F}_f + \mathbf{F}_a$ is in the saturation region then $\mathbf{F}_f$ and $\mathbf{F}_a$ acting simultaneously will produce a flux less than $\mathbf{\Phi}_f + \mathbf{\Phi}_a$ and an induced emf less than $\mathbf{E}_f + \mathbf{E}_a$.

For greatly lagging power factor currents, *i.e.*, for large α; $\mathbf{F}_f$ and $\mathbf{F}_a$ are nearly in phase opposition, and therefore in opposition in space too. The resultant air gap mmf $\mathbf{F}_r$ is reduced in magnitude, *i.e.*, F_r becomes less than F_f. The air gap flux is also reduced, and is reduced proportionately. So, are reduced E_r and V_t. This is the demagnetizing effect of the lagging armature currents. The machine is not saturated. However, to maintain the terminal voltage, excitation will have to be considerably increased, taking the machine into saturation.

For greatly leading currents, *i.e.*, for large $-\alpha$, $\mathbf{F}_f$ and $\mathbf{F}_a$ are nearly in phase, or are aligned in space. The resultant air gap mmf $\mathbf{F}_r$ is increased in magnitude, *i.e.*, F_r becomes greater than F_f. The machine, therefore, is very likely to be saturated. The air gap flux is also increased, but not proportionately. So are increased E_r and V_t. This is the magnetizing effect of the leading armature currents. However, to maintain the terminal voltage, excitation will have to be decreased but the machine still remains saturated. The rated terminal voltage V_t, and hence the resultant air gap emf E_r $(E_r \approx V_t)$, is such that the necessary F_r is in the saturation region.

For constant terminal voltage operation, the field winding current and the armature current have always to be such that F_r is maintained very nearly constant.

Because of saturation, x_{ar}, and hence x_s is not constant but decreases with the extent of saturation.

32.6 OCC AND SCC

Open-circuit characteristic (**OCC**), which is replica of the magnetization characteristic, is a plot of open circuit armature emf E_f per phase against field current I_f, obtained at rated speed. The plot of E_r against F_r (if we could measure it) will have the same shape as OCC. Initially, the curve is a straight line; but as saturation sets in, the line bends downwards a little. The extended initial straight line portion, which is the same thing as the tangent drawn from the origin to OCC, is the ***air gap line***, *i.e.*, if there were no saturation E_f vs I_f plot would be this straight line.

Short-circuit characteristic (**SCC**) is a plot of short circuited armature current I_a against the field current I_f, obtained at rated speed. The plot is a straight line all through. For short-circuit condition, $\mathbf{E}_r = \mathbf{I}_a (r_a + j\,x_l) = \mathbf{I}_a \mathbf{z}_l$. As z_l is constant and relatively very small, E_r remains very small, even for armature currents much above the rated value. This in turn means $\mathbf{\Phi}_r$ remains very small. The machine is therefore, not at all saturated. It is operating on the linear portion of the *B-H* curve.

We may write $\mathbf{E}_r = K_E\,\mathbf{\Phi}_r$. Since, the machine is unsaturated, we can write $\mathbf{\Phi}_r = K_{\Phi r}\,\mathbf{F}_r$.

$$\mathbf{\Phi}_r = \mathbf{K}_{\Phi r}\,(\mathbf{F}_f + \mathbf{F}_a) = K_{\Phi r}\,(K_{Ff}\,\mathbf{I}_f + K_{Fa}\,\mathbf{I}_a)$$

Thus, $$\mathbf{E}_r = K_E\,K_{\Phi r}\,K_{Ff}\,\mathbf{I}_f + K_E\,K_{\Phi r}\,K_{Fa}\,\mathbf{I}_a$$

Also, $$\mathbf{E}_r = \mathbf{I}_a\,\mathbf{z}_l$$

Therefore, $$\mathbf{I}_a = K_E\,K_{\Phi r}\,K_{Ff}\,\mathbf{I}_f / (\mathbf{z}_l - K_E\,K_{\Phi r}\,K_{Fa})$$

It shows that I_a is linearly proportional to I_f, and this is confirmed by the corresponding plot obtained in the SC test.

If $K_{\Phi r}$ were not a constant and it decreased with F (as would be the case under saturation), I_a also would have not increased proportionately with I_f but would have increased at a slower rate.

We note that under SC condition, even at large values of I_f and I_a, the net mmf acting in the machine is quite small; therefore the armature mmf must be in direct opposition to the field mmf. This further implies that the armature current flowing is greatly lagging, as if E_f is sending a current into an inductor. This strengthens modeling the armature reaction by an inductive reactance (x_{ar}).

The ratio of the field current required to develop rated voltage under open circuit condition as found from OCC to that required to circulate rated current in the armature under short-circuit condition is known as the ***short-circuit ratio***, SCR of the machine. It equals the reciprocal of per unit value of the working synchronous impedance.

32.7 VOLTAGE REGULATION, UNSATURATED AND SATURATED SYNCHRONOUS IMPEDANCE

Synchronous impedance is an important parameter of the machine as it enables us to determine its performance. For some aspects of the performance, or for greater accuracy in the results, saturation is given due consideration.

In finding the ***voltage regulation*** of a synchronous generator, we need the value of the voltage which the armature terminals acquire when the full load on it is thrown off; field current remaining unchanged. This value is E_f. E_f may be found using the equation $\mathbf{E}_f = \mathbf{V}_t + \mathbf{I}_a\,\mathbf{z}_s$; which is known as the ***emf method or the synchronous impedance method*** of finding the voltage regulation. The correctness of the results depends upon the correctness of the value of $\mathbf{z}_s$, and this in turn depends on the saturation conditions in the machine.

If we express $\mathbf{E}_f = \mathbf{V}_t + \mathbf{I}_a\,\mathbf{z}_s$, and set $\mathbf{V}_t = 0$, then z_s for a particular field current is E_f/I_a. As the field current I_f is increased beyond a certain point, the emf E_f instead of increasing proportionately increases slowly, but I_a continues to increase proportionately. Therefore, from that point onwards, z_s decreases somewhat. If OCC were the air gap line, z_s would have remained the same all through, and that value is called the unsaturated synchronous impedance. Thus, slow increase in E_f with I_f is taken into account by decreasing value of z_s.

Apart from the *OCC*, and the air gap line we can have the ***working line***, which is a straight line joining the origin and the rated voltage point on the *OCC*. Synchronous impedance as found from this straight line is the ***working synchronous impedance***, and includes the effect of saturation partially.

Effect of saturation can be more elaborately included by using the ***saturation factor k***.

The saturation factor k depends on the conditions in the air gap of the machine.

For the value of E_r in the machine, we find $F_{r,\,ag}$, and $F_{r,\,occ}$ from the air gap line, and from the *OCC* respectively. Then $k = F_{r,\,occ} / F_{r,\,ag}$ (k>1). For rated terminal voltage conditions, k may be around 1.25.

k is thus, a function of E_r, and beyond E_r equal to about 0.5 pu, it increases substantially as E_r increases.

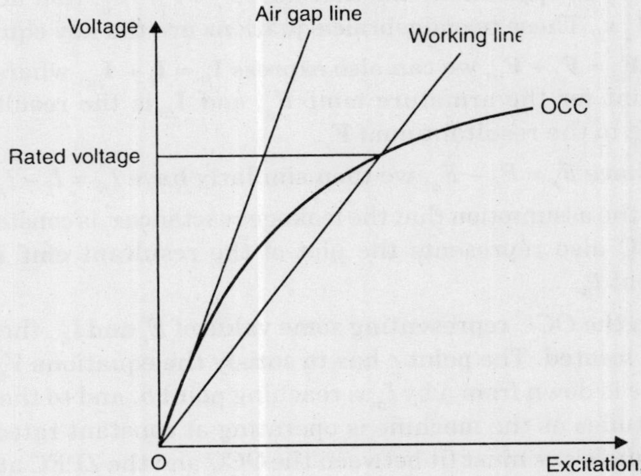

FIGURE 32.3. Open-circuit characteristic, air-gap line, and the working line of a synchronous machine.

We say that $x_{ar,\,sat} = x_{ar,\,ag}\,/k = (x_{s,ag} - x_l)/k$

Therefore, $x_{s,\,sat} = (\,x_{s,ag} - x_l)/k + x_l$

If x_l is neglected, then $x_{s,\,sat} = x_{s,ag}\,/k$

Knowing x_s correctly, one can find E_f correctly.

The other way of accounting for saturation and finding the correct value of E_f is that we first find E_f and I_f from the knowledge of V_t, I_a, and the power factor using unsaturated synchronous impedance and the air gap line; add to this I_f, a correction term which depends on air gap voltage; and then for this corrected value of the field current, find the new value of E_f from the *OCC*, which is taken as the correct value of the excitation emf. The correction term is the difference between the field currents as found from the *OCC*, and from the air gap line for the value of the air gap voltage E_r.

Another method of finding the VR is the ***mmf method***. In this method, we take $\mathbf{E}_r$ equal to $\mathbf{V}_t$, *i.e.*, $\mathbf{z}_l$ is considered negligible. For this E_r, find the corresponding field current I_{fr} from the *OCC*. For the given armature current I_a, find the corresponding I_{fa} from the *SCC*. The excitation I_f is found by finding the resultant of two vectors—I_{fr} and I_{fa}, at an angle (90° + φ), where φ is the angle between $\mathbf{E}_r$ (*i.e.*, $\mathbf{V}_t$) and $\mathbf{I}_a$, *i.e.*, it is the load power factor angle. For this I_f, we find the corresponding E_f from the *OCC*.

The synchronous impedance method results in somewhat higher value of the voltage regulation; while the mmf method yields a somewhat lower value.

Another way of determining the excitation emf is to find the leakage reactance and the armature mmf equivalent field current. This is possible through another characteristic, known as ***zero power factor characteristic*** (ZPFC), which is a plot of the armature terminal voltage against field current, the armature supplying rated current at zero lagging power factor. The curve shall have the $V_t = 0$ point also, which belongs to SCC as well. This point o' is located on the I_f-axis, and its distance from the origin o equals the field current required at the rated armature current in the SC test. For zero lagging power factor current, the armature mmf and the field mmf are nearly in opposition, and therefore $F_r = F_f - F_a$. And neglecting the winding resistance, $E_r = V_t + I_a x_l$. These two algebraic equations are the key equations.

As we express $\mathbf{F}_r = \mathbf{F}_f + \mathbf{F}_a$, we can also express $\mathbf{I}_{fr} = \mathbf{I}_f + \mathbf{I}_{fa}$, where $\mathbf{I}_{fa}$ is the equivalent field current to account for the armature mmf $\mathbf{F}_a$, and $\mathbf{I}_{fr}$ is the resultant equivalent field current corresponding to the resultant mmf $\mathbf{F}_r$.

For ZPFC, we have $F_r = F_f - F_a$, we then similarly have $I_{fr} = I_f - I_{fa}$.

Further, under the assumption that the leakage reactance x_l is constant and the resistance is negligible; the OCC also represents the plot of the resultant emf E_r against resultant equivalent field current I_{fr}.

For a point a on the *OCC* representing some value of E_r and I_{fr}, the corresponding point c on the ZPFC can be located. The point c has to satisfy the equations $V_t = E_r - I_a x_l$, and $I_f = I_{fr} + I_{fa}$, *i.e.*, the point c is down from a by $I_a x_l$ reaching point b, and to the right by I_{fa}. $I_a x_l$ and I_{fa} are of fixed magnitudes as the machine is operating at constant rated current. Therefore, such a right angle triangle abc must fit between the *OCC* and the ZPFC at all points, including the point o'.

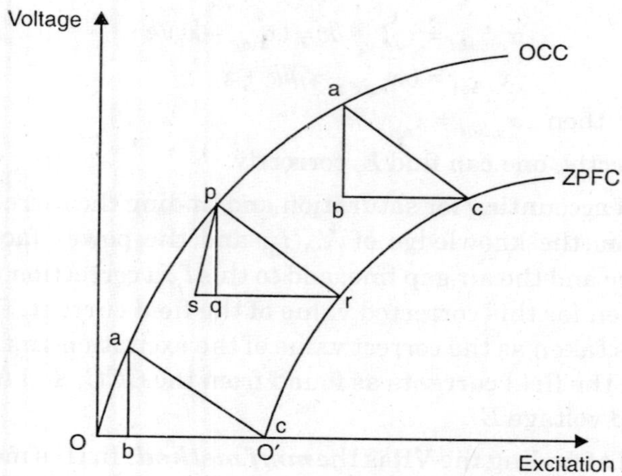

FIGURE 32.4. OCC, ZPFC and the Potier triangle.

When we slide the triangle *abc* down so that *bc* is on the I_f-axis, *c* coinciding with *o′* (*a* remaining on the OCC), we see a bigger triangle *aoo′*. *ao* is a straight section because the initial part of the OCC is linear. The right angle triangle *abc* is contained in the triangle *aoo′*.

Thus, the ZPFC is nothing but the OCC displaced horizontally to the right by distance *oo′*. Therefore, for any point *r* on the ZPFC, a horizontal line *rs* equal to *o′o* and towards left is drawn. From the point *s*, a line parallel to the initial part of the OCC is drawn to cut the OCC at *p*. From point *p*, perpendicular is dropped to cut *sr* at *q*. The right angle triangle *pqr* is obtained which is identical to the right angle triangle *abc*. The triangle *pqr* is the ***Potier triangle***. *pq* is $I_a x_p$, and *qr* is I_{fa}. x_p is the Potier reactance, which is only slightly greater than the leakage reactance x_l.

We have thus found the leakage reactance x_l, and the armature reaction equivalent field current I_{fa} for the rated armature current. Using proportionality, I_{fa} for any other value of armature current can be found.

For the given terminal conditions of the generator, we build E_r by applying $\mathbf{V}_t = \mathbf{E}_r - \mathbf{I}_a (r_a + j\,x_l)$, find corresponding I_{fr} from the OCC, and find I_{fa} for the given I_a. $\boldsymbol{I}_{fr}$ leads $\mathbf{E}_r$ by 90°, and $\mathbf{I}_{fa}$ is along $\mathbf{I}_a$. $\mathbf{I}_f$ is found using $\mathbf{I}_{fr} = \mathbf{I}_f + \mathbf{I}_{fa}$. Then for this I_f we find E_f from the OCC, and hence the voltage regulation can be found.

This method gives better results as compared to the synchronous impedance method, and the mmf method.

32.8 LOSSES

Mechanical losses P_m, which comprises friction and windage losses, can be found by measuring the mechanical power input when the machine is run at synchronous speed with the armature circuit, and the field circuit open.

Rotational losses P_r, which comprise the mechanical losses, and the iron (core) losses, can be found by measuring the mechanical power input when the machine is run at synchronous speed with the armature circuit open, and the field winding excited so as to develop rated voltage.

Iron losses P_i, are given by $P_r - P_m$.

Short-circuit load loss $P_{sc,}$ which comprises copper losses in the armature windings and very small core loss, can be found by measuring the mechanical power input when the machine is run at synchronous speed with the armature circuit short-circuited, and the field windings excited so as to circulate rated armature current, and subtracting from it the mechanical losses P_m.

Copper losses P_c, in the armature winding can be calculated as $3 \times I_a^2 r_a$, where I_a is the rated armature current, and r_a is the per phase dc resistance of the armature winding calculated at 75° C.

Stray load loss P_s, is given by $P_{sc} - P_c$.

Effective resistance of the armature $r_{a,\ \text{effective}}$ is calculated as follows:

$$\mathbf{r}_{\mathbf{a,\ effective}} = P_{sc} / 3 I_a^2$$

32.9 POWERS

We take all the quantities on per phase basis. The machine terminal voltage $\mathbf{V}_t$ is taken as the reference phasor, the armature current $\mathbf{I}_a = I_a \angle -\phi$, and the synchronous impedance is $\mathbf{z}_s = z_s \angle\theta$. The angle θ is quite close to 90°, *i.e.*, r_a is very nearly negligible. We introduce the angle α such that $\theta + \alpha = 90°$. α is quite small.

Generator

For a generator, the current $\mathbf{I}_a$ is the output current, and the excitation emf $\mathbf{E}_f = E_f \angle\delta$

$$\text{Active power output } P_{out} = \frac{E_f V_t}{z_s}\cos(\theta - \delta) - \frac{V_t^2}{z_s}\cos\theta$$

$$\text{Reactive power output } Q_{out} = \frac{E_f V_t}{z_s}\sin(\theta - \delta) - \frac{V_t^2}{z_s}\sin\theta$$

$$\text{Active power input } P_{in} = -\frac{E_f V_t}{z_s}\cos(\theta + \delta) + \frac{E_F^2}{z_s}\cos\theta$$

$$\text{Reactive power input } Q_{in} = -\frac{E_f V_t}{z_s}\sin(\theta + \delta) + \frac{E_f^2}{z_s}\sin\theta$$

Input power is the prime mover power less the rotational losses, *i.e.*, the air gap power converted into electrical power.

If we express θ as $(90° - \alpha)$, the above relations become

$$\text{Active power output } P_{out} = \frac{E_f V_t}{z_s}\sin(\delta + \alpha) - \frac{V_t^2}{z_s}\sin\alpha$$

$$\text{Reactive power output } Q_{out} = \frac{E_f V_t}{z_s}\cos(\delta + \alpha) - \frac{V_t^2}{z_s}\cos\alpha$$

$$\text{Active power input } P_{in} = \frac{E_f V_t}{z_s}\sin(\delta - \alpha) + \frac{E_f^2}{z_s}\sin\alpha$$

$$\text{Reactive power input } Q_{in} = -\frac{E_f V_t}{z_s}\cos(\delta - \alpha) + \frac{E_f^2}{z_s}\cos\alpha$$

For various other forms, one may use the relations

$$r_a = z_s \cos\theta = z_s \sin\alpha, \quad \text{and} \quad x_s = z_s \sin\theta = z_s \cos\alpha$$

If $\theta = 90°$, the relations become very simple.

$$P_{out} = \frac{E_f V_t}{x_s}\sin\delta$$

$$Q_{out} = \frac{E_f V_t}{x_s}\cos\delta - \frac{V_t^2}{x_s} = \frac{V_t^2}{x_s}\left(\frac{E_f}{V_t}\cos\delta - 1\right)$$

$$P_{in} = \frac{E_f V_t}{x_s}\sin\delta$$

$$Q_{\text{in}} = \frac{E_f V_t}{x_s}\cos\delta - \frac{V_t^2}{x_s} = \frac{E_F^2}{x_s}\left(1 - \frac{V_t}{E_f}\cos\delta\right)$$

Motor

For a motor, the current $\mathbf{I}_a$ is the input current, and the excitation emf $\mathbf{E}_f = E_f \angle -\delta$.

$\mathbf{E}_f$ and $\mathbf{V}_t$ get interchanged in the above relations. For example,

$$\text{Reactive power input } Q_{\text{in}} = -\frac{E_f V_t}{z_s}\sin(\theta+\delta) + \frac{V_t^2}{z_s}\sin\theta$$

$$= -\frac{E_f V_t}{z_s}\cos(\delta-\alpha) + \frac{V_t^2}{z_s}\cos\alpha$$

$$\text{Active power output } P_{\text{out}} = \frac{E_f V_t}{z_s}\cos(\theta-\delta) - \frac{E_f^2}{z_s}\cos\theta$$

$$= \frac{E_f V_t}{z_s}\sin(\delta+\alpha) - \frac{E_f^2}{z_s}\sin\alpha$$

Output power is the shaft power plus the rotational losses, *i.e.*, the air gap power converted into mechanical power.

With resistance neglected, $P_{\text{in}} = P_{\text{out}} = \dfrac{E_f V_t}{x_s}\sin\delta$, for the generator as well as for the motor.

In all the above power relations, if the voltages are line-to-line voltages in kV, we obtain the three-phase power (reactive power) in MW (MVAr).

32.10 V-CURVES

The expression $P = \dfrac{E_f V_t}{x_s}\sin\delta$ is the most commonly used expression for power, and is called the power-angle ($P - \delta$) equation.

For a generator, the power out; and for a motor, the power in is also given by $V_t I_a \cos\varphi$.

$$\text{Thus,} \quad V_t I_a \cos\varphi = \frac{E_f V_t}{x_s}\sin\delta$$

For constant terminal voltage operation, $I_a x_s \cos\varphi$ should always adjust to be equal to $E_f \sin\delta$; and both of these should equal $(P\,x_s)/V_t$. The minimum value of I_a which occurs at $\varphi = 0$, is P/V_t; and the minimum value of E_f which occurs at $\delta = 90°$, is $(P\,x_s)/V_t$.

For a fixed value of P, $I_a \cos\varphi$ is constant; if φ is large (both for +ve, and –ve values), I_a also is large, and vice versa. Similarly, $E_f \sin\delta$ also has to be constant; if δ is large E_f need be small, and vice versa. The angle δ has to remain between 0° and 90°.

From the phasor diagram of a generator, we can find that

$$\tan\delta = \frac{P x_s}{V_t^2 + P x_s \tan\phi}$$

For fixed values of P, V_t, and x_s we note that if φ is large, δ will have to be small, and vice-versa. If φ goes on decreasing from near about 90° through 0 to the negative value given by $\tan^{-1}(-V_t^2/Px_s)$, δ will go on increasing from zero to 90°. Beyond this value of φ, δ will become greater than 90°, which makes the machine fall.

Thus, when δ has the highest value (nearly 90°), I_f has the lowest value, φ is negative (*i.e.*, the *pf* is leading) and has a value of $\tan^{-1}(-V_t^2/Px_s)$, and I_a has quite a large value. As δ decreases, I_f increases, φ becomes less negative, and I_a decreases; till a point is reached when φ becomes zero, and I_a has its lowest value. For the given value of the power, the value of the excitation at this point is the normal excitation, and prior to this, it was under-excitation. As δ decreases further, I_f continues to increase (this is now over-excitation), φ becomes positive (*i.e.*, the *pf* becomes lagging) and increases, and therefore, I_a now onwards goes on increasing.

The plot of I_a against I_f, at constant power, will therefore, look like V ; and the plot is therefore, named ***V-curve***. The points on this curve are for constant power and for varying power factor.

For a motor, in the under-excitation region, the power factors is lagging; and in the over-excitation mode, it is leading.

Thus, a synchronous machine, irrespective of the fact whether it is operating as a motor or as a generator; when over-excited supplies reactive power to the bus bars, and when under-excited absorbs reactive power from the bus bars.

It may be explicitly stated that if a synchronous machine is connected to constant voltage bus bars, then any change in the excitation has no effect on the terminal voltage, rather it has an effect on the machine power factor.

If mechanical power is changed, then $I_a \cos\varphi$, as well as $E_f \sin\delta$ will have to change accordingly.

32.11 SYNCHRONOUS PHASE MODIFIER

It has been noted above that a synchronous motor when over-excited draws leading current, and when it is under-excited, it takes lagging current. An unloaded, and over-excited synchronous motor therefore, acts as a capacitor; and when under-excited, it behaves as an inductor. In power systems, the loads are of lagging power factor, and varying too. Such a machine is used at substations, or in large industries, to correct the phase by adjusting its excitation so that all the apparatus behind is relieved of the reactive component of the current of the load. The shaft of such a machine is relatively weak, as it is not to take any load torque.

32.12 SYNCHRONIZATION

Only the dark and bright lamp method of determining the correct instant of closing the paralleling switch is given. Let the busbar terminals of frequency f be R, Y, and B; and those of the incoming machine of frequency f' be R', Y', and B'. The phase sequences are R-Y-B, and R'-Y'-B'. Lamps are connected between R and R'(lamp1), Y and B'(lamp 2), and B and Y' (lamp 3). It can be found by drawing the phasor diagram of the voltages that if the sequence of dimming of the lamps is 1-2-3, then the incoming machine is fast; and if the sequence is 1-3-2, then the incoming machine is slow. The frequency of lighting of each lamp is absolute (f-f'). And if lamp1 remains dark, and the other two remain full bright; then the frequency of the incoming machine is the same as that of the bus bars, and the paralleling switch can be closed.

32.13 TWO ALTERNATORS OPERATING IN PARALLEL

If two alternators operating in parallel are supplying a load at a constant voltage, then

(*i*) the mechanical input to the two machines should be such that the sum of the active components of the currents of the two equals the active component of the load current, and

(*ii*) the excitation of the two machines should be such that the sum of the reactive components of the currents of the two equals the reactive component of the load current, and the terminal voltage remains constant.

If the input to alternator 1 is reduced, then the input to alternator 2 has to be increased, so as to satisfy (*i*). The current, power factor, and the load angle of machine 1 decrease, and those of machine 2 increase.

If the excitation of alternator 1 is reduced, then that of alternator 2 has to be increased, so as to satisfy (*ii*). For machine 1, the current decreases, but the power factor and the load angle increase, and for machine 2 the opposite effects occur.

Apart from above effects, the speed of a generator falls as the load on it is increased. This change in speed from no load condition to full load condition may be of the order of 3%. This occurs on account of the turbine speed governor characteristics, wherein as the machine load changes, the input to the machine changes automatically.

The frequency, or the electrical speed of both the machines operating in parallel, has to be the same. This forces the load (active power) sharing to be as per the following relations.

$$\frac{\Delta P_1}{\Delta P_2} = \frac{\tan\alpha_2}{\tan\alpha_1},$$

where $\tan\alpha = \dfrac{\Delta f}{R}$; Δf being the change in frequency from no load to full load, and R, the rating of the machine. Δf gives the droop characteristic of the machine. Further, $\Delta P_1 + \Delta P_2 = \Delta P_L$.

The change in frequency due to the additional load ΔP_L can simply be found from $\Delta f = \Delta P_1 \tan\alpha_1 = \Delta P_2 \tan\alpha_2$.

32.14 SYNCHRONIZING POWER

An isolated synchronous machine can even be operated at speeds different from the synchronous speed, though it is desirable to operate it at synchronous speed. But if several machines are operating in parallel, it is necessary that all the machines run at the same electrical speed, and this speed should be the synchronous speed. Whenever a sudden change on the system occurs; like large change in the load, a fault, major change in the system configuration, then the machines have to readjust their power angles to satisfy the new operating condition. During this short period of readjustment, one or more machines may fall out of step. This may happen if the ***synchronizing power*** of the machine is not enough to ride over the disturbance. The synchronizing power of the machine is its capability or the power to restore synchronism. If the machine is operating at power angle δ_o, then the synchronizing power for a departure of $\Delta\delta$ (in radians) in the power angle is given by

$$P_{\text{synchro}} = \frac{d}{d\delta}(P_{\max} \sin \delta)\Big|_{\text{at } \delta = \delta_o} \times \Delta\delta = P_{\max} \cos \delta_o \times \Delta\delta$$

$P_{\max} \cos \delta_o$ is called the synchronizing power coefficient.

Thus, SPC depends on $P_{\max}$, and the initial loading reflected by δ_o.

32.15 HUNTING

Whenever any imbalance occurs between the mechanical power input and the electrical power output of a generator, the rotor speed becomes different from the synchronous speed. The rotor reaches steady state, *i.e.*, it reacquires steady synchronous speed after undergoing oscillations in the speed. The oscillation in the rotor speed is known as ***hunting***. The rotor relative speed w.r.t. the synchronous speed oscillates between a small positive maximum, and the negative maximum. The duration of oscillations will depend on the degree of mechanical, and electrical damping present in the machine system. The oscillations are governed by the so-called swing equation

$$P_{\text{mech}} - P_{\max} \sin \delta = M \frac{d^2\delta}{dt^2},$$

where M = moment of inertia of the rotating mass of the machine, I × synchronous speed, ω_s. If powers are in MWs, M is in MJ *s* per electrical degree, δ also being in electrical degrees. In deriving this equation, damping has been neglected.

To reduce hunting, short-circuited damper bars, similar to those in a cage rotor of an induction machine, are provided on the pole faces of the synchronous machine. Because of the relative speed of the rotor, the synchronously rotating armature mmf flux induces an emf in the bars, the magnitude of which is proportional to the relative speed of the rotor. The resulting currents and the synchronously rotating armature mmf flux interact to develop a torque which opposes the very cause of it, *i.e.*, the relative speed of the rotor, and hence the oscillations are reduced.

In a synchronous motor, the damper bars serve the additional function of developing the starting torque.

32.16 WINDINGS INDUCTANCES IN SALIENT-POLE SYNCHRONOUS MACHINES

In a ***salient-pole synchronous machine***, the field poles are of projecting type; and therefore, the air gap length is not uniform. As such, most of the windings inductances depend on the position of the rotor. Suitable general expressions for the inductances and the flux linkages of the windings are given below. Also, because of the non-uniform air gap, the synchronously rotating armature mmf flux path finds different reluctances for different operating conditions. This situation is resolved by resolving the armature mmf along two axes, the reluctances of the magnetic circuits along which are fixed. These two axes are the *d*-axis, and the *q*-axis. The reluctance of the magnetic path along the *q*-axis is greater than that along the *d*-axis. It further means that equal mmf acting along the two axes shall produce smaller flux along the *q*-axis than that produced along the *d*-axis.

The rotor self inductance:

$L_{ff} = L_{ff}$, *i.e.*, it is independent of the rotor position.

It may be recalled that the rotor angle with phase *a*-axis is $\theta = \omega t + \theta_o = \omega t + \pi/2 + \delta$

The stator to rotor mutual inductances:

$$L_{af} = M_f \cos\theta$$
$$L_{bf} = M_f \cos(\theta - 120)$$
$$L_{cf} = M_f \cos(\theta - 240)$$

The stator self inductances:

$$L_{aa} = L_{aao} + L_{g2}\cos 2\theta$$
$$L_{bb} = L_{aao} + L_{g2}\cos(2\theta + 120)$$
$$L_{cc} = L_{aao} + L_{g2}\cos(2\theta - 120)$$

where $L_{aao} = L_{al} + L_{go}$, $L_{go} = N_a^{\,2}(P_{gd} + P_{gq})/2 = N_a^{\,2} P_{\text{mean}}$; N_a being the number of turns of phase *a*; P_{gd}, and P_{gq} being the permeance of magnetic flux path along the *d*-axis, and *q*-axis respectively,

$$L_{g2} = N_a^{\,2}(P_{gd} - P_{gq})/2 = N_a^{\,2} P_{diff}$$

The stator to stator mutual inductances:

$$L_{ab} = L_{ba} = -0.5\,L_{go} + L_{g2}\cos(2\theta - 120)$$
$$L_{bc} = L_{cb} = -0.5\,L_{go} + L_{g2}\cos 2\theta$$
$$L_{ca} = L_{ac} = -0.5\,L_{go} + L_{g2}\cos(2\theta + 120)$$

It can be found that the mutual inductance between phase *i* and phase *j* is given by

$$L_{ij} = N_i N_j (P_{\text{mean}}\cos\alpha_{ij} + P_{diff}\cos(2\theta - \alpha_i - \alpha_j));$$

where N_i, α_i are the number of turns and the angle of phase *i*-axis from the reference phase, and $\alpha_{ij} = \alpha_i - \alpha_j$.

If the permeance of the air gap of the magnetic circuit is same everywhere, *i.e.*, ***if the machine is a cylindrical rotor machine***, then

$$P_{gd} = P_{gq} = P_g$$
$$L_{go} = N_a^{\,2} P_g,\; L_{g2} = 0$$
$$L_{aa} = L_{bb} = L_{cc} = L_{aao} = L_{al} + L_{go} = L_{al} + N_a^{\,2} P_g = L_s\text{(say)}$$
$$L_{ab} = L_{bc} = L_{ca} = -0.5\,L_{go} = -0.5\,N_a^{\,2} P_g = -M_s\text{(say)} \approx -0.5\,L_s$$

32.17 FLUX LINKAGES

$$\begin{bmatrix} \lambda_a \\ \lambda_b \\ \lambda_c \\ \lambda_f \end{bmatrix} = \begin{bmatrix} L_{aa} & L_{ab} & L_{ac} & L_{af} \\ L_{ba} & L_{bb} & L_{bc} & L_{bf} \\ L_{ca} & L_{cb} & L_{cc} & L_{cf} \\ L_{fa} & L_{fb} & L_{fc} & L_{ff} \end{bmatrix} \begin{bmatrix} i_a \\ i_b \\ i_c \\ i_f \end{bmatrix}$$

The terminal voltage of any winding , the machine acting as a generator, is given by

$$v = -\frac{d\lambda}{dt} - r\,i = -p\lambda - r\,i$$

32.18 TRANSFORMATIONS TO *d*-AND *q*-AXES VARIABLES

The mmf of any phase can be resolved along d-axis, and along q-axis. For phase a, it is

$$F_{da} = N_a\, i_a \cos\theta$$

$$F_{qa} = N_a\, i_a \cos(\theta - 90°) = N_a\, i_a \sin\theta$$

It may be reminded that θ is the angle between the rotor axis (d-axis), and the phase a-axis, and the angle between the q-axis and the phase a-axis is $\theta - 90°$, as it is behind d-axis by 90°.

Similar expressions for the other two phases can be written easily.

As stated earlier, the advantage is that the d-axis mmf, and the q-axis mmf act on their respective magnetic circuits, which are fixed and their reluctances are not dependent on the rotor position.

The net mmf's due to the three phases are

$$F_{ds} = N_a\, [i_a \cos\theta + i_b \cos(\theta - 120°) + i_c \cos(\theta - 240°)]$$

$$F_{qs} = N_a\, [i_a \sin\theta + i_b \sin(\theta - 120°) + i_c \sin(\theta - 240°)]$$

Accordingly, we may set

$$i_d = \sqrt{(2/3)}\, [i_a \cos\theta + i_b \cos(\theta - 120°) + i_c \cos(\theta - 240°)]$$

and $$i_q = \sqrt{(2/3)}\, [i_a \sin\theta + i_b \sin(\theta - 120°) + i_c \sin(\theta - 240°)]$$

A fictitious current i_o is introduced.

$$i_o = \sqrt{(1/3)}\, [i_a + i_b + i_c]$$

In some literature, 2/3 instead of $\sqrt{(2/3)}$; and (1/3) instead of $\sqrt{(1/3)}$is used.

In matrix form, we can write

$$\begin{bmatrix} i_d \\ i_q \\ i_o \end{bmatrix} = \sqrt{\frac{2}{3}} \begin{bmatrix} \cos\theta & \cos(\theta - 120°) & \cos(\theta - 240°) \\ \sin\theta & \sin(\theta - 120°) & \sin(\theta - 240°) \\ 1/\sqrt{2} & 1/\sqrt{2} & 1/\sqrt{2} \end{bmatrix} \begin{bmatrix} i_a \\ i_b \\ i_c \end{bmatrix}$$

and $$\begin{bmatrix} i_a \\ i_b \\ i_c \end{bmatrix} = \sqrt{\frac{2}{3}} \begin{bmatrix} \cos\theta & \sin\theta & 1/\sqrt{2} \\ \cos(\theta - 120°) & \sin(\theta - 120°) & 1/\sqrt{2} \\ \cos(\theta - 240°) & \sin(\theta - 240°) & 1/\sqrt{2} \end{bmatrix} \begin{bmatrix} i_d \\ i_q \\ i_o \end{bmatrix}$$

or $$i_{dqo} = P\, i_{abc} \quad \text{and} \quad i_{abc} = P^{-1} i_{dqo} = P^t\, i_{dqo},$$

where, P is above 3 × 3 transformation matrix, called Park's transformation matrix.

If we let $i_a = I_m \cos(\omega t + \delta - \alpha)$, $i_b = I_m \cos(\omega t + \delta - \alpha - 120°)$, and $i_c = I_m \cos(\omega t + \delta - \alpha - 240°)$, *i.e.*, i_a, i_b, and i_c are balanced three-phase currents, and setting $\theta = \omega t + \pi/2 + \delta$,

We find $i_d = -kI_m \sin\alpha$, $i_q = kI_m \cos\alpha$, and $i_o = 0$, where $k = \sqrt{(3/2)}$.

We note that for the balanced three-phase currents, i_d and i_q are constant currents. This is quite interesting. These fictitious currents are supposed to be flowing in fictitious concentrated windings whose axes are the d-axis, and the q-axis and which rotate along with the rotor. In the air gap, these produce the same effect as the three stationary but time-varying armature currents taken together.

It may be seen that when the d-axis is along the axis of phase a, the phase a carries current $-I_m \sin\alpha$ (at $t = 0$, $\theta = 0$).

32.19 BASIC MACHINES RELATIONS IN dq0 VARIABLES

Flux linkages

$$\lambda_f = L_{ff}\, i_f + k\, M_f\, i_d$$
$$\lambda_d = kM_f\, i_f + L_d\, i_d$$
$$\lambda_q = L_q\, i_q$$
$$\lambda_o = L_o\, i_o$$

The ***d-q-0 inductances*** are

$$L_d = L_{al} + 3/2(L_{go} + L_{g2}) = L_s + M_s + \frac{3}{2} L_{g2};$$

where $$L_s = L_{al} + L_{go},\ M_s = 0.5\, L_{go}$$

$$L_q = L_{al} + 3/2(L_{go} - L_{g2}) = L_s + M_s - \frac{3}{2} L_{g2}$$

$$L_o = L_{al} = L_s - 2M_s$$

The ***f-d-q-0 terminal voltages*** (for generating action) are given by

$$v_f = p\, \lambda_f + r_f i_f$$
$$v_d = -p\, \lambda_d - \omega\, \lambda_q - r_a\, i_d$$
$$v_q = -p\, \lambda_q + \omega\, \lambda_d - r_a\, i_q$$
$$v_o = -p\, \lambda_o - r_a\, i_o$$

The ***electromagnetic torque*** (due to all the three phases) is

$$T = \frac{\text{poles}}{2}(\lambda_d i_q - \lambda_q i_d)$$

Under ***balanced steady state conditions***, i_f, i_d, and i_q are constant, and therefore, the flux linkages λ_f, λ_d, and λ_q are all constant.

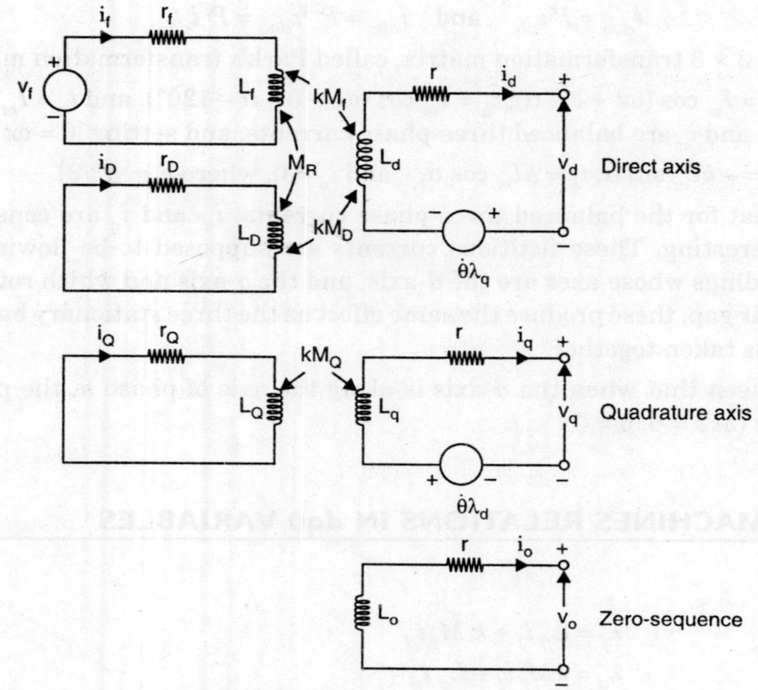

FIGURE 32.5. The equivalent two-axis model of a synchronous machine. (Damper circuits are also included).

We then have
$$v_f = r_f i_f$$
$$v_d = -\omega\lambda_q - r_a i_d = -\omega L_q i_q - r_a i_d = -x_q i_q - r_a i_d$$
$$v_q = \omega\lambda_d - r_a i_q = \omega k M_f i_f + \omega L_d i_d - r_a i_q = \omega k M_f i_f + x_d i_d - r_a i_q$$

x_d, and x_q are respectively the direct-axis synchronous reactance, and the quadrature-axis synchronous reactance.

As
$$i_a = \sqrt{(2/3)}\,(i_d \cos\theta + i_q \sin\theta)$$
$$= \sqrt{(2/3)}\,(i_d \cos(\omega t + \theta_o) + i_q \sin(\omega t + \theta_o))$$
$$= \sqrt{(2/3)}\,(i_d \cos(\omega t + \theta_o) + i_q \cos(\omega t + \theta_o - 90°)),$$

it can be transformed into the ***phasor equation*** as

$$\mathbf{I}_a = (\mathbf{I}_d + \mathbf{I}_q), \text{ where } \mathbf{I}_d = \frac{i_d}{\sqrt{3}}\angle\theta_o, \text{ and } \mathbf{I}_q = \frac{i_d}{\sqrt{3}}\angle(\theta_o - 90°) = \frac{i_q}{\sqrt{3}}\angle\delta$$

For lagging currents, i_d has negative value and therefore, $\mathbf{I}_d$ is along $-d$-axis.

Similarly, v_a or v_t can also be expressed as a phasor in terms of $\mathbf{I}_d$ and $\mathbf{I}_q$.

$$v_t = \sqrt{(2/3)}\,(v_d \cos(\omega t + \theta_o) + v_q \cos(\omega t + \theta_o - 90°))$$

$$\mathbf{V}_t = (\mathbf{V}_d + \mathbf{V}_q), \text{ where } \mathbf{V}_d = \frac{v_d}{\sqrt{3}}\angle\theta_o, \text{ and } \mathbf{V}_q = \frac{v_q}{\sqrt{3}}\angle(\theta_o - 90°) = \frac{v_q}{\sqrt{3}}\angle\delta$$

$$\frac{v_d}{\sqrt{3}}\angle\theta_o = (-x_q i_q \angle\theta_o - r_a i_d \angle\theta_o)/\sqrt{3} = -jx_q\mathbf{I}_q - r_a\mathbf{I}_d$$

Thus, $\mathbf{V}_d = -j\,x_q\,\mathbf{I}_q - r_a\,\mathbf{I}_d$

$$\frac{v_q}{\sqrt{3}}\angle(\theta_o - 90^\circ) = (\omega k M_f\, i_f \angle(\theta_o - 90^o) + x_d\, i_d \angle(\theta_o - 90^o) - r_a\, i_q \angle(\theta_o - 90^o))/\sqrt{3}$$

$$= \mathbf{E}_f - jx_d\,\mathbf{I}_d - r_a\,\mathbf{I}_q$$

where $\mathbf{E}_f = E_f\angle\delta$, and $E_f = \omega M_f\, i_f/\sqrt{2}$

Thus $\mathbf{V}_q = \mathbf{E}_f - jx_d\,\mathbf{I}_d - r_a\,\mathbf{I}_q$

Therefore, $\mathbf{V}_t = -j\,x_q\,\mathbf{I}_q - r_a\,\mathbf{I}_d + \mathbf{E}_f - j\,x_d\,\mathbf{I}_d - r_a\,\mathbf{I}_q$

$$= \mathbf{E}_f - r_a(\mathbf{I}_d + \mathbf{I}_q) - j\,x_q\,\mathbf{I}_q - jx_d\,\mathbf{I}_d = \mathbf{E}_f - r_a\,\mathbf{I}_a - j\,x_q\,\mathbf{I}_q - jx_d\,\mathbf{I}_d$$

$$\mathbf{E}_f = \mathbf{V}_t + r_a\mathbf{I}_a + j\,x_q\,\mathbf{I}_q + jx_d\,\mathbf{I}_d$$

For a motor, E_f and V_t get interchanged.

$$\mathbf{V}_t = \mathbf{E}_f + r_a\,\mathbf{I}_a + j\,x_q\,\mathbf{I}_q + jx_d\,\mathbf{I}_d$$

The phasor diagrams should satisfy these equations.

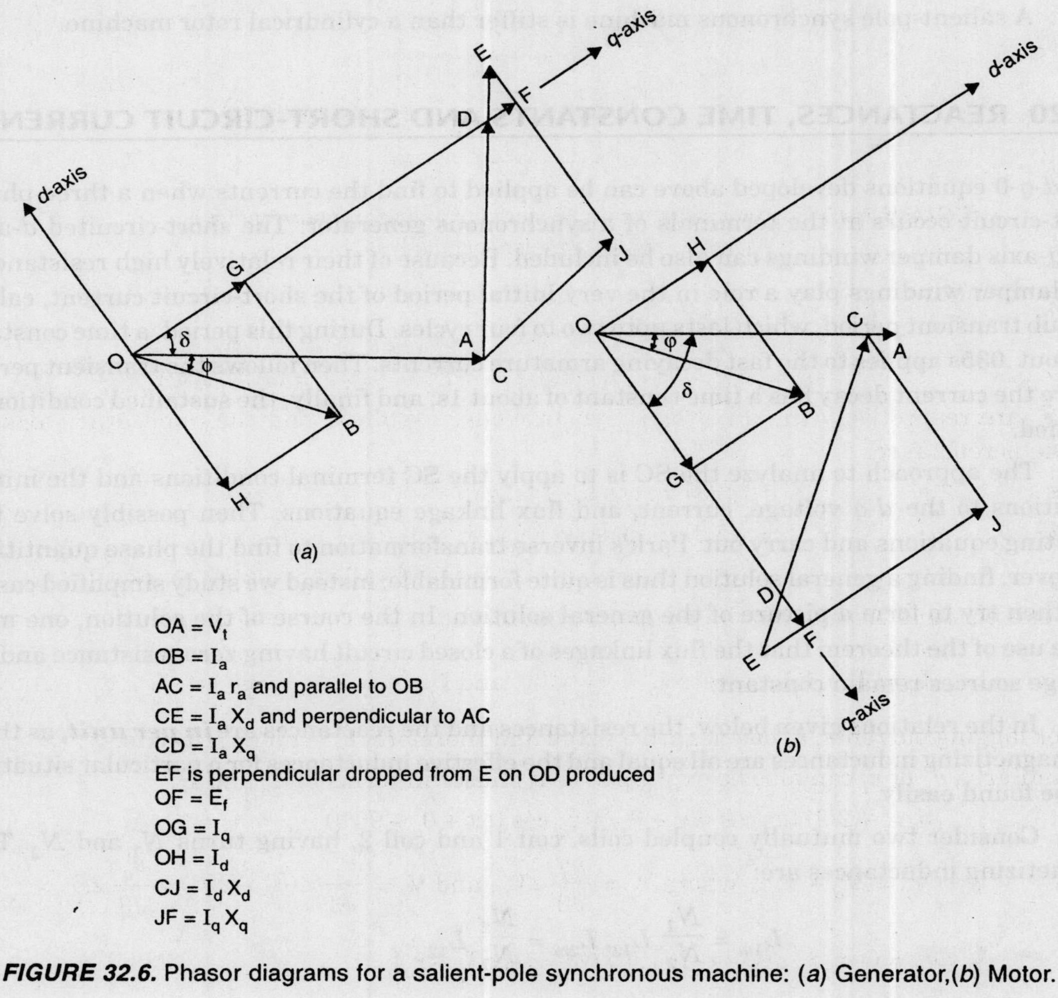

FIGURE 32.6. Phasor diagrams for a salient-pole synchronous machine: (*a*) Generator,(*b*) Motor.

One can deduce from the phasor diagram, with resistance neglected, that the power is given by

$$P = \frac{E_f V_t}{x_d} \sin \delta + V_t^2 \frac{x_d - x_q}{2x_d x_q} \sin 2\delta$$

The second term in this expression is the reluctance power of the machine; it being due to the saliency of the rotor.

P acquires maximum value at angle δ greater than 90°, the angle δ being given by

$$\frac{\cos \delta}{1 - 2\cos^2 \delta} = \frac{V_t (x_d - x_q)}{E_f x_q}$$

The synchronizing power coefficient is

$$\frac{E_f V_t}{x_d} \cos \delta + V_t^2 \frac{x_d - x_q}{x_d x_q} \cos 2\delta$$

A salient-pole synchronous machine is stiffer than a cylindrical rotor machine.

32.20 REACTANCES, TIME CONSTANTS AND SHORT-CIRCUIT CURRENTS

The d-q-0 equations developed above can be applied to find the currents when a three-phase short-circuit occurs at the terminals of a synchronous generator. The short-circuited d-axis and q-axis damper windings can also be included. Because of their relatively high resistances, the damper windings play a role in the very initial period of the short-circuit current, called the sub transient period, which lasts upto two to four cycles. During this period, a time constant of about .035s applies to the fast decaying armature currents. Then follows the transient period where the current decay has a time constant of about 1s; and finally, the sustained condition is reached.

The approach to analyze the SC is to apply the SC terminal conditions and the initial conditions to the d-q voltage, current, and flux linkage equations. Then possibly solve the resulting equations and carry out Park's inverse transformation to find the phase quantities. However, finding a general solution thus is quite formidable; instead we study simplified cases, and then try to form a picture of the general solution. In the course of the solution, one may make use of the theorem that the flux linkages of a closed circuit having zero resistance and no voltage sources remain constant.

In the relations given below, the resistances and the reactances are ***in per unit***, as then the magnetizing inductances are all equal and the effective inductances for a particular situation can be found easily.

Consider two mutually coupled coils, coil 1 and coil 2, having turns N_1 and N_2. The magnetizing inductances are:

$$L_{1m} = \frac{N_1}{N_2} L_{12}, \; L_{2m} = \frac{N_2}{N_1} L_{12}$$

The base values of inductances L_{B1} and L_{B2} for the two coils are related as $\dfrac{L_{B1}}{L_{B2}} = \left(\dfrac{N_1}{N_2}\right)^2$.

The per unit values of the magnetizing inductances are

$$L_{1mpu} = \frac{N_1}{N_2}\frac{L_{12}}{L_{B1}},\ L_{2mpu} = \frac{N_2}{N_1}\frac{L_{12}}{L_{B1}}\left(\frac{N_1}{N_2}\right)^2 = L_{1mpu} = L_{mpu}$$

Thus, in *pu* system, the magnetizing inductances of any number of coupled windings are all equal and the *pu* self inductance of a winding is the sum of its *pu* leakage inductance and the *pu* magnetizing inductance.

Several resistances, reactances, and time constants are involved in the complete expression of short-circuit armature current. We have already met many of these. A more elaborate set of the parameters is given below. The resistances and inductances/reactances are **in pu**.

Resistances and leakage inductances

Armature winding, field winding, *d*-axis damper winding, *q*-axis damper winding:

$$r_a,\ r_f,\ r_{kd},\ r_{kq}\ ;\ L_{al},\ L_{fl},\ L_{kdl},\ L_{kql}$$

Corresponding **leakage** reactances $x_{al} = \omega L_{al},\ x_{fl} = \omega L_{fl},\ x_{kdl} = \omega L_{kdl},\ x_{kql} = \omega L_{kql}$

d-axis inductances and reactances

Magnetizing inductance and reactance $L_{ad},\ x_{ad} = \omega L_{ad}$

Synchronous inductance and reactance $L_d = L_{ad} + L_{al},\ x_d = \omega(L_{ad} + L_{al}) = x_{ad} + x_{al} = x_d$

Transient reactance $$x'_d = \left(\frac{1}{x_{ad}} + \frac{1}{x_{fl}}\right)^{-1} + x_{al}$$

Subtransient reactance $$x''_d = \left(\frac{1}{x_{ad}} + \frac{1}{x_{fl}} + \frac{1}{x_{kdl}}\right)^{-1} + x_{al}$$

q-axis inductances and reactances

Magnetizing inductance and reactance

$$L_{aq},\ x_{aq} = \omega L_{aq}$$

Synchronous inductance and reactance

$$L_q = L_{aq} + L_{al},\ x_q = \omega(L_{aq} + L_{al}) = x_{aq} + x_{al} = x_q$$

Transient reactance $x'_q = x_q$

Subtransient reactance:

$$x''_q = \left(\frac{1}{x_{aq}} + \frac{1}{x_{kql}}\right)^{-1} + x_{al}$$

d-axis Time Constants

Open-circuit transient time constant

$$T_{do}' = \frac{x_{ad} + x_{fl}}{r_f} = \frac{L_{ff}}{r_f}$$

Short-circuit transient time constant

$$T_d' = \frac{x'_d}{x_d} T_{do}' = \frac{1}{r_f}\left(\left(\frac{1}{x_{ad}} + \frac{1}{x_{al}}\right)^{-1} + x_{fl}\right)$$

Open-circuit subtransient time constant

$$T_{do}'' = \frac{1}{r_{kd}}\left(\left(\frac{1}{x_{ad}} + \frac{1}{x_{fl}}\right)^{-1} + x_{kdl}\right)$$

Short-circuit subtransient time constant

$$T_d'' = \frac{1}{r_{kd}}\left(\left(\frac{1}{x_{ad}} + \frac{1}{x_{fl}} + \frac{1}{x_{al}}\right)^{-1} + x_{kdl}\right)$$

q-axis Time Constants

Open-circuit sub-transient time constant

$$T_{qo}'' = \frac{1}{r_{kq}}(x_{aq} + x_{kql})$$

Short-circuit sub-transient time constant

$$T_q'' = \frac{1}{r_{kq}}\left(\left(\frac{1}{x_{aq}} + \frac{1}{x_{al}}\right)^{-1} + x_{kql}\right)$$

Armature short-circuit time constant

$$T_a = \frac{2}{r_a}\left(\frac{1}{x_d''} + \frac{1}{x_q''}\right)^{-1}$$

Following relations amongst the time constants hold.

$$\frac{T_d''}{T_{do}''} = \frac{x_d''}{x_d'} \quad \text{and} \quad \frac{T_q''}{T_{qo}''} = \frac{x_q''}{x_q'}$$

It may be observed that the various reactances are found looking into the d-axis or the q-axis armature windings terminals, with one or more of the other circuits open/closed on itself, depending on the reactance being found. These reactances govern the magnitude of the armature current under steady state, transient state, and subtransient state.

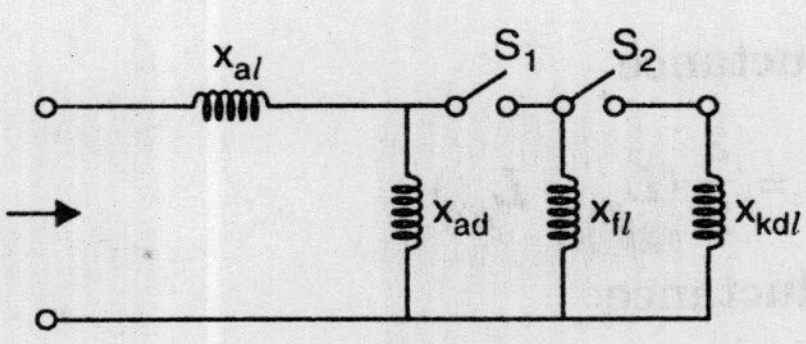

(*a*)

For synchronous reactance, S_1 and S_2 are both open
For transient reactance, S_1 is closed, S_2 open
For sub transient reactance, S_1 and S_2 are both closed

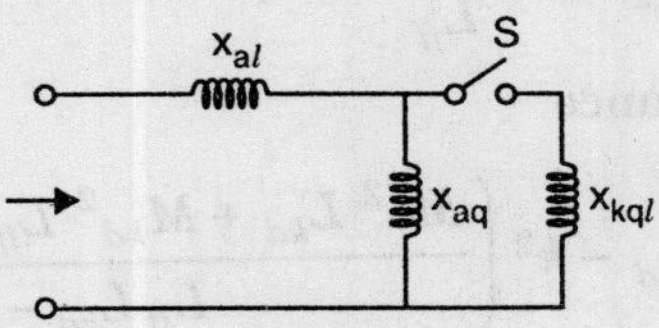

(*b*)

For synchronous reactance, S is open
For transient reactance, S is open (same as above)
For sub transient reactance, S is closed

FIGURE 32.7. Equivalent circuit to find reactances: (*a*) *d*-axis reactances, (*b*) *q*-axis reactances.

T_{do}' governs the decay of i_f with armature circuit open. T_d' governs the decay of i_f with armature short circuited. For both the cases, we should find the reactances while looking into the field winding with armature circuit open/shorted.

Similarly in determining T_{do}'' and T_d'', which govern the decay of the transient in the *d*-axis damper winding, we should find the reactances while looking into the *d*-axis damper winding circuit with field winding circuit shorted and the armature circuit open/shorted .

Similar comments apply to the *q*-axis time constants.

T_a governs the decay of the dc and the second harmonic components of the armature current.

Given below next are the ***actual inductances***, rather than the *pu* inductances, and the relations.

Mutual inductance between *d*-axis armature winding and the field winding–KM_f.

Mutual inductance between *d*-axis armature winding and *d*-axis damper winding–KM_{kd}.

Mutual inductance between *q*-axis armature winding and *q*-axis damper winding–KM_{kq}.

Mutual inductance between the rotor winding and the *d*-axis damper winding–M_{kf}.

Self inductances of the *d*-axis, *q*-axis damper windings – L_{kd} , L_{kq}

$$k = \sqrt{\frac{3}{2}}$$

Armature winding inductances:

leakage inductance $\qquad L_{al}$

d-axis magnetizing inductance

$$L_{ad} = \frac{3}{2}(L_{go} + L_{g2})$$

d-axis synchronous inductance

$$L_d = L_{ad} + L_{al}$$

d-axis transient inductance

$$L_d' = L_d - k^2 \frac{M_f^2}{L_{ff}}$$

d-axis subtransient inductance

$$L_d'' = L_d - k^2\left(\frac{M_f^2 L_{kd} + M_{kd}^2 L_{ff} - 2M_f M_{kd} M_{kf}}{L_{ff}L_{kd} - M_{kf}^2}\right)$$

q-axis magnetizing inductance

$$L_{aq} = \frac{3}{2}(L_{go} - L_{g2})$$

q-axis synchronous inductance

$$L_q = L_{aq} + L_{al}$$

q-axis subtransient inductance

$$L_q'' = L_q - k^2 \frac{M_{kq}^2}{L_{kq}}$$

The expressions for various ***short circuit*** currents are as under.

(*a*) No damper windings, all resistances, and transformer emf's neglected:

$$i_d = -\sqrt{\frac{3}{2}}\frac{M_f i_{fo}}{L'_d} = -\sqrt{\frac{3}{2}}\frac{\sqrt{2}E_{fo}}{x'_d}$$

$$i_q = 0$$

$$i_f = i_{fo} + \frac{3}{2}\frac{M_f}{L_{ff}}\frac{M_f i_{fo}}{L'_d} = i_{fo}\left[1 + \frac{x_d - x'_d}{x'_d}\right]$$

$$i_a = \frac{\sqrt{2}E_{fo}}{x'_d}\sin(\omega t + \delta)$$

(*b*) No damper windings, all resistances except r_f, and transformer emf's neglected:

$$i_d = -\sqrt{\frac{3}{2}}\frac{\sqrt{2}E_{fo}}{x_d} - \sqrt{\frac{3}{2}}\sqrt{2}E_{fo}\left(\frac{1}{x_d'} - \frac{1}{x_d}\right)e^{-t/T_d'}$$

$$i_f = i_{fo} + \frac{3}{2}\frac{M_f}{L_{ff}}\frac{\sqrt{2}E_{fo}}{x'_d}e^{-t/T_d'} = i_{fo}\left[1 + \frac{x_d - x'_d}{x_d'}e^{-t/T_d'}\right]$$

$$i_a = \frac{\sqrt{2}E_{fo}}{x_d}\sin(\omega t + \delta) + \sqrt{2}E_{fo}\left(\frac{1}{x_d'} - \frac{1}{x_d}\right)e^{-t/T_d'}\sin(\omega t + \delta)$$

(c) No damper windings, all resistances neglected, and transformer emf's considered :

$$i_d = -\sqrt{\frac{3}{2}}\,\frac{\sqrt{2}E_{fo}}{x'_d}(1 - \cos\omega t)$$

$$i_q = \sqrt{\frac{3}{2}}\,\frac{\sqrt{2}E_{fo}}{x_q}\sin\omega t$$

$$i_f = i_{fo}\left[1 + \frac{x_d - x'_d}{x'_d}(1 - \cos\omega t)\right]$$

$$i_a = \frac{\sqrt{2}E_{fo}}{x'_d}\sin(\omega t + \delta) - \frac{\sqrt{2}E_{fo}}{2}\left(\frac{1}{x_d'} + \frac{1}{x_q}\right)\sin\delta - \frac{\sqrt{2}E_{fo}}{2}\left(\frac{1}{x_d'} - \frac{1}{x_q}\right)\sin(2\omega t + \delta)$$

(d) Damper windings, all resistances, and transformer emf's considered:

$$i_a = \sqrt{2}\,E_{fo}\left[\frac{1}{x_d} + \left(\frac{1}{x_d'} - \frac{1}{x_d}\right)e^{-t/T_d'} + \left(\frac{1}{x_d''} - \frac{1}{x_d'}\right)e^{-t/T_d''}\right]\sin(\omega t + \delta)$$

$$- \sqrt{2}\,E_{fo}\left[\frac{x_d'' + x_q''}{2x_d''\,x_q''}e^{-t/T_a}\sin\delta + \frac{x_d'' - x_q''}{2x_d''\,x_q''}e^{-t/T_a}\sin(2\omega t + \delta)\right]$$

If the transformer emf's were neglected, the second term, which consists of dc and second harmonic components would not be present.

Here, $E_{fo} = \omega M_f i_{fo}$ and i_{fo} is the field current before the occurrence of the short-circuit.

For a salient pole machine typical values of various reactances in *pu* are as under:

x_d	x_q	x_d'	x_d''	x_q''
0.8 to 1.5	0.5 to 1.0	0.25 to 0.50	0.15 to 0.30	0.20 to 0.30

The synchronous reactance found from OCC and SCC is the *d*-axis synchronous reactance, x_d.

To find the rms value of the symmetrical fundamental component of the SC subtransient, transient, steady state current of a loaded alternator supplying before the fault, current $\mathbf{I}_a$ at terminal voltage $\mathbf{V}_t$; we can simply first find the appropriate voltage $\mathbf{E}''$, $\mathbf{E}'$, $\mathbf{E}$ behind the corresponding reactance x_d'', x_d', x_d; and then solve the circuit using such voltage and the reactance.

$$\mathbf{E}'' = \mathbf{V}_t + j\,\mathbf{I}_a\,x_d'' \qquad \mathbf{E}' = \mathbf{V}_t + j\,\mathbf{I}_a\,x_d' \qquad \mathbf{E} = \mathbf{V}_t + j\,\mathbf{I}_a\,x_d$$

Here, $\mathbf{E}$ is $\mathbf{E}_{fo}$. $E'' < E' < E$.

If the generator was unloaded prior to the fault, all the three voltages equal $\mathbf{E}_{fo}$.

It has been assumed all through that on the occurrence of the SC, the speed ω of the rotor does not change.

32.21 DETERMINING x_q

x_q is found from the slip test, or the maximum lagging current test.

In the ***slip test***, a reduced voltage is applied to the armature with the field circuit open, and the machine is driven at a speed close to synchronous speed, *i.e.*, at a very small slip. The rotating armature mmf meets the rotor at slip speed. When these are in quadrature with each other, the effective reactance is x_q; the armature current (rms value) is maximum and the voltage minimum. The ratio of the minimum value of the phase voltage to the maximum value of the phase current gives the value of x_q. The oscillations in the current are relatively quite large; and those in the voltage are quite small, and are due to the variation in the source impedance voltage drop with variation in the armature current.

In the ***maximum lagging current test***, the machine is run as a synchronous motor at no load. The machine may be over-excited, or normal-excited. Its field current is then gradually reduced and in the process the armature current also varies. At some point the current is maximum possible lagging and has relatively a large value. Such maximum stable value of the armature current is noted. The phase voltage divided by this value of the armature current gives the value of x_q.

Section 8 Power Systems

Chapter **33** GENERATION

33.1 GENERAL

Energy is neither available, nor is utilized in electric form. Yet the naturally available energy is converted to Electric form (*ac*). It is so because it is a very neat form of energy and can be sent to any nook or corner, near or far. It is the most desired form of energy.

The conventional method of generating electric energy or power is to rotate the electric generator rotor by means of a suitable turbine, which in turn may be driven by steam power, gas power, hydro power, wind power etc. Very high pressure steam is generated utilizing the heat of fossil fuels, nuclear fuels, or directly the solar heat. Several unconventional methods of electric power generation like geothermal, tidal, photovoltaic, fuel cells are also being adopted, but these do not generate bulk electric power and have only limited applications.

Some information of interest follows.

The installed generating capacity in India in the year 1947 was merely 1362 MW. Now, it is 13825 MW.

India has as much as 17% of world's population.

Per capita electric energy consumption per year in India = 606 kWh

Per capita electric energy consumption per year in USA = 18000 kWh

Power position in India as in Nov. 2007 :

The installed generating capacity	= 13825	MW	
Coal-fired thermal power	= 73495	MW	= 53.16%
Diesel	= 1201	MW	= 0.87%
Gas	= 14580	MW	= 10.55%
Hydropower	= 34680	MW	= 25.08%
Nuclear power	= 4120	MW	= 02.98%
Wind power	= 7760	MW	= 05.61%
Others	= 2415	MW	= 01.75%

The length of our transmission and distribution lines stands at 6497727 circuit-km. we have 765-kV lines also, though being operated at 400 kV. The losses in our power system are really electric! Transmission, distribution, transformation, and unaccounted energy losses total 32.15%.

A target of adding 41110 MW of power during tenth 5-year plan (2002-07) was set. This comprises 25417 MW of thermal power, 14393 MW of hydro-power, and 1300 MW of nuclear power. The installed generating capacity as in Nov. 2007 stands at 138250 MW, of which 73495 MW is coal based, 14580 MW is gas based, 120 MW is Diesel based, and 4120 MW is nuclear based. For the 11th 5-year plan (2007–12), a target of adding 78000 MW has been set. It is estimated that by the year 2012, India's need of power will stand at 212000 MW.

1 MW is equivalent to	= 24000 units (kWh) per day
1 Lac units per day is equivalent to	= Av power of 4.0167 MW
Per capita installed capacity in India	= 0.125 kW
Per capita installed capacity in USA	= 3 kW

33.2 VARIOUS GENERATION RELATED TERMS AND FACTORS

Consistent units for power and energy are assumed, *i.e.*, kW and kWh; or MW and MWh.

Average load or power, P_{av} at a station or a bus bar over a certain cycle period (day, week, or a year) = (total energy consumption at the point during the cycle period)/ (cycle duration in hours).

Maximum load or power, P_{max} at a station or a bus bar is the maximum of the powers during the cycle being considered. It is not the momentary maximum, but the maximum of averages of powers in successive short intervals, of say half an hour.

Load factor of the station or the plant for the cycle being considered,

$$\text{LDF} = \frac{\text{Average load}}{\text{Maximum load}} = \frac{P_{av}}{P_{max}}$$

The LDF is less than or equal to 1.

A high load factor is desirable as it lowers the effective cost per unit of energy.

Connected load, P_{cnctd} at a point means the sum of the continuous ratings of all the loads or the power absorbers fed from that point.

The ***Demand factor*** (usually over a year) at a point

$$\text{DMF} = \frac{\text{Maximum demand seen or put up by the consumer, whichever is greater}}{\text{Connected load}}$$

$$= \frac{P_{max}}{P_{cnctd}}$$

The DMF is less than or equal to 1.

If the demand factor at a substation is low, a relatively larger number of consumers can be served by the substation.

The ***diversity factor*** at a point

$$\text{DVF} = \frac{\text{Sum of the maximum demands of all the consumers at that point}}{\text{Maximum demand at that point}}$$

$$= \frac{\sum_i P_{\max, i}}{P_{\text{Max}}}$$

The DVF is equal to or greater than 1.

A high diversity factor is desirable as it lowers the capital investment.

The ***plant factor*** or the plant capacity factor

$$\text{PCF} = \frac{\text{Average annual load on the plant, } P_{av}}{\text{Rated capacity of the plant, } P_{rated}}$$

$$= \frac{\text{Energy produced over the year by the plant}}{\text{Rated capacity of the plant, } P_{rated} \times 8760}$$

The plant factor is similar to the load factor, and it should also be high.

The ***plant use factor*** PUF

$$= \frac{\text{Average annual load on the plant on the basis of the duration the plant has been in operation, } P_{av,\,op}}{\text{Rated capacity of the plant, } P_{rated}}$$

$$= \frac{\text{Energy produced over the year by the plant}}{\text{Rated capacity of the plant, } P_{rated} \times \text{number of hours the plant has been in operation}}$$

The plant factor is less than the plant use factor.

33.3 VARIOUS ENERGY GENERATION RELATED CURVES AND COSTS

Load curve. The load curve at a point (a station bus bar, for instance) is the plot of the load at that point against time (in chronological order) over the cycle period of interest. A point on this curve is the load on the system at the corresponding point of time.

A daily load curve is quite often considered in power system operation. It begins at 00.00 hour (midnight) and ends at the next 00.00 hour. The curve is characterized by a valley, and a peak. Usually, the load at the two ends of the curve is the same. The load is divided into three classes – the ***base load*** which is the lowest value of the load, the ***intermediate load*** which is the load having values between the average load and the base load, and the ***peaking load*** which is the load having values between the peak load and the average load.

A base load plant should have low operating cost; while a peak load plant should be able to start quickly, though it may have relatively high operating cost. A base load plant supplies almost constant power throughout and thus operates at a high load factor; while the load on a peak load plant varies and it operates at relatively low load factor. Hydro stations and Gas turbine stations are particularly suited to supply peak loads as these can be started within about 5 to 10 minutes. But from the operating cost point of view, the hydro stations suit to supply base load as well. Gas turbine stations have relatively very high operating cost.

Load duration curve. It is the load curve replotted showing the loads in descending order (*i.e.*, the highest load at the origin). A point (P, t) on this curve tells that the system load is equal to or greater than P MW for a duration of t hours; and less than or equal to P MW for a duration of $(24 - t)$ hours.

Mass curve. It is the curve obtained by integrating the load curve. A point (E, t) on this curve tells that upto time t, energy used up is E. It is a rising curve.

Integrated duration curve. It is the plot of the total energy produced or consumed corresponding to a particular power; E is taken along x-axis and P along y-axis. A point (P, E) on this curve tells that a total energy E is used up by all such loads which are less than or equal to P. It is also a rising curve.

Firm capacity of a plant is the power which is always available at the plant busbars, including the emergency conditions.

Spinning reserve at a point of time is the capacity available (ready to be supplied) at the bus bars but not being used.

Hot reserve at a point of time is the boiler capacity (MW) available (ready to run the prime movers) for the generators but not being used.

Cold reserve at a point of time is the capacity of the boilers and the generators taken together available for generation but not being used.

Spinning reserve ≤ Hot reserve ≤ Cold reserve

Capital cost of a plant is the cost which is incurred once; and comprises the costs of land, design, consultancy, planning, building construction, equipment and facilities, erection, commissioning, testing etc.

Annual fixed cost comprises interest on the capital cost, depreciation, insurance, taxes, fixed managerial and maintenance costs etc.

Annual variable cost comprises cost of the fuel, oil, stores, repairs and maintenance, salaries and wages of operating personnel, and any other cost to be incurred to generate energy from the plant.

The annual operating cost is approximated as $A \times kW$ capacity + $B \times kWh$ generated

Depreciation fund

Depreciation to be set aside every year

(*i*) on the basis of straight line method is given by

$$DF = \frac{\text{capital cost}, C - \text{salvage value}, S}{\text{number of years of useful life}, n}$$

(*ii*) on the basis of sinking fund method is given by

$$DF = (\text{ capital cost}, C - \text{salvage value}, S) \times \frac{r}{(1+r)^n - 1},$$

r being the annual percent rate of interest

(*iii*) and on the basis of declining balance method is given by

$$DF_i = C\,(1 - x/100)^i\, x/100 \text{ at the end of } i^{\text{th}} \text{ year}, x \text{ found from } C(1 - x/100)^n = S.$$

33.4 THERMAL PLANTS

Approximate values of some data of interest:

Capital cost = Rs 3.5 to 4 crores per MW

Efficiency = 30% to 40%

Starting time = 3 to 4 hours

Calorific value of commonly used coal in thermal plants = 27000 kJ/kg = 6430 kCal/kg

1 kWh of energy generated thermally (coal- fired) = 2500 kCal

Coal requirement per MW of power = 10 tons per day

Ash released per MW of power = 2.5 tons per day

Cooling water requirement per MW of power = 3.5×10^6 litres per day

Height, and the diameter at the base of a cooling tower = 150 m each

Pressure in the boiler = 350 atmospheres

Temperature in the boiler = 650° C

An ***economizer*** utilizes the heat in the flue gasses in heating the feed water.

An ***air preheater*** further utilizes the heat from the flue gasses leaving the economizer in heating the air used in combustion.

A ***forced draft fan*** is installed near the base of the boiler, while an ***induced draft fan***, near the base of the chimney.

33.5 HYDRO PLANTS

$$P = \frac{735.5}{75} q\, h\, \eta$$

P is the hydro power generated in kW

q is the water discharge in m^3 per second

h is the water head in m

η is the overall fractional efficiency of the turbine-generator set

Specific speed of a turbine is the speed at which it shall develop one metric horsepower under a head of one metre.

$$N_s = 4.45\, N \frac{\sqrt{P_t}}{h^{1.25}}$$

N_s is the turbine specific speed in rpm, and N is the turbine actual speed in rpm at which the turbine develops P_t metric horsepower under a head of h m.

Turbine to be employed for running the generator is decided on the basis of head available.

Head	*Turbine type*	*Specific speed*
Between 0 and 30 m	Propeller and Kaplan	300 – 1000
Between 30 m and 200 m	Francis	60 – 300
Above 200 m	Pelton	10 – 60

Very small hydro-power generation is classified as

Micro	Less than 1 MW
Mini	1 to 5 MW
Small	5 to 15 MW

Hydro stations have very low operating cost, and also these can be started within about 5-10 minutes.

33.6 NUCLEAR PLANTS

In a nuclear power plant, the necessary heat energy to heat water into steam is derived through nuclear reaction of a suitable ***radioactive*** material; rather than through chemical reaction as in a coal-fired plant.

The part of the plant where the nuclear reaction takes place is known as ***reactor***. It is the heart, or the core of the plant. The heat generated in the reaction is carried away by the circulating ***coolant*** to the ***heat exchanger*** where water is heated into steam. Water, heavy water, liquid sodium, liquid potassium, and carbon dioxide are the commonly used coolants. Apart from heat, the nuclear reaction emits very harmful radiations. To guard the people and the surroundings against the radiations, the reactor is very strongly shielded with thick layers of lead, concrete, steel etc. The reactor is extremely heavy, and as such, the load bearing capacity of the soil has to be far greater than that required for other plants.

Radioactive materials are characterized by their disintegration rate or ***decay rate***, also called activity, which is given by $-\frac{dN}{dt} = \lambda N$; N is the number of nuclei at time t, and λ is the decay constant of the material. This equation leads to ***half life***, $T_{half} = 0.693/\lambda$, and ***average life***, $T_{av} = 1/\lambda$.

One disintegration per second has been given the name one becquerel (*bq*), 10^6 integrations per second is called one rutherford (*rf*), and 3.7×10^{10} disintegrations per second, one curie (*ci*).

A commonly used nuclear fuel is $_{92}U^{235}$, *i.e.*, uranium of atomic number 92, and of mass number 235. The atomic number is the number of electrons revolving around the nucleus, and the mass number is the number of protons plus the number of neutrons in the nucleus. $_{92}U^{235}$ has 235 – 92 = 143 neutrons. The mass of the nucleus is less than the total mass of 92 protons and 143 neutrons. This difference in mass is the ***mass defect***. This mass defect is contained in the nucleus in the form of binding energy as per the well known mass-energy relation $\mathbf{E = mc^2}$.

The nucleons masses are often expressed in atomic mass units (***amu***). One amu equals 1/12 times the mass of the carbon atom. Further, one amu = 1.6604×10^{-27} kg. The mass of a proton, m_p = 1.007 276 amu = 1.672 52 × 10^{-27} kg, and that of a neutron, m_n = 1.008 665 amu = 1.674 82 × 10^{-27} kg.

When a nuclear reaction (***fission***) occurs, the original atom splits into more than one particle, and the total mass of the products is less than the total mass of the reactants before the reaction. This difference in mass appears as energy in the form of heat and electromagnetic radiations. One ***amu*** mass defect is equivalent to **931.1 MeV** of energy.

A nuclear reaction can be caused artificially, by subjecting nuclei to ***neutron bombardment***.

When $_{92}U^{235}$ material is bombarded with a thermal (speed 2.2 km/s) neutron, a nuclear reaction

$$_{92}U^{235} + {}_{o}n^{1} = {}_{56}\text{Ba}^{137} + {}_{36}\text{Kr}^{97} + 2\ {}_{o}n^{1} + \gamma \text{ rays can occur.}$$

The other possible reaction is

$$_{92}U^{235} + {}_{o}n^{1} = {}_{57}La^{148} + {}_{35}Br^{85} + 3\ {}_{o}n^{1} + \gamma \text{ rays}$$

By carrying out the additions of masses of the elements on the two sides, one can find the mass difference and hence the energy released.

We also see that on the average, 2.5 neutrons per fission of $_{92}U^{235}$ are released. These released neutrons have very high speed, of the order of $c/20$. These have to be slowed down to thermal speed for causing further fission. Slowing down of the neutrons can be achieved by making them collide with molecules of small atomic mass, so that only the required loss of energy of the neutron occurs in the collision. The speed ratio, and the energy ratio; after, and before the collision are given by

$$\frac{v}{v_0} = \left[\frac{A-1}{A+1}\right], \frac{E}{E_0} = \left[\frac{A-1}{A+1}\right]^2$$

where A is atomic mass of the nucleus with which the neutron collides.

Energy loss is also measured as $Ln\frac{E_0}{E}$, denoted by ξ. For A large, the average logarithmic decrement in energy is $\xi \approx 2/(A + 2/3)$.

The material used for slowing down the fast neutrons is called the ***moderator***. ***Heavy water*** and ***graphite*** are two of the most commonly used moderator materials.

A thermal speed neutron can bombard another $_{92}U^{235}$ atom, thereby releasing more energy, and more neutrons. A ***chain reaction***, thus, sets in. This chain reaction has to be appropriately controlled, if the released energy is to be tapped and utilized.

If chain reaction is to be sustained, at least one neutron should be emitted in each fission.

We define ***multiplication factor k*** as the ratio of the number of neutrons at a particular stage to the number of neutrons at the preceding stage.

k has to be maintained very close to 1.04 for controlled reaction, *i.e.*, the number of neutrons produced is only slightly greater than the number of neutrons absorbed/used.

Fuel elements are gradually loaded into the reactor till the neutron production rate acquires a constant value, and then we say that the reactor has gone ***critical***.

The neutron flux in the reactor can be controlled by inserting in it suitable neutron absorbing material; ***boron*** and ***cadmium*** being two such materials. More is the insertion of neutron absorbing material in the core, more is the absorption of neutrons. These are in the form of rods, and as such are called ***control rods***.

Only about **0.7%** of the naturally found uranium is ${}_{92}U^{235}$. However, even ${}_{92}U^{238}$ which is about 99.3%, can be transformed into a fissionable material.

$$ {}_{92}U^{238} + {}_{o}n^{1} = {}_{92}U^{239} + \gamma \text{ rays} $$

$$ {}_{92}U^{239} + {}_{-1}e^{0} = {}_{93}Np^{239} $$

$$ {}_{93}Np^{239} + {}_{-1}e^{0} = {}_{94}Pu^{239} + \beta \text{ particles} $$

${}_{94}\mathbf{Pu}^{239}$, plutonium, is a fissionable material.

Thus, if natural uranium is used, ${}_{92}U^{235}$ is used up as fuel, and simultaneously ${}_{92}U^{238}$ gets used up to produce fissionable plutonium, ${}_{94}Pu^{239}$. We could perhaps produce more ${}_{94}Pu^{239}$ than the amount of ${}_{92}U^{235}$ got consumed in energy generation. This is called ***breeding***.

Another material, which like ${}_{92}U^{238}$, is not fissionable but can produce fissionable material is thorium, ${}_{90}\mathbf{Th}^{232}$.

$$ {}_{90}Th^{232} + {}_{o}n^{1} = {}_{90}Th^{233} + \gamma \text{ rays} $$

$$ {}_{90}Th^{233} + {}_{-1}e^{0} = {}_{91}Pa^{233} $$

$$ {}_{91}Pa^{233} + {}_{-1}e^{0} = {}_{92}U^{233} $$

${}_{92}\mathbf{U}^{233}$ is a fissile material, and can be used in a reactor.

${}_{92}U^{238}$ and ${}_{90}Th^{232}$ have been classed as fertile materials; and ${}_{92}U^{235}$, ${}_{92}U^{233}$ and ${}_{94}Pu^{239}$ are fissile materials.

In context with the conversion of a fertile material into a fissile material, a parameter r is defined as

$$ r = \frac{\text{number of new fissile atoms formed}}{\text{number of original fuel atoms consumed in the fission capture process}} $$

over one fuel cycle, *i.e.*, number of new fissile atoms per original fuel atom.

For a breeder reactor, $r > 1$, and it is called the breeding ratio. The period over which the amount of new fuel produced equals the amount of original fuel placed in the reactor, is called the doubling time of the breeder, which may be several years. If $0 < r < 1$, then r is called the conversion ratio.

If N_o is the number of fuel atoms at $t = 0$, then by the end of n fuel cycles, the total number of new fuel atoms produced; which is the same thing as the total number of fertile atoms which got converted, is

$$ N_o(r + r^2 + \ldots + r^n) $$

For $r < 1$, and n very large, the number of new fuel atoms becomes $N_o\, r/(1 - r)$; which is the maximum possible conversion.

Reactors can be classified on the basis of several considerations; but these have mostly been named on the basis of the coolant, or moderator, or fuel used; like pressurized heavy water reactor, advanced gas cooled reactor, boiling water reactor, and liquid metal fuelled reactor. Fast breeder reactor is so named as it also produces nuclear fuel using fast neutrons. India's reactors are mostly PHWRs. We have BWR and FBR as well.

To meet the future power needs of India, nuclear power generation has to be given priority. It is planned to have nuclear power of 24000 MW, and of 50000 MW by the years 2020 and 2030 respectively. As already stated, today it stands at merely 2770 MW.

The radiation emitted has harmful effects on living beings. ***Radiation dose*** is measured as energy received per kg by the absorbent from the decay of the radionuclide, its SI unit being **gray** (*Gy*). One *Gy* is 1 J of energy per kg of the absorbent. ***RAD*** is another unit, which equals 0.01 J per kg. Yet another commonly used unit is ***REM***—roentgen equivalent man. REMs = ***RBE*** × RADs, where *RBE* is the relative biological effectiveness.

The general public, and the plant operating personnel must not be subjected to an average annual dose greater than 0.5, and 5*REMs* respectively. Thus, the maximum permissible whole body dose to plant personnel of age *y* years is given by

$$D = 5(y - 18)\ REMs.$$

33.7 WIND POWER PLANTS

When wind blowing with velocity v strikes a surface of area A normally, it imparts to it an energy of $0.5\ \rho\ v^2\ A$ per second; ρ being the density of air (= 1.2 kg/m^3). The surface can be made to turn a rotor, and hence can generate electric power.

Wind turbines have very large blades; the sweep diameter being of the order of 20*m*. These are installed in areas where an annual average wind speed of about 20km/h exists. The capital investment is around Rs 3 crores per MW and the efficiency of wind turbine generators is around 30% only.

In India, wind farms have been created in the states of Maharashtra, Tamilnadu, Gujarat, and Orissa. Our country has made good progress in wind power generation. We now stand 4th in the world, surpassing even Denmark. 1100 MW of wind power was added in the year 2004-05 alone, out of which 200 MW was added in Tamilnadu. Wind farms occupy large land area. At Satara in Maharashtra, the 225 MW wind farm is scattered over an area of 35 km × 5 km.

33.8 SOLAR POWER PLANTS

India's plains receive an average of about 200 W/m^2 of solar energy. It can be utilized for many useful purposes including electric power generation. For bulk solar power generation, only solar power tower type of plants can be installed; wherein the solar heat is focused on to a high tower, steam is raised, and power generated. An extensive area is required to install focusing mirrors which direct the solar heat to the top of the tower. In India, the Thar Desert in North-west part holds good promise for a solar power tower plant.

As of today, only photovoltaic cells are used in India, to develop small *kWs* of electric power. A solar cell is basically a semiconductor diode which can develop a voltage of 0.5 -1 volt and can deliver 20–40 mA/cm^2. Several solar cells are suitably assembled to form a module. Modules are assembled to form an array. The maximum possible output of a solar array is about 250 W/m^2. The efficiency of a solar cell is about 10-14%.

Any renewable energy source requires some kind of storage system to store the excess energy available and utilize the stored energy when the actual source is not available.

There are hosts of other non-conventional methods of power generation which utilize tidal energy, geo-thermal energy, fuel cells, MHD etc. These have not yet been harnessed to any significant extent in India.

Chapter 34 SUPPLY SYSTEMS AND LINE PARAMETERS

34.1 GENERAL

The bulk electric power is usually generated at locations quite far away from the load centers. There is a network of lines and transformers between the generation point and the power consumption point. Between the two, there may be primary (hv) and secondary (lv) transmission network; and primary and secondary distribution network. Following are the typical voltage levels at various stages.

Level	*Voltage, kV line-to-line*
Generation	6.6, 11, 33
Primary transmission	110, 132, 220, 400, 765, and ± 500 kV dc
Secondary transmission	33, 66, 110
Primary distribution	3.3, 6.6, 11, 33
Secondary distribution	0.440 and 3-wire ± 220 V dc

Transmission lines are much longer than the distribution lines. As such, a large amount of line conductor material is needed. If other considerations permit, we use high voltages for lines to economize in conductor material.

The per phase line conductor volume in m^3 is given by $\dfrac{\rho P^2 l^2}{pV^2 \cos^2 \varphi}$, where ρ is the resistivity of the conductor material in ohm-m, P- the per phase power in MW at the receiving end, l- the length of the line in m, p- the per phase I^2R loss in MW in the line, V- the line to neutral voltage in kV, and cos φ is the power factor of the balanced load. We may put it as $\dfrac{\rho Pl^2}{kV^2 \cos^2 \varphi}$ where k is the I^2R loss fraction, *i.e.*, $k = p/P$. If the transmission voltage is doubled, the conductors volume required reduces to ¼ of the former. Though the line efficiency remains the same, the current density δ in the conductor has doubled, and there will be greater temperature rise as the temperature rise is proportional to $\delta^2 r$.

We can put it this way also that for a given line, if the voltage is doubled (assuming that the line can stand that voltage), transmitted power becomes 4 times (assuming that there is no undue temperature rise). Percentage power loss remains the same.

For a resistive circuit, per unit voltage drop (= IR/V) and per unit power loss (= I^2R/IV) is one and the same thing.

34.2 COMPARISON

Following table gives the comparison of the line conductors volumes for fixed ρ, P, l, and p for various supply systems. Balanced load is assumed. The first basis of comparison applies to the overhead lines, while the other applies to cables.

S.No.	*Supply system*	*Conductor volume required in comparison to dc 2-wire, mid point earthed system on the basis of*	
		Same value of maximum potential difference between a line conductor and earth (applies to overhead lines)	*Same value of maximum potential difference between any two line conductors (applies to multicore cables)*
1.	dc, 2-wire, mid point earthed	unity	unity
2.	ac, 1-phase, 2-wire, mid point earthed	$2/\cos^2\varphi$	$2/\cos^2\varphi$
3.	ac, 2-phase, 3-wire, neutral point earthed	$5.82/\cos^2\varphi$	$2.91/\cos^2\varphi$
4.	ac, 3-phase, 3-wire, neutral point earthed	$2/\cos^2\varphi$	$1.5/\cos^2\varphi$
5.	ac, 3-phase, 4-wire, neutral point earthed, neutral wire of the same size as the phase wire	$2.67/\cos^2\varphi$	$2/\cos^2\varphi$

34.3 CHOICE OF TRANSMISSION VOLTAGE

For deciding the transmission voltage, transmission cost C_{trmsn} is worked out. The sum of the costs of transformers, switchgear and lightning arrestors may be expressed as $(A + BV)P$, that of insulators and supports as $(C + DV)l$, and that of the lines ElP/V. Here P is the power which the system has to carry. A, B, C, D, E, are constants. The transmission cost becomes, $C_{trmsn} = (A + BV)P + (C + DV)l + ElP/V$. The voltage at which C_{trmsn} is minimum is chosen; which is given by $V = [ElP/(BP + Dl)]$.

An empirical formula to select a suitable transmission voltage is

$V_{L\text{-to-}L}$ in kV = 5.5 [(l/1.6) + 10P], l is the distance of transmission in km, and P- the 3-phase power in MW.

34.4 KELVIN'S LAW

Kelvin's law enables one to choose the current density in a line conductor on the criterion of least annual financial burden due to the cost of the energy wasted by way of I^2R loss in the line conductors, and the line depreciation and insurance costs. It is unconcerned with the temperature rise of the lines, or other physical factors.

The annual cost due to the lines can be expressed as $C_1 + C_2\,a + C_3/a$; which gives the best value of a as (C_3/C_2). It is simple to fix C_2. C_3 works out to be $8.76\, n\, c\, \rho\, l\, k^2 L^2\, I_{max}^2$; where

n is the number of phases

c is the cost of energy per kWh

k is the form factor $= \dfrac{\text{rms current over the year}}{\text{mean current over the year}} = \dfrac{I}{I_{mean}}$

L is the annual load factor for the line

$$= \frac{\text{mean current over the year}}{\text{maximum current over the year}} = \frac{I_{mean}}{I_{max}}$$

Thus, the current density

$$\delta = \frac{I_{max}}{a} = \sqrt{\frac{C_2}{8.76\, nc\rho l}}\ \frac{1}{kL}$$

It is independent of power P and the voltage V of the lines.

L and k are related as under :

L	0.1	0.2	0.3	0.4	0.5	0.7	1.0
k	2.2	1.7	1.45	1.3	1.2	1.08	1.0

The value of ρ should be at the operating temperature of the conductor. And this may be further increased by 1–2% (depending on the conductor size) to take into account the skin effect under ac conditions.

For overhead lines, though there is no such problem as damage to the conductor insulation, the conductors cannot be let to operate at high temperature. At high temperatures, annealing and creeping cause loss in tensile strength of the conductors and hence fast aging. Further, a non-uniform part of the conductor may experience self-reinforcing cycle of increasing temperature, increasing resistance, decreasing cross-sectional area under the weight of the conductor itself, and eventually snapping of the conductor at that point.

The lines therefore, should be operated within their thermal limits. The lines of lengths upto about 100 km can be loaded upto their thermal limits; but for longer lines stability considerations prohibit loading to such limits.

34.5 LINE PARAMETERS

It is assumed that the line conductors are homogenous, straight, cylindrical, and are parallel to each other. The radius of ith conductor is denoted by r_i. The distances are all perpendicular distances from/to the conductor axis. The axis-to-axis perpendicular distance between conductor i and conductor j is D_{ij}. And $D_{ij} = D_{ji}$. In the cross-sectional view, the conductors appear as circles, and we may say centre-to-centre distances.

34.6 INDUCTANCE

Ampere's circuital law is applied to find flux linkages of a conductor or a line due to all the currents. Knowing the flux linkages, inductance is found as flux linkages per unit self current. It may be noted that the inductance thus found ***includes the mutual effects*** also.

It is also seen that $\frac{\mu_0}{2\pi}$ equals 0.2×10^{-6} H/m = 0.2 mH/km.

The inductance of a conductor due to its internal flux linkages, $\boldsymbol{L_{int} = 0.05\ mH/km}$.

It is independent of the conductor size.

The inductance of a conductor due only to the flux contained between points P_1 and P_2,

$$L_{12} = 0.2 \text{ Ln } \frac{D_2}{D_1} \text{ mH/km, } D_2 > D_1.$$

Inductance of a ***single-phase, two-wire line*** $= 2 \times 0.2 \text{ Ln } \dfrac{D}{\sqrt{r_1' r_2'}}$ mH/km, $r' = 0.7788\, r$.

The factor 0.7788 appears to account for the internal inductance.

The ***flux linkages*** of conductor i in a group of n conductors, carrying currents I_1, I_2, I_n such that $\sum_{k=1}^{k=n} I_k = 0$, is

$$\lambda_i = 0.2 \sum_{k=1}^{k=n} I_k Ln \frac{1}{D_{ik}} \text{ mWb/km; } D_{ii} = 0.7788\, r_i.$$

If the net exponent of D's is not zero, then D's should be in meters, *e.g.*, in the above expression D should be in meters ; and in $\frac{D_j}{D_k}$, D's can be in any same units.

The currents may be instantaneous currents, or dc currents, or currents in phasor form. λ will also be accordingly instantaneous flux linkages, constant flux linkages, or flux linkages in phasor form.

34.7 GEOMETRIC MEAN DISTANCES (GMD)

To afford ease in handling the conductors, these are stranded, *i.e.*, there are several thinner sub-conductors formed into one. When all the strands are of the same radii, the number of strands in layer x is given by $6(x - 1)$, $x > 1$; and the total number of strands is given by $3x^2 - 3x - 1$. The overall radius of the conductor $r_c = r(x + 1)$, $x > 1$.

The ***self GMD*** of a conductor x consisting of m sub-conductors is

$$D_{s,x} = \sqrt[m^2]{\prod_{k=1}^{k=m} D_{1k} \prod_{k=1}^{k=m} D_{2k} \cdots \prod_{k=1}^{k=m} D_{mk}}$$

The self GMD values for a few cases are given below.

Number of strands	3	7	19	37	61	91	solid
Self GMD	$0.677\ r_c$	$0.726\ r_c$	$0.758\ r_c$	$0.768\ r_c$	$0.722\ r_c$	$0.774\ r_c$	$0.7788\ r_c$

The ***ACSR*** has aluminium strands to carry the current, and a few central strands of steel to provide mechanical strength to the conductor. The steel strands carry only a very small fraction of the conductor current.

The ***mutual GMD*** between two conductors x and y ; the conductors having m and m' sub-conductors is

$$D_{m,\,x-y} = \sqrt[mm']{\prod_{k'=1'}^{m'} D_{1k'} \prod_{k'=1'}^{m'} D_{2k'} \cdots \prod_{k'=1'}^{m'} D_{mk'}}$$

Bundling is splitting the original conductor into two or more sub-conductors, suitably spaced from each other. Primarily, bundling of conductors is resorted to reduce the electric field at the surface of the conductors so that corona is reduced/eliminated. With bundling, for the same total area of cross-section of the conductor, the effective radius of the conductor increases and hence the electric field at the surface reduces. It reduces the inductance also, and hence improves the line's power-carrying capacity as well.

The inductance of a single-phase line consisting of ***composite conductors*** x and y is

$$2 \times 0.2 \left(\text{Ln} \frac{D_{m,x-y}}{\sqrt{D_{s,x}\, D_{s,y}}} \right) \text{mH//km}$$

For a ***three-phase line (transposed if unsymmetrical)***, average inductance per phase over the transposition cycle is

$$L = 0.2\ \text{Ln}\ \frac{D_{m'}\ \text{eqt.}}{D_{s},\ \text{eqt.}}\ \text{mH/km},$$

where $D_{m,\,\text{eqt}} = \sqrt[3]{D_{m,a-b}\, D_{m,b-c}\, D_{m,c-a}}$, and $D_{s,\text{eqt}} = \sqrt[3]{D_{s,a}\, D_{s,b}\, D_{s,c}}$

evaluated in any one part of the transposition cycle.

It is assumed that in a phase, the current divides equally amongst all its sub-conductors.

For a ***symmetrical*** line $D_{m,\,eqt}$ is D, and $D_{s,\,eqt}$ is r'.

For a symmetrical ***double-circuit***, transposed three-phase line having identical circuits distant $h/2$ apart on the either side of the tower, and conductors in a circuit laid vertically one above the other with spacing d between the adjacent phase conductors, the per phase inductance averaged over the transposition cycle is

$$L = 0.02\ \text{Ln}\left[2^{1/6}\left(\frac{d}{r'}\right)^{1/2}\left(\frac{g}{f}\right)^{1/3}\right] \text{ mH/km, where} \quad f^2 = 4\,d^2 + h^2 \text{ and } g^2 = d^2 + h^2$$

The conductors of the two circuits are so arranged that if b and b' are h apart, then a and a'; and c and c' are $(4\,d^2 + h^2)^{1/2}$ apart.

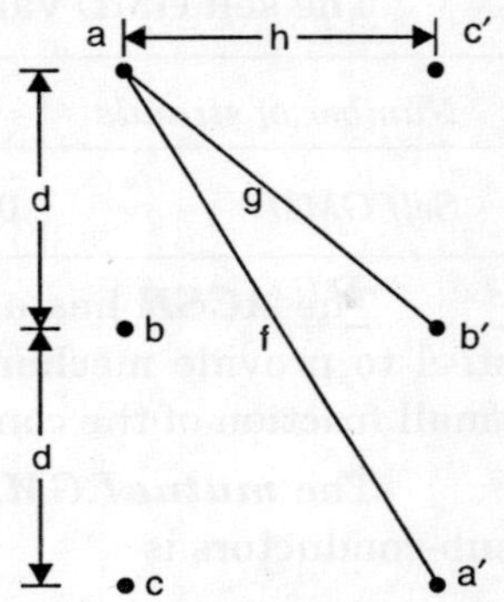

FIGURE 34.1. A typical arrangement of conductors in a double-circuit line.

For high voltage lines, L may be of the order of ***1 mH/km.***

The current-carrying lines induce electromagnetic emf in the neighboring circuits, if any.

The ***emf induced*** by a current-carrying conductor in a circuit running parallel to it, is

$$\mathbf{E} = 0.2 \times j\omega\,\mathbf{I}\ \text{Ln}\,\frac{D_2}{D_1} \text{ mV/km of the circuit, where } D_1 \text{ and}$$

D_2 are the distances of the two sides of the circuit from the emf inducing conductor.

The total emf induced is the phasor sum of the emf's induced by all the conductors.

34.8 SKIN EFFECT AND THE RESISTANCE

When the current flowing in a cylindrical conductor is alternating, the current density over the cross-section of the conductor is not uniform. The current density decreases as we move from the edge towards the centre. This causes an increase in the effective resistance of the conductor under ac conditions.

We can imagine the conductor to be composed of infinitely large number of thin, coaxial, tubular conductors of increasing radii. Because of the circular flux lines within the conductor also, the inner tubular conductors will have higher flux linkages as compared to the tubular conductors away from the centre, or the axis. This means inner tubes have comparatively higher inductance and therefore, higher reactance. On this count the current density is smaller in the centre, and it increases towards the edge.

The effect depends on the conductor size, conductivity and permeability, and the frequency. The effect accentuates with an increase in all the four quantities.

The ratio ac resistance/dc resistance is given by a combination of Bessel's functions of the variable $mr = 0.05013\sqrt{\dfrac{\mu_r f}{R_0}}$, where R_0 is the dc resistance of the wire in ohms per km. The ratio increases with mr, though not in a linear manner.

It can be shown that due to skin effect, L_{int} decreases with mr.

The ***dc resistance*** is found using the relation $R = \dfrac{\rho l}{a}$. The resistivity ρ should be at the operating temperature. The resistivity, or the dc resistance R_2, at temperature t_2 ^0C, can be found using the relation $\dfrac{R_2}{R_1} = \dfrac{T+t_2}{T+t_1}$. T has values of 234.5, 241, and 228 respectively for annealed copper, hard-drawn copper, and hard-drawn aluminium.

34.9 PROXIMITY EFFECT

The flux linkages of the strands, and hence the ac resistance of a conductor is also affected if there is a current carrying conductor close by. Let there be two parallel conductors a and b, carrying currents in the same direction. Imagine this time that the conductors are composed of several strands. The flux lines due to the current in conductor b link the strands of conductor a also. Those strands of conductor a which are nearer to conductor b will have greater flux linkages in comparison to those strands which are relatively away from b. Under ac conditions, the reactance of these nearer strands is higher, and therefore, the current density smaller. The effect will be opposite if the conductors carry currents in opposite directions. However, in both the cases the effective resistance increases.

The difference in flux linkages of the various strands is insignificant if the conductors themselves are not quite close to each other. In a multi-core cable, the conductors are very close to each other, and hence the proximity effect is pronounced ; while in overhead lines it is negligible.

34.10 CAPACITANCE

Conductors are all cylindrical and parallel. We assume that a conductor k has a uniformly distributed charge q_k per unit length. This assumption is slightly invalid as in the presence of other conductors, the charge distribution is not exactly uniform.

The potential difference between any two points 1 and 2 due to such a conductor is

$$V_{12} = \frac{q_k}{2\pi\varepsilon_0} \text{Ln} \frac{D_2}{D_1}.$$

The variables may be instantaneous, dc, or ac phasors.

We can write the expression for the potential of conductor i w.r.t. some point P due to all the conductors as

$$V_{iP} = \frac{1}{2\pi\varepsilon_0} \sum_{k=1}^{k=n} q_k \operatorname{Ln} \frac{D_{kP}}{D_{ki}} . \; D_{ii} \text{ is } r_i.$$

If P goes to infinity, and $\sum q_k$ is zero, then

$$V_i = \frac{1}{2\pi\varepsilon_0} \sum_{k=1}^{k=n} q_k \operatorname{Ln} \frac{1}{D_{ki}}.$$

For a single-phase, two-wire line, the capacitance to neutral, per unit length is

$$C_n = \frac{2\pi\varepsilon_0}{\operatorname{Ln} \dfrac{D}{\sqrt{r_a r_b}}} \text{ F/m}$$

and the capacitance between the lines is half of it.

$$C_n = \frac{1/18}{\operatorname{Ln} \dfrac{D}{r}} \mu\text{F/km}, \quad \text{where } r = \sqrt{(r_a r_b)}.$$

The exact expression for the capacitance of equal radii line is

$$C_n = \frac{1/18}{\operatorname{Ln} (D/2r + \sqrt{D^2/4r^2 - 1})} \mu\text{F/km}.$$

All the capacitance formulas are of $C_n = \dfrac{1/18}{\operatorname{Ln} \dfrac{D}{r}}$ μF/km form only.

For a composite conductor line, D is replaced by $D_{m,\,a-b}$, and r by $\sqrt{(D_{s,a} D_{s,b})}$.

For a ***three-phase line***, D is replaced by $D_{m,eqt}$ and r by $D_{s,eqt}$. The capacitance obtained is the capacitance per phase averaged over the transposition cycle.

$$C_n = \frac{1/18}{\operatorname{Ln} \dfrac{D_{m,\,eqt}}{D_{s,\,eqt}}} \mu\text{F/km}$$

It might have been noted that in the capacitance formulas, r, rather than r' appears. There is no such factor as 0.7788.

34.11 EFFECT OF EARTH ON CAPACITANCE

If the line conductors are close to the earth surface, the capacitance is affected. The effect can be easily examined by applying the method of images to the charged lines and the earth. For a single-phase line, the capacitance expression becomes

$$C_n = \frac{1/18}{\text{Ln}\,\dfrac{D}{r} - \text{Ln}\sqrt{\dfrac{H_{ab'}\,H_{ba'}}{H_{aa'}\,H_{bb'}}}}\ \mu\text{F/km}.$$

The dashed letters denote the image conductory.

For a transposed three-phase line, the expression is similar.

$$C_n = \frac{1/18}{\text{Ln}\,\dfrac{D_{m,\,eqt}}{D_{s,\,eqt}} - \text{Ln}\sqrt[3]{\dfrac{H_{ab'}\,H_{ac'}H_{bc'}}{H_{aa'}\,H_{bb'}H_{cc'}}}}\ \mu\text{F/km}$$

We see that the capacitance increases due to the presence of the earth. The effect vanishes as the ratio slant distance/vertical distance of conductors, like $H_{ab'}/H_{aa'}$ approaches unity ; *i.e.*, the lines are high above the earth's surface.

Charged lines induce ***electrostatic emf*** on the neighboring conductors, or lines, if any.

The emf induced is given by $V_T = V_P \dfrac{\text{Ln}\,\dfrac{h}{d}}{\text{Ln}\,\dfrac{h}{r}}$, where V_P is the potential of the charged conductor P w.r.t. earth, h its height from earth, r its radius, and d the distance between conductor P and conductor T.

If the image of conductor P in earth is taken into account, the induced potential is

$$V_T = V_P \frac{\text{Ln}\,\dfrac{d'}{d}}{\text{Ln}\,\dfrac{2h}{r}},$$

where d' is the distance between the image conductor P' and the conductor T.

34.12 THE PRODUCT LC

We see that $L = \dfrac{\mu_0}{2\pi}\,Ln\,\dfrac{D_{m,\,eqt}}{D_{m,\,eqt}}$ H/m, and $C = \dfrac{2\pi\varepsilon_0}{Ln\,\dfrac{D_{m,eqt}}{D_{s,eqt}}}$ F/m

We also have noted that $\dfrac{D_{m,\,eqt}}{D_{s,\,eqt}}$ in the expression for L has a value slightly greater than what it has in the expression for C. Thus, the product LC is only slightly greater than $\mu_0\varepsilon_0$. $\mu_0\varepsilon_0$ is $1/c^2$ *i.e.*, $10^{-16}/9$. Therefore, if we know L, or C, the other parameter can be found.

$$LC \approx 1/c^2;\ L = (1/C)(\,1/c^2) = 10^{-16}/(9C),\ \text{or}\ C = (1/L)(\,1/c^2) = 10^{-16}/(9L).$$

Chapter 35 TARIFFS AND POWER FACTOR IMPROVEMENT

35.1 TARIFFS

A tariff is simply a way of charging the consumers for supplying electricity to them. The 'way' has to be so devised that the annual fixed costs on account of the capital investment in the plant, transmission system, and the distribution system; and the annual operating cost to generate energy are recovered from the consumers in an equitable manner.

The fixed charges are directly linked to the plant size, which in turn depend on the sum of the maximum demands of all the consumers being served, and the diversity factor of the group of those consumers. Higher the diversity factor smaller is the plant size to meet the same sum of the maximum demands; and hence less is the annual fixed cost per kW imposed on the consumers.

The cost of energy generation, *i.e.*, the cost per kWh generated, reduces if the load factor of the group being served is high.

Further, the tariff may have the provision of giving incentives to those whose peak demands are staggered in time, or to those whose loads are of higher power factors. Such provisions make the tariff a little complex, and also detailed metering is necessitated.

The tariffs are simplified, or made sophisticated, to suit a group of a particular type of load.

A tariff can be of the form $c = a\,x + b\,y$; where x is the maximum kW demand put by the consumer, and y is the energy (kWh) consumed over the billing period. This is actually the ***two- part tariff***.

In the ***flat rate tariff***, the energy charges are merged in the demand charges on the assumption that all the consumers of that group use electricity in proportion to the demand put up by them. Such is the case when the loads are for agricultural purposes. The consumers are charged in proportion to their respective demands. Flat rates are different for different types of consumers.

In the ***block rate tariffs***, different rates of energy charges are applied for different blocks of energy units. Usually, first block of a certain number of units is charged at a high rate, and the subsequent blocks of units are charged at lower rates. However, there being perennial shortage of electricity, the charging pattern is opposite to it. First block of units is charged at low rates, and the subsequent blocks of units are charged at higher rates.

The ***maximum demand tariff*** is the same as the two part tariff, except that instead of estimating the maximum demand from the connected load or the premises size, it is actually metered.

Similarly, there are a few ***power factor related tariffs***. In the maximum kVA demand tariff, kVA rather than kW is measured and the fixed charges are realized based on this kVA. In yet another tariff, the kVArh, and kWh are measured separately and are charged on different rates, the rate for kVArh being only a fraction of that for kWh. We can also have a mechanism in which the low power factor load is charged at a higher rate in comparison to the high power factor load consumer.

35.2 POWER FACTOR IMPROVEMENT

Any device which needs ac magnetizing current for its functioning lowers the power factor of the system. Under-excited synchronous machines also draw reactive current from the mains. For a given *k* Wattage and conductor sizes, low power factor results in poor voltage regulation and poor efficiency. It reflects loss of energy, and hence financial burden.

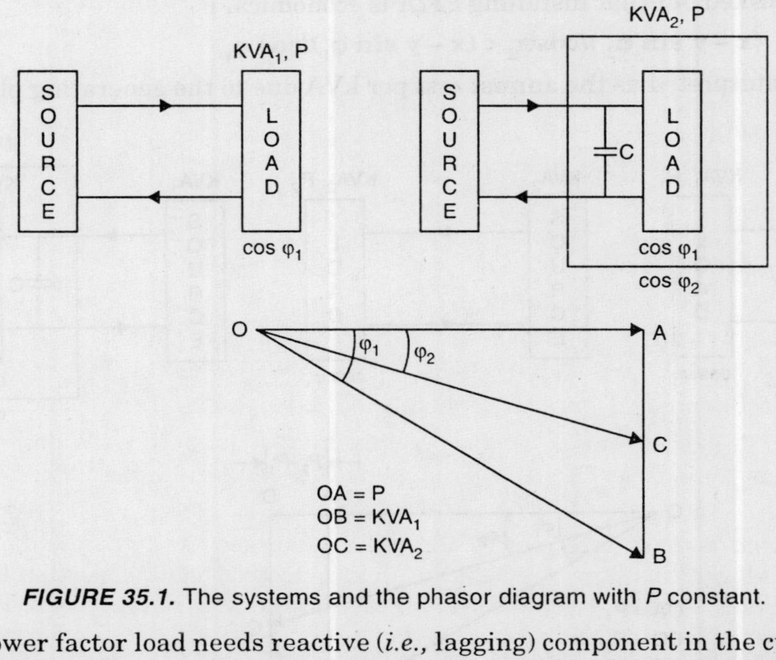

FIGURE 35.1. The systems and the phasor diagram with *P* constant.

A low power factor load needs reactive (*i.e.*, lagging) component in the current; and this wattless component flows in all the equipment and apparatus behind the load. Close to the load, if we install a device that draws reactive current of opposite phase (*i.e.*, leading) so that the reactive component of the current needed by the load is nearly cancelled, all the equipment and apparatus behind the device is then relieved of the wattles current, and the power factor of the system is improved. Usually the static capacitors or the overexcited synchronous machines are the power factor improving apparatus.

However, installing any power factor improving apparatus requires investment. The economics of power factor improving apparatus depend on the prevailing costs of such apparatus, and the rates of demand charges.

We let φ_1 be the power factor angle (lagging) of the load, or the initial power factor angle; and φ_2 be the power factor angle (lagging) of the system after installing the capacitors, or the new power factor angle ($\varphi_2 < \varphi_1$). After installing the *PFIA*, the kVA or the MVA demand of the load for the same power required reduces.

Further, let x be the demand charge per kVA per annum, and y be the annual cost per kVAr by way of depreciation, insurance etc., due to the power factor improving apparatus.

If we assume that the MW is constant at P, the MVAr of the PFIA required is P (tan φ_1 – tan φ_2), and the capacitance in farads is given by $P(\tan\varphi_1 - \tan\varphi_2)/(V^2\omega)$; where V is the voltage in kV. For three phase loads, P is the three phase power in MW, and V is the line to line voltage in kV. We assume the system is balanced.

Maximum saving in annual cost occurs if $\mathbf{\sin\varphi_2 = y/x}$.

Another situation connected with power factor improvement is supplying additional load either by installing additional generating plant and keeping the power factor the same, or by improving the power factor of the system and keeping the kVA to be the same.

Clearly, the first option requires higher kVA, and the second option needs kVAr.

It can be worked out that installing *PFIA* is economical if

$$\mathbf{(x - y\sin\varphi_2)/\cos\varphi_2 < (x - y\sin\varphi_1)/\cos\varphi_1}$$

Now, x is interpreted as the annual cost per kVA due to the generating plant.

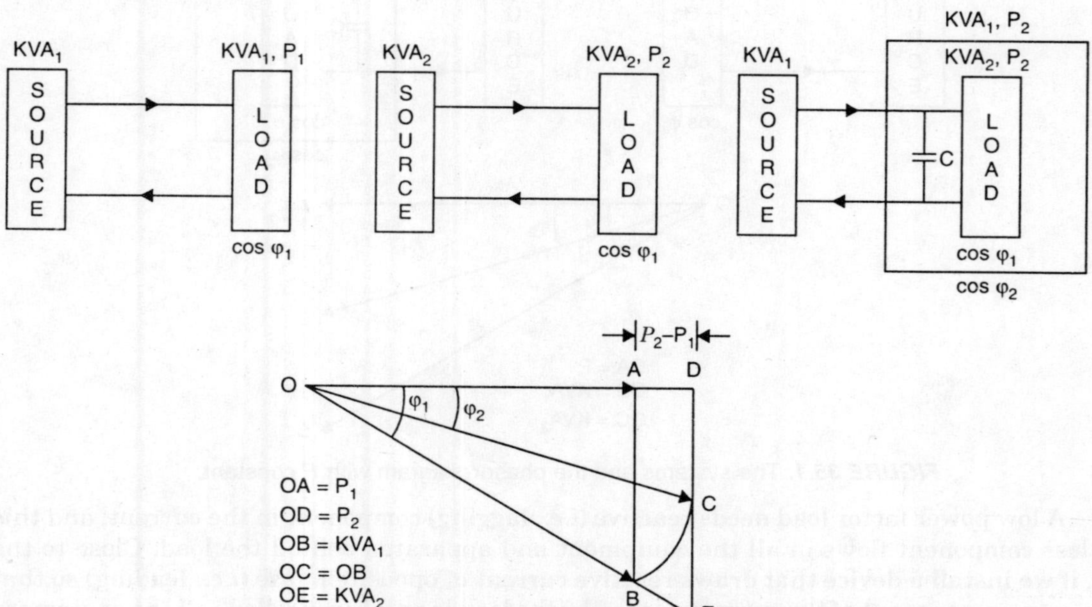

FIGURE 35.2. The systems and the phasor diagram to supply additional load by improving the power factor with kVA constant.

Chapter 36

SAG CALCULATIONS, INSULATORS AND CORONA

SAG CALCULATIONS

36.1 GENERAL

Consider that a transmission line conductor having weight per unit length as w is freely supported at two support points, P_1 and P_2; P_1 being on left of P_2. P_1 and P_2 may be at the same level, or at different levels. The horizontal distance between P_1 and P_2 is L. Because of the weight of the line, the line is in tension, and acquires the shape of a catenary. The coordinate axes are chosen to lie in the plane of the line with lowest point on the line as the origin O. The horizontal tangent to the line at point O and pointing to right is the x-axis and the vertical line through O is the y-axis. Let the coordinates of a point P on the line be (x, y), and the line length (curve) from O to P be s. The coordinates of the support points P_1 and P_2 are (x_1, y_1) and (x_2, y_2) respectively. The tension at O is H, and that at P is T.

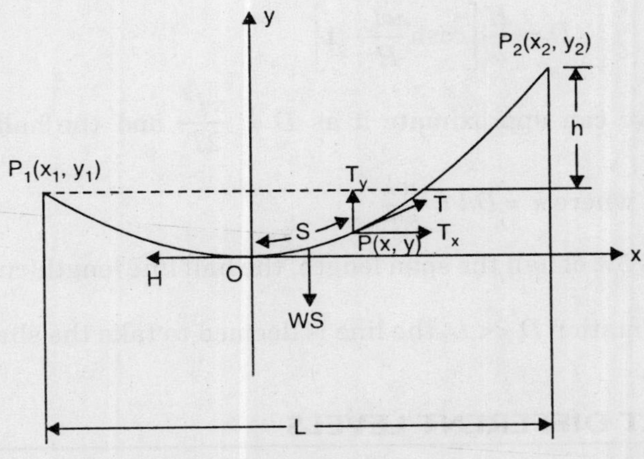

FIGURE 36.1. A freely supported line supported at points P_1 and P_2. The coordinate axes and the various forces are also shown.

By setting up the horizontal and vertical force balance equations acting on the segment OP of the line, one can find the various relations.

$$\frac{T_y}{T_x} = \frac{ws}{H}, \quad \frac{dy}{dx} = \sqrt{1+\left(\frac{ws}{H}\right)^2}, \quad x = \frac{H}{w}\sinh^{-1}\left(\frac{ws}{H}\right),$$

or

$$\sinh\frac{wx}{H} = \frac{ws}{H} \quad \text{or} \quad s = \frac{H}{w}\sinh\frac{wx}{H}$$

$$\mathbf{y} = \frac{\mathbf{H}}{\mathbf{w}}\left[\mathbf{cosh}\frac{\mathbf{wx}}{\mathbf{H}} - \mathbf{1}\right], \quad \mathbf{T} = \mathbf{H}\,\mathbf{cosh}\,\frac{\mathbf{wx}}{\mathbf{H}}$$

The tension is highest at the support points. H is the lowest tension.

For $\frac{wx}{H}$ small, $T \simeq H$, and $y = \frac{wx^2}{2T}$.

36.2 SUPPORTS AT THE SAME LEVEL

Let the support points be at the same level. The half span length $l = L/2$. It means $x_2 = l$, and $x_1 = -l$.

$$y_2, \text{ or } y_1 = \frac{H}{w}\left[\cosh\frac{wl}{H} - 1\right]$$

Thus, the lowest point O on the line is down by y_2 from the support level line, P_1P_2. This is the ***sag*** of the mid point.

The sag of a point $P(x, y)$ on the line may be defined as the difference between the ordinate y of the point P, and the ordinate y' of the point P' on the vertical through P cutting the straight line P_1P_2 at P'. With level supports, the maximum sag is at the lowest point as found above.

$$D = \frac{H}{w}\left[\cosh\frac{wl}{H} - 1\right]$$

When $D << L$, we can approximate it as $D = \frac{wl^2}{2T}$ and the half line length (curve) $= l + \frac{w^2l^3}{6T^2} = l\left[1+\frac{2}{3}k^2\right]$, where $k = D/l = \frac{wl}{2T}$.

Thus, if the sag is 5% of half the span length, the half line length(curve) is only app 0.2% greater than l.

With the approximation $D << L$, the line is deemed to take the shape of a parabola.

36.3 SUPPORTS AT DIFFERENT LEVELS

When the supports are at different levels such that $y_2 = y_1 + h$, we have

$$x_2 = \frac{L}{2} + \frac{Th}{wL}, \quad \text{and} \quad -x_1 = \frac{L}{2} - \frac{Th}{wL}$$

36.4 EFFECT OF ICE AND WIND

Ice and wind, both add to the effective weight per unit length of the line.

Ice

If d is the diameter of the conductor alone, and d' the diameter with the ice coating of thickness t, then the additional weight on the conductor due to ice is

$$w_i = (\pi/4)\,(d'^2 - d^2) \times 915.$$

The effective weight becomes

$$w_{eff} = w + (\pi/4)\,(d'^2 - d^2) \times 915.$$

Wind

If p is the wind pressure in kgf/m^2, assumed to be acting horizontally, then the wind force on the line per unit length is

$$w_w = (2/3) \times (d \text{ or } d', \text{ as the case may be}) \times p$$

$$w_{eff} = [w_w{}^2 + (w + w_i)^2]^{1/2}$$

We use w_{eff} in place of w to find the sag D. With wind pressure included in w_{eff}, the sag obtained is the slant sag. The vertical sag is obtained as Slant sag $\times \dfrac{W + W_i}{W_{eff}}$, which is the same as would be obtained by using $w_{eff} = (w + w_i)$ in place of w in the sag relation.

The sag and the tension are affected with the temperature. As the temperature increases, the sag increases, while the tension in the conductor decreases.

Typical span lengths of lines:

Line voltage (kV)	*Spans (m)*
66	240, 250, 275
110/132	315, 325, 336
220	336, 360, 375
400	400
765	400, 450

Typical conductor spacing for *EHV* lines:

Line voltage (kV)	*220*	*400*	*765*
Phase to phase	5.5	8	15
Phase to tower	2.13	3	5.1
Width of right of way	30	50	64

INSULATORS

36.5 GENERAL

The lines are supported on the steel cross arms of the towers. Between the line and the cross arm, there has to be adequate insulation. This is provided by insulators. The material of the insulators should possess all the properties desired from any insulating material like high breakdown strength, high resistivity, little affected by temperature variation, no trapped air packets etc. In addition, it should possess good mechanical strength, should be impervious to gasses and liquids, should have a small coefficient of expansion, and should not have any pores, and so on. The commonly used material is ***porcelain***, though ***toughened glass*** is also used. The approximate *BD* strengths of the two are 60kV/cm, and 140kV/cm respectively. A demerit of glass insulators is that on glass surfaces, moisture condenses.

The insulator surface should be so shaped that it is nearly an ***equipotential surface***, and the length of the ***leakage current path*** from the line to grounded metallic parts, is large; of course, keeping in mind the economic aspect as well.

The line insulators are in the form of strings of units, or also called, discs. Each disc has a metallic part; above and below it, so that these can be fitted one after the other to form the string of discs. The number of discs to be used in a string depends on the voltage of the conductor to earth.

Usually, the strings are suspended from the cross arms and the conductors are secured to the lower most units. These strings are then called ***suspension insulators***.

Because of the metallic links, each disc is a small capacitor; the dielectric being porcelain. The whole string acts as a set of capacitors in series; the capacitances then called link-to-link, or pin-to-pin, or ***mutual capacitances***. Capacitances are also formed by the intervening metallic links and the tower; the dielectric being air. These are pin-to-earth, or ***shunt capacitances***. An n-unit insulator string can, therefore, be modeled as a string of n mutual capacitances, and (n-1) shunt capacitances between the junction points of the units, and the earth.

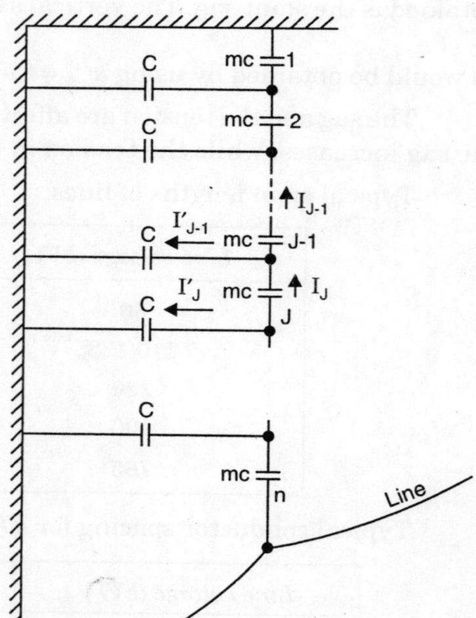

FIGURE 36.2. String insulator model.

Both the capacitances are in the range of pico farads; the mutual capacitance being about ten times the shunt capacitance. For ac lines, very small currents flow through these capacitances. Because of the presence of shunt capacitances, the current in the string is different in different units and as a result, the voltage distribution across the string becomes non-uniform. The unit nearest the line has the highest voltage across it. For dc lines there cannot be any current through the capacitances, and the voltage distribution is uniform.

Only the line unit experiences the rated voltage, all other units having progressively lower voltages. Thus, those units are under utilized. This leads to the term—'**string efficiency**'.

$$\text{Per cent string efficiency} = \frac{\text{Voltage of the conductor w.r.t. earth}}{n \times \text{Voltage across the line unit}} \times 100$$

36.6 UNIT VOLTAGE EVALUATION

The voltage across the various units can be evaluated by applying *KCL* at the junction points.

The units are numbered serially, top unit being unit 1, and line unit being unit n. The mutual capacitance is mC, C being the shunt capacitance. For unit j, we can write

$$I_j = I_{j-1} + I'_{j-1}$$

$$I_{j-1} = mC\omega V_{j-1}$$

$$I'_{j-1} = C\omega \sum_{k=1}^{k=j-1} V_k, \ V_k \text{ is the voltage across unit } K$$

$$V_j = V_{j-1} + \frac{1}{m} \sum_{k=1}^{k=j-1} V_k \qquad j = 2, 3, \ldots, n$$

We can also find that

$$V_j = D_j V_1; \quad \text{where} \quad D_j = \frac{1}{m}\left[(1+m)D_{j-1} + \sum_{k=1}^{k=j-2} D_k\right] \text{ with } D_1 = 1$$

All unit voltages are thus expressed in terms of the top unit voltage V_1.

The voltage of the conductor w.r.t. earth $= \sum_{k=1}^{k=n} V_k$, and therefore, V_1 can be found out.

We can obtain the expressions for a few of the D's.

$$D_1 = 1$$

$$D_2 = \frac{m+1}{m}$$

$$D_3 = \frac{m^2 + 3m + 1}{m^2}$$

$$D_4 = \frac{m^3 + 6m^2 + 5m + 1}{m^3}$$

$$D_5 = \frac{m^4 + 10m^3 + 15m^2 + 7m + 1}{m^4}$$

When ***n is large***, we can use the following formula for string efficiency.

$$\eta = \frac{\sinh \dfrac{n}{\sqrt{m}}}{2n \sinh \dfrac{1}{2\sqrt{m}} \cosh \dfrac{n - 1/2}{\sqrt{m}}}$$

This relation can be derived by applying the distributed parameter line analysis to the string model.

36.7 UNIT VOLTAGE EQUALIZATION

(*a*) Grading of units

The string is assembled with different units. As such, their capacitances also are different. The unit capacitances for uniform voltage distribution are found from the following relation.

$$C_j = C_{j-1} + (j-1)C \qquad j = 2, 3, \ldots, n; \qquad \text{and} \quad C_1 = mC$$

(*b*) Static shielding

With the help of suitably shaped shield connected to the line, additional capacitances between the line and the junction points are created. For uniform voltage distribution, the capacitance between the junction point just above the jth unit and the shield should be

$$C_j' = \frac{(j-1)}{n-(j-1)} C$$

Typical number of 11-kv insulator discs used for lines of various voltages:

Line voltage(kV)	66	132	220	400	765
Number of discs	4–5	9–11	15–17	22–24	40–42

CORONA

36.8 GENERAL

Whenever the electric field in air exceeds its breakdown strength (30 kV peak/cm), the air over there breaks down and an ***electric discharge*** is initiated. If the field is maintained, the discharge too is maintained. Depending on the field strength, the breakdown is accompanied with low hissing sound and luminous glow around the conductor. The electric discharge, hissing sound, and the glow—is ***corona***. Such conditions may be found at the surfaces of high voltage transmission lines.

36.9 ELECTRIC FIELD

The electric field at the surface of the conductor is dependent on the conductor radius, the spacing between the conductors and the voltage difference between the two. The electric field at the surface of the line conductor for a two-wire line is given by

$$E = \frac{V'}{2r\, Ln \frac{D}{r}},$$ where D is the axis-to-axis distance between the two identical cylindrical conductors each of radius r, and V' is the voltage difference between the two conductors.

It has been assumed that $r << D$. We can equally well say that V, equal to $V'/2$, is the potential difference between the line conductor and the neutral wire (at ground potential), and as such

$$E = \frac{V}{rLn\dfrac{D}{r}}.$$

Now this relation is applicable to three-phase lines as well.

The discharge initiation depends on the atmospheric temperature, pressure and humidity, and the condition of the conductor surface.

36.10 DISRUPTIVE VOLTAGE

The ***disruptive gradient*** at NTP, denoted by E_o, is 21.1 kV(rms)/cm. At a pressure of b cm of Hg, and temperature t °C, it is δE_o; δ being the ***air density factor*** given by $\dfrac{3.92b}{273+t}$.

The ***surface irregularity factor*** is denoted by m. The disruptive gradient then is

$$E_d = m\ \delta E_o.$$

This enables us to find line to neutral ***disruptive voltage*** in kV, as

$$V_d = m\ \delta E_o\ r\ Ln\ (D/r);\ r \text{ in cm.}$$

The term $r\ Ln\ (D/r)$ increase with r until $D/2r$ decreases to e, thereafter it increase with r. For the overhead lines, $D/2r >> e$, and therefore, increasing r raises the value of V_d. Using high value of r is one of the ways to reduce corona.

For rainy, sleety, and dusty weather, corona starts at lower voltages and we take

$$V_{d,\,bad} = 0.8\ V_d.$$

36.11 VISUAL CORONA VOLTAGE

The above expression gives the conductor voltage at which the corona occurs, but the corona will not be visible. For the corona to be ***visible***, still higher gradient is necessary. An additional factor of $(1 + 0.3/(\delta r))$ is applied.

$$V_{\text{visual}} = m\ \delta E_o\ r\ (1 + 0.3/(\delta r))\ Ln\ (D/r)$$

36.12 SURFACE FACTOR VALUES

The ***value of m*** depends on the surface condition, and the type of corona being considered.

For normal corona:

Perfectly smooth surface	$m = 1$
Weathered surface	$m = 0.93$ to 0.98
7-strand conductor surface	$m = 0.83$ to 0.87
n-strand conductor surface	$m = 0.90$, n larger

For local visual corona:

Perfectly smooth surface $m = 1$

Stranded conductor surface $m = 0.72$

For general visual corona:

Perfectly smooth surface $m = 1$

Stranded conductor surface $m = 0.82$

Corona causes continuous power loss and radio interference. Two ***loss formulas*** are in use.

36.13 PEEK'S FORMULA

$$P_c = 244\,\frac{f+25}{\delta}\sqrt{\frac{r}{D}}\,(V - V_d)^2 \times 10^{-5} \text{ kW per phase per km; } V > V_d$$

For bad weather conditions, we should use $V_{d,\,bad}$ in place of V_d. As already stated, $V_{d,\,bad} = 0.8\,V_d$.

When corona losses are small, or V/V_d is less than 1.8, we use Peterson formula.

36.14 PETERSON FORMULA

$$P_c = \frac{21\times 10^{-6} fV^2}{\left(Lg\dfrac{D}{r}\right)^2} \times F \text{ kW per phase per km.}$$

The voltage ratio factor F is found from the following table.

V/V_d	0.6	0.8	1.0	1.2	1.4	1.6	1.8	2.0	2.2
F	0.012	0.018	0.05	0.08	0.3	1.0	3.5	6.0	8.0

Due to the ionization of air being caused by the field even at voltages less than V_d, there is very small power loss for V/V_d less than 1.

To ***reduce corona***, we use bundled conductors, or hollow conductors.

From corona considerations, the line conductor diameter d in cm is found from $\dfrac{V_{L-L}\,(kV)}{86.6}$.

The fair weather corona loss may be around ***0.25 kW per phase per km***; while for foul weather conditions this figure is about 4 kW.

The energy contained in a high voltage wave traveling over the line gets somewhat dissipated by way of corona. On reaching the terminal equipment, the wave may not be able to cause any damage to the equipment. Thus, the corona phenomenon offers an ***advantage*** too.

Chapter 37

PER UNIT SYSTEM

37.1 GENERAL

The numerical value of any quantity expressed as a fraction of base (reference) value of that quantity, is its per unit (*pu*) value.

$$\textbf{\textit{Per unit value}} \text{ of a quantity} = \frac{\text{Actual numerical value of the quantity}}{\text{Base value of that quantity}}$$

For example, an ohmic value of 10 ohms on a base of 100 ohms is 0.1 *pu*. Though pu values are dimensionless, we may associate with a *pu* value the unit of the quantity, just to distinguish different *pu* values from each other. We could call the 0.1 *pu* resistance as 0.1 *pu* ohm.

37.2 BASE VALUES

We come across various types of quantities ; like voltage, current, impedance, power etc. The various quantities are inter-related. It so happens that we need to choose independently the base values for any two electrical quantities, and the base values for all other quantities get fixed due to the inter-relations. The phase (*i.e.*, the angle) does not occur in any of the base values, or the per unit values.

The quantities which are dimensionally identical have the same base value. Resistance, reactance, and impedance have a common base value ; likewise, active power, reactive power, apparent power, and complex power also have the same base value.

We shall use the suffixes *b*, *a*, and *pu* to denote the base, actual, and the *pu* values respectively. As usual ; V, I, Z, S etc., denote voltage, current, impedance, apparent power etc. All the base values, and the actual values are assumed to be in SI units, unless stated otherwise.

It is usual to choose independently the base values for the apparent power, and the voltage.

37.3 SINGLE-PHASE CIRCUITS

Base value of current in amperes

$$= \frac{\text{Base value of the apparent power in volt amperes}}{\text{Base value of the voltage in volts}}$$

$$I_b = \frac{S_b}{V_b}.$$

This very relation holds if S_b, and V_b are in kVA, and kV respectively

$$I_b = \frac{S_{b,\,\text{kVA}}}{V_{b,\,\text{kV}}}$$

$$Z_b = \frac{V_b \text{ in volts}}{I_b \text{ in amperes}} = \frac{V_b^{\,2}}{S_b}.$$

This very relation holds if S_b, and V_b are in MVA, and kV respectively.

$$Z_b = \frac{V_{b,\,kV}^{\;2}}{S_{b,\,\text{MVA}}}$$

Having fixed the base values, per unit values can be found out.

37.4 PU VALUE OF AN IMPEDANCE ON A NEW PAIR OF BASE VALUES

If $Z_{pu,\,\text{given}}$ is on base values of $S_{b,\,\text{given}}$, and $V_{b,\,\text{given}}$; then $Z_{pu,\,\text{new}}$ of that impedance on base values of $S_{b,\,\text{new}}$, and $V_{b,\,\text{new}}$ is given by the following easily verifiable relation.

$$Z_{pu,\,\text{new}} = Z_{pu,\,\text{given}} \times \frac{S_{b,\,\text{new}}}{S_{b,\,\text{given}}} \times \left(\frac{V_{b,\,\text{given}}}{V_{b,\,\text{new}}}\right)^2$$

37.5 EFFECT OF A SINGLE-PHASE TRANSFORMER ON PU VALUES

If a transformer is present in a given circuit, the base voltages on the two sides of the transformer are in the ratio of the transformer nominal voltages. The volt-ampere base is always the same throughout the whole of the circuit. The current base, and the impedance base on the two sides can, therefore, be determined. Let t be the nominal hv/lv ratio. Then

$$Z_{b,\,\text{hv}} = t^2 Z_{b,\,lv}.$$

$V_{a,\,lv}$, $I_{a,\,lv}$ and $Z_{a,\,lv}$ on the lv side appear as $tV_{a,\,lv}$, $I_{a,\,lv}/t$, and $t^2 Z_{a,\,lv}$ on the hv side. The pu values of any of these quantities on the two sides can be seen to be identical.

Thus, the way the base values are decided on the two sides of the transformer, the *pu* value of a quantity remains the same on both the sides.

37.6 THREE-PHASE CIRCUITS

The above relations hold for a three-phase system also, provided the base voltage is line to neutral voltage, and the base volt ampere is per phase volt ampere.

However, if the base voltage is line to line voltage, and base volt ampere is the three-phase volt ampere, the relations are changed as under.

$$I_b = \frac{S_b}{\sqrt{3}\,V_b}.$$

This very relation holds if S_b, and V_b are in kVA, and kV respectively, *i.e.*,

$$I_b = \frac{S_{b,\mathrm{kVA}}}{\sqrt{3}\,V_{b,\mathrm{kV}}}$$

$$Z_b = \frac{V_b^{\,2}}{S_b}.$$

This very relation holds if S_b, and V_b are in MVA, and kV respectively, *i.e.*,

$$Z_b = \frac{V_{b,\mathrm{kV}}^{\;2}}{S_{b,\mathrm{MVA}}}.$$ Thus, the relation for Z_b remains unchanged.

37.7 EFFECT OF A THREE-PHASE TRANSFORMER ON PU VALUES

If a three-phase transformer is present in a given circuit, the base line voltages on the two sides of the transformer are in the ratio of the transformer line to line (or line to neutral) nominal voltages. The volt-ampere base is always the same throughout the whole of the circuit. The current base, and the impedance base on the two sides can, therefore, be determined. Let t be the nominal *hv*/*lv* ratio. Then

$$Z_{b,\,hv} = t^2 Z_b,\ lv.$$

The line $V_{a,\,lv}$ and $I_{a,\,lv}$ on the *lv* side are seen as $tV_{a,\,lv}$ and $I_{a,\,lv}/t$ on the *hv* side. $Z_{a,\,lv}$ on per phase basis (*Y*) on the *lv* side is reflected as $t^2 Z_{a,\,lv}$ on per phase basis (*Y*) on the *hv* side. If the actual value of a quantity is given, say on *lv* side, its *pu* values can be found on the two sides, and can be checked to be identical. Thus, *pu* values remain unaltered.

The type of connection of the three-phase transformer has no effect on the per unit values. Also, whether the neutral is grounded, or ungrounded is insignificant.

37.8 COMMON BASE VALUES

For various equipment, the *pu* values specified on the equipment are on the base of the equipment rating itself. When solving a problem, involving equipment and apparatus of different ratings, it is necessary to first convert all the given *pu* values to a common MVA and kV base. On changing the MVA and kV base, $Z_{pu,\,\mathrm{new}}$ is found by using the same relation as given in case of single-phase circuits, viz.

$$Z_{pu,\text{new}} = Z_{pu,\text{given}} \times \frac{S_{b,\text{ new}}}{S_{b,\text{ given}}} \times \left(\frac{V_{b,\text{ given}}}{V_{b,\text{ new}}}\right)^2$$

37.9 ADVANTAGES OF WORKING IN *PU* SYSTEM

One may solve a problem using actual values, or per cent values, or *pu* values. Working in terms of *pu* values is most preferred.

(*i*) One does not have to deal with too small, or too large numbers.

(*ii*) When two quantities in per cent are multiplied, the result has to be divided by 100 to get the correct % age figure. Similarly, when two quantities in per cent are divided, the result has to be multiplied by 100 to get the correct % age figure. No such division or multiplication is required when working in terms of *pu* values.

(*iii*) The actual values of a quantity are different on the two sides of a transformer, whereas the *pu* values are identical.

(*iv*) The type of connection of a three-phase transformer does not affect the *pu* values.

(*v*) The *pu* values of a particular quantity of an equipment of the same kind but of ratings of a wide range are nearly the same. For example, the *pu* impedance of synchronous generators of different ratings may be in the range of 1*pu* to 1.5 *pu* ; whereas the actual ohmic values are widely different.

Chapter 38 UNDERGROUND CABLES

38.1 GENERAL

When in an area, the population density is high and the overhead wires become a potential hazard, or the overhead lines are not acceptable to the area as it mars its good looks, we resort to the use of underground cables for distribution, and in some cases for transmission too; even though cables are several times more expensive than the overhead lines.

Voltage, kV	0.4	11	33	66	132	220	400
Cost $\frac{\text{Cable}}{\text{OH lines}}$	2	3	5	7	9	13	24

Also, sometimes a short length of the cable is used between the overhead line and the station to reduce the intensity of the incoming transient voltage wave, and thereby prevent damage to the terminal equipment.

Cables may be single-core, three-core, four-core, or even five-core cables. These are classified on the basis of the voltage classes also. Essentially, the cable consists of a conductor or conductors to carry the load current; suitable insulation like impregnated paper and cotton covering on the conductors, and a few more coverings like a metallic (usually lead) sheath to prevent the ingress of moisture, steel armoring to protect the cable from mechanical injuries, and jute covering to facilitate handling them.

Properties of a few of the cable insulations are given below.

Tan δ is considered in the latter part of this chapter.

Material	*Breakdown strength kV/mm*	*Relative permittivity*	*Tan δ*
Butyl rubber	13	3.2	0.015
Ethylene propylene rubber (EPR)	15	2.8	0.009
Polyvinyl chloride (PVC)	17	5	0.045
Cross-linked polyethylene (XLPE)	18	2.5	0.008
Polyethylene thermoplastic	21	2.3	0.0004
Oil impregnated paper	21	3.6	0.008

It is important that the insulation is free from voids (tiny packets of air or vacuum), the electric gradient in the insulation does not exceed the safe limit of the stress, and the temperature of the insulation does not exceed the thermal limit, usually 65 °C. The voids are the seats of cable breakdown as the field in the void is ε_r times that in the insulation and the *BD* strength of air is much lower than that of the insulation. We will analyze single-core cables for electric stress evaluation. The radius of the conductor is denoted by r, and the external radius of the insulation by R. The voltage of the conductor w.r.t. earth is taken as V. The permittivity of the insulation or the dielectric is $\varepsilon = \varepsilon_r \varepsilon_o$. The electric field or the stress or the gradient is denoted by E.

38.2 ELECTRIC GRADIENT IN THE INSULATION

It is already known, that the electric field at a normal distance x from the axis of the cylindrical conductor having charge per unit length q is given by

$$E = \frac{q}{2\pi\varepsilon x}$$

For a cable, this leads to $V = \dfrac{q}{2\pi\varepsilon} \operatorname{Ln} \dfrac{R}{r}$. Therefore, the electric field, in terms of V is

$$E_x = \frac{V}{x \operatorname{Ln} \frac{R}{r}}$$, x greater than or equal to r, and less than or equal to R.

E_x is independent of the dielectric constant, as the expression is in terms of V. For given values of V, r, R, the ***maximum stress*** occurs at $x = r$, *i.e.*, at the surface of the conductor. Thus,

E_r, or the ***maximum stress*** $= \dfrac{V}{r \operatorname{Ln} \frac{R}{r}}$

$r \operatorname{Ln}(R/r)$ increases with r till the ratio R/r falls to e, and thereafter, it decreases with r.

Thus, at the conductor surface, the stress is least possible if R/r equals e.

The minimum stress occurs at $x = R$, *i.e.*, at the outer surface of the insulation.

E_R, or the ***minimum stress*** $= \dfrac{V}{R \,\text{Ln}\, \dfrac{R}{r}}$

Thus, $$\frac{E_{\max}}{E_{\min}} = \frac{R}{r}$$

38.3 ECONOMIC DIMENSIONS

We note that $$R = r\, e^{\frac{V}{rE_r}}$$

For given values of V and E_r, we can find the minimum value of R w.r.t. r, *i.e.*, the minimum insulation, and therefore, the most economic size.

The relation obtained is $\text{Ln}\dfrac{R}{r} = 1$, *i.e.*, $\boldsymbol{R = e\, r}$.

38.4 DIMENSIONS FOR THE MINIMUM VALUE OF THE STRESS AT THE SURFACE

We can also find that for given values V and R, the minimum value of E_r, *i.e.*, the stress at the conductor surface, is obtained if the same condition is satisfied, *i.e.*, $\boldsymbol{R = e\, r}$. If R/r is different from e, the stress at the surface is higher than the minimum found as above. For $R/r < e$, the cable operation may become ***unstable***, *i.e.*, a partial discharge in the insulation, if any, will further cause an increase in the electric stress at the surface, and eventually breakdown of the cable.

38.5 GRADING OF INSULATION

We note that the dielectric remains underutilized because the maximum stress occurs at the conductor surface only and as we move away from the surface, the stress progressively decreases. With a view to better utilize the dielectric, we may resort to grading of dielectric, *i.e.*, use of more than one (usually 2 or 3) dielectric. It is usual to set maximum value of stress in each layer to be the same. Each progressive dielectric will then be of decreasing dielectric constant.

Layer 1—outer radius r_1, dielectric constant $\varepsilon_{1r}\varepsilon_o$, maximum stress $E_{1\max}$, *i.e.*, the value of the stress at the conductor surface itself.

Layer 2—outer radius r_2, dielectric constant $\varepsilon_{2r}\varepsilon_o$, maximum stress $E_{2\max}$. Similarly,

Layer 3—outer radius r_3, dielectric constant $\varepsilon_{3r}\varepsilon_o$, maximum stress $E_{3\max}$.

Now we again consider the electric field expression in terms of charge per unit length, q, on the conductor.

$$E = \frac{q}{2\pi\varepsilon x}$$

We find that the potential of the conductor is given by

$$V = \frac{q}{2\pi\varepsilon_o}\left[\frac{1}{\varepsilon_{3r}}\text{Ln}\frac{R}{r_2} + \frac{1}{\varepsilon_{2r}}\text{Ln}\frac{r_2}{r_1} + \frac{1}{\varepsilon_{1r}}\text{Ln}\frac{r_1}{r}\right]$$

$$q = \frac{2\pi\varepsilon_o V}{\frac{1}{\varepsilon_{3r}}\text{Ln}\frac{R}{r_2} + \frac{1}{\varepsilon_{2r}}\text{Ln}\frac{r_2}{r_1} + \frac{1}{\varepsilon_{1r}}\text{Ln}\frac{r_1}{r}}$$

$$E_{1\max} = \frac{q}{2\pi\varepsilon_{1r}\varepsilon_o r} = \frac{K}{\varepsilon_{1r} r}, \text{ where } K = \frac{q}{2\pi\varepsilon_o} = \frac{V}{\frac{1}{\varepsilon_{r3}}\text{Ln}\frac{R}{r_2} + \frac{1}{\varepsilon_{r2}}\text{Ln}\frac{r_2}{r_1} + \frac{1}{\varepsilon_{r1}}\text{Ln}\frac{r_1}{r}}$$

Similarly,
$$E_{2\max} = \frac{q}{2\pi\varepsilon_{2r}\varepsilon_o r_1} = \frac{K}{\varepsilon_{2r} r_1}, \text{ and}$$

$$E_{3\max} = \frac{q}{2\pi\varepsilon_{3r}\varepsilon_o r_2} = \frac{K}{\varepsilon_{3r} r_2}$$

Any of the $E_{\max}$'s can be found in terms of V; for instance,

$$E_{2\max} = \frac{1}{\varepsilon_{2r} r_1}\frac{V}{\frac{1}{\varepsilon_{3r}}\text{Ln}\frac{R}{r_2} + \frac{1}{\varepsilon_{2r}}\text{Ln}\frac{r_2}{r_1} + \frac{1}{\varepsilon_{1r}}\text{Ln}\frac{r_1}{r}} = \frac{V/r_1}{\frac{\varepsilon_{2r}}{\varepsilon_{3r}}\text{Ln}\frac{R}{r_2} + \text{Ln}\frac{r_2}{r_1} + \frac{\varepsilon_{2r}}{\varepsilon_{1r}}\text{Ln}\frac{r_1}{r}}$$

We can express V in terms of $E_{\max}$'s.

$$V = E_{1\max}\, r\,\text{Ln}\frac{r_1}{r} + E_{2\max}\, r_1\,\text{Ln}\frac{r_2}{r_1} + E_{3\max}\, r_2\,\text{Ln}\frac{R}{r_2}$$

We see that $E_{1\max}\varepsilon_r = K/r$, $E_{2\max}\varepsilon_{2r} = K/r_1$, and $E_{3\max}\varepsilon_{3r} = K/r_2$

Since, $r < r_1 < r_2$,

$$E_{1\max}\varepsilon_{1r} > E_{2\max}\varepsilon_{2r} > E_{3\max}\varepsilon_{3r}$$

If all the $E_{\max}$'s are equal,

$$\varepsilon_{1r}\, r = \varepsilon_{2r}\, r_1 = \varepsilon_{3r}\, r_2, \quad \frac{\varepsilon_{3r}}{\varepsilon_{2r}} = \frac{r_1}{r_2}, \quad \frac{\varepsilon_{3r}}{\varepsilon_{1r}} = \frac{r}{r_2}, \quad \frac{\varepsilon_{2r}}{\varepsilon_{1r}} = \frac{r}{r_1}, \quad \text{and} \quad \varepsilon_{1r} > \varepsilon_{2r} > \varepsilon_{3r}.$$

$$V = E_{\max}\left[r\,\text{Ln}\frac{r_1}{r} + r_1\,\text{Ln}\frac{r_2}{r_1} + r_2\,\text{Ln}\frac{R}{r_2}\right]$$

Further, if $r_1/r = r_2/r_1 = R/r_2 = \alpha$, then $V = E_{\max}\,\text{Ln}\,\alpha\,[r + r_1 + r_2]$

If we examine the minimum stress relations and let all the minimum stresses be equal, the resulting condition obtained is

$$\varepsilon_{1r} r_1 = \varepsilon_{2r} r_2 = \varepsilon_{3r} R, \frac{\varepsilon_{3r}}{\varepsilon_{2r}} = \frac{r_2}{R}, \frac{\varepsilon_{3r}}{\varepsilon_{1r}} = \frac{r_1}{R}, \frac{\varepsilon_{2r}}{\varepsilon_{1r}} = \frac{r_1}{r_2}$$

The sets of conditions for all the E_{max}'s to be equal, and for all the E_{min}'s to be equal are different.

However, if we let $r_1/r = r_2/r_1 = R/r_2 = \alpha$, the two sets of conditions become identical and

$$V = E_{min} \text{ Ln } \alpha \, [r_1 + r_2 + R]$$

38.6 USE OF INTERSHEATHS

We can achieve the same effect as above by using single dielectric along with intersheaths. The intersheaths are thin coaxial metallic cylindrical tubes maintained at suitable potential, and interposed in the dielectric. Let there be two intersheaths, one of radius r_1 at potential V_1, the other of radius r_2 at potential V_2; $r < r_1 < r_2 < R$, and $V > V_1 > V_2$.

Now let $K = \dfrac{q}{2\pi\varepsilon}$

For $r < x < r_1$

$$E_1 = \frac{V - V_1}{x \text{ Ln } \frac{r_1}{r}}, \qquad E_{1max} = \frac{V - V_1}{r \text{ Ln } \frac{r_1}{r}}, \qquad E_{1min} = \frac{V - V_1}{r_1 \text{ Ln } \frac{r_1}{r}}$$

For $r_1 < x < r_2$

$$E_2 = \frac{V_1 - V_2}{x \text{ Ln } \frac{r_2}{r_1}}, \qquad E_{2max} = \frac{V_1 - V_2}{r_1 \text{ Ln } \frac{r_2}{r_1}}, \qquad E_{2min} = \frac{V_1 - V_2}{r_2 \text{ Ln } \frac{r_2}{r_1}}$$

For $r_2 < x < R$

$$E_3 = \frac{V_2}{x \text{ Ln } \frac{R}{r_2}}, \qquad E_{3max} = \frac{V_2}{r_2 \text{ Ln } \frac{R}{r_2}}, \qquad E_{3min} = \frac{V_2}{R \text{ Ln } \frac{R}{r_2}}$$

Let $\dfrac{R}{r_2} = \dfrac{r_2}{r_1} = \dfrac{r_1}{r} = \alpha > 1$, then

$$E_{1max} = \alpha E_{1min}, E_{2max} = \alpha E_{2min}, E_{3max} = \alpha E_{3min}$$

Further, let $E_{1max} = E_{2max} = E_{3max} = E_{max}$, then

$$E_{1min} = E_{2min} = E_{3min} = E_{min} = E_{max} / \alpha$$

We see that $V_2 = r_2 E_{max} \text{ Ln } \alpha$, $V_1 = r_1 E_{max} (1 + \alpha) \text{ Ln } \alpha$, and $V = r E_{max} (1 + \alpha + \alpha^2) \text{ Ln } \alpha$

$$E_{max} = \frac{V}{\frac{r}{3}(1 + \alpha + \alpha^2)\left(\text{Ln } \frac{R}{r}\right)}$$

As $\alpha > 1$, $\dfrac{V}{\dfrac{r}{3}(1+\alpha+\alpha^2)\left(\text{Ln}\,\dfrac{R}{r}\right)}$ is less than $\dfrac{V}{r\left(\text{Ln}\,\dfrac{R}{r}\right)}$, the value of E_{max} when no intersheath is used. Thus, we see that with the intersheaths we can use a dielectric of lower *BD* strength, and also the material is better utilized.

The intersheaths voltages in terms of the conductor voltage are given by

$$V_2 = \frac{V}{1+\dfrac{1}{\alpha}+\dfrac{1}{\alpha^2}}, \quad \text{and} \quad V_1 = \frac{V\left(1+\dfrac{1}{\alpha}\right)}{1+\dfrac{1}{\alpha}+\dfrac{1}{\alpha^2}} = V_2\left(1+\frac{1}{\alpha}\right)$$

Implementing the use of intersheaths in practice is difficult, however.

38.7 INSULATION RESISTANCE OF A SINGLE-CORE CABLE

The insulation resistance of a cable of length l is $R_{ins,\,l} = \dfrac{\rho}{2\pi l}\,\text{Ln}\,\dfrac{R}{r}$, where ρ is the resistivity of the insulating material. It may be seen that longer the cable, smaller the insulation resistance; whereas obviously, the conductor resistance increases with cable length.

38.8 CAPACITANCE OF A SINGLE-CORE CABLE

$$C = \frac{1/18}{\text{Ln}\,\dfrac{R}{r}}\ \mu\text{F/km}$$

Compared with overhead lines, the cables have higher capacitances, and lower inductances. As a consequence, the L/C ratio for a cable is much smaller. $\sqrt{(L/C)}$ is called the characteristic resistance, and for the cables it is of the order of 40–100 Ω.

38.9 POWER FACTOR OR TAN δ OF A CABLE

Cos φ, or tan δ (δ + φ = 90°) of a cable, at angular frequency ω is $(\omega\, C_l\, R_{ins,l})^{-1}$; the capacitance, and the insulation resistance are for a length l of the cable, and in consistent units. For instance, capacitance may be in μF, and the insulation resistance in $M\Omega$. It is desired that the power factor, or tan δ be as small as possible. It is of the order of 10^{-4}. Since, tan δ is small, tan δ is approximately equal to δ in radians. δ is called the loss angle, or the loss factor.

Tan δ increases with electric stress, and also with working temperature. Beyond a certain point, the increase is very rapid. Increase in tan δ means more heating of the cable. The cables should, therefore, be operated within the safe stress and temperature limits, as already stated in the beginning.

38.10 CHARGING kVA

For a single-phase cable, charging kVA = $V^2C\,\omega\,l \times 10^{-3}$; V in kV, C in μF/km, and the cable length l in km.

For a three-phase cable, total charging kVA = $V^2C\,\omega\,l \times 10^{-3}$; V is the line-to-line voltage in kV, C is the capacitance per phase in μF /km, and the cable length l in km.

The cables draw a large amount of charging kVA or MVA. For instance, a three-phase, 132–kV cable may draw a charging kVA of 1250 per km. Because of this problem, cables for longer distances are not used.

38.11 FINDING THE CAPACITANCE OF A THREE-PHASE CABLE

Let C_c, and C_s respectively be the core-to-core capacitance, and the core-to-sheath capacitance. Then the effective capacitance of any core to sheath/neutral/earth is $C_n = 3C_c + C_s$.

Two measurements are made. Bunch all the cores together, and find the capacitance between the bunch and the sheath. This is seen to be $C_1 = 3C_s$. Next, bunch two cores together and connect them to the sheath as well. Now find the capacitance between this bunch and the third conductor. This capacitance is $C_2 = 2C_c + C_s$. These give $C_s = C_1/3$, and $C_c = C_2/2 - C_1/6$. This further leads to $C_n = (9C_2 - C_1)/6$.

38.12 CURRENT CARRYING CAPACITY OF A CABLE ON THERMAL/ TEMPERATURE RISE CONSIDERATIONS

We consider a three-core cable. Let R be the ac resistance of any of the cores. To account for sheath losses and dielectric losses, the effective resistance is taken as $R(1 + \lambda)$. The power loss which appears as heat is $3I^2R(1 + \lambda)$. The heat has to flow through the dielectric layer, bedding layer, serving layer, and soil or the ground to the surface of the ground and finally to the ambient. All these intervening layers offer thermal resistances to the flow of heat from the cores to the atmosphere. Taking the maximum temperature to be 65°C, and that of the ambient as 25 °C, the current rating of the three-phase cable can be evaluated from the relation

$$I = \sqrt{\frac{40}{3R(S_d + (1+\lambda)S_{b-s-g})}},$$

where S_d is the thermal resistance of a dielectric layer, and

S_{b-s-g} is the sum of the thermal resistances of the bedding, serving, and the ground.

Finding thermal resistances accurately is difficult.

Based on the maximum conductor temperature of 80° C and ground temperature of 15°C, following are the current carrying capacities of underground cables of voltages of about 1000 V.

Conductor area , mm²	4.52	9.35	14.52	25.80	38.71	64.52	96.77	161.29	322.58
Current carrying capacity, A	51	77	100	135	170	225	280	375	530

If the ground temperature is 30ºC, the current capacity reduces by a factor of 0.88.

For voltages of 3.3 & 6.6 kV, 11 kV, and 22 kV, the reduction factors are 0.95, 0.9, and 0.8 respectively. Further, if 2 or 3 cables are laid side by side, at a spacing of 15 cm, the multiplying factors are 0.84 and 0.75; for a spacing of 30 cm, these are 0.88 and 0.80, and for a spacing of 60 cm, the factors are 0.92 and 0.85 respectively.

Chapter 39

LINE PERFORMANCE

39.1 GENERAL

Transmission lines are the highways of electric power. Large blocks of power glide along the line from one point to the other. As such one need to understand the behaviour of the lines under various operating conditions. Here, first we wish to examine the steady-state performance of a line under sinusoidal excitation and balanced conditions. We study the terminal behavior of the line, *i.e.*, the relations between the voltages, currents, powers at the two ends of the line. The variation of the quantities of interest-like voltage and current with distance along the line is also taken up. The voltages and the currents are represented by their corresponding phasors. The impedances and admittances are, in general, complex quantities. The equations, therefore, have to be manipulated in phasor algebra. For ease, **no boldfacing** has been applied to phasors, or complex quantities. Magnitudes of the complex quantities are indicated with vertical bars as $|V|$.

A transmission line is a distributed parameter circuit, *i.e.*, the line's series resistance and inductance, and shunt capacitance and conductance—are all spread uniformly along the length of the line. For finding the exact performance, this fact has to be incorporated in the analysis.

We will consider the line to be a three-phase line, and analysis will be carried out on per phase basis.

39.2 NOTATION

The quantities are assumed to be in their SI units, unless stated otherwise. The parameters of the line are ***on per unit length*** basis.

Following notation is adopted.

$R, L, \omega L, z = R + j\omega L$	series resistance, inductance, reactance, impedance
$G, C, \omega C, y = G + j\omega C$	shunt conductance, capacitance, susceptance, admittance. Usually, $\mid y \mid << \mid z \mid$
l	the length of the line
Z	$z\,l$, the total series impedance
Y	$y\,l$, the total shunt admittance
x	the distance of a point P from the receiving end (R-end) of the line, *i.e.*, the end where the load is connected. $x = 0$, at the R-end; and $x = l$, at the sending end (S-end), *i.e.*, the end where the source is connected.
V, I	the voltage, current at point P
V_R, I_R	the voltage, current at the R-end
V_S, I_S	the voltage, current at the S-end
Z_L, Z_S	the load impedance, the impedance seen by the source
$Z_c = (z/y)$	the characteristic impedance of the line, which is approximately equal to 250 to 500 ohms, and slightly capacitive. The line designs are such that Z_c is smaller for higher voltages lines because of the larger self GMD of the phase conductors of higher voltages lines.
$\gamma = \alpha + j\beta = \sqrt{(yz)}$	the propagation constant of the line; α- the attenuation constant, and β-the phase shift constant. β is nearly the same for all the lines.

Some of the subject matter presented here, is also appearing in the chapter on 'Transmission Lines', in the subject of 'Fields and Waves'.

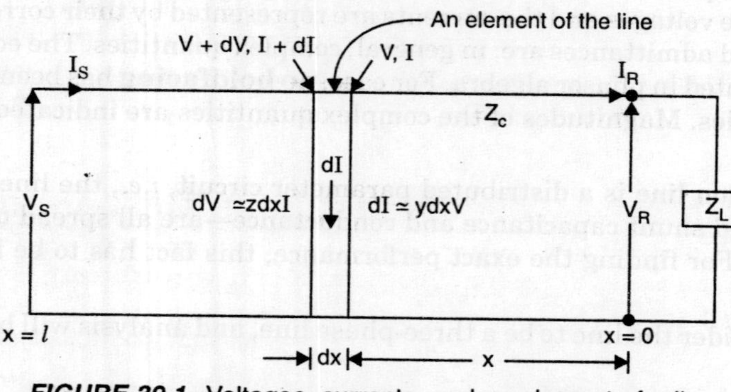

FIGURE 39.1. Voltages, currents, and an element of a line.

39.3 V, I, Z EVALUATION

The governing differential equations of the line are

$$\frac{d^2V}{dx^2} = yz\,V, \quad \text{and} \quad \frac{d^2I}{dx^2} = yzI$$

Their solutions, after incorporating the R-end conditions, are

$$V = \frac{V_R + I_R Z_c}{2} e^{\gamma x} + \frac{V_R - I_R Z_c}{2} e^{-\gamma x}$$

$$I = \frac{V_R / Z_c + I_R}{2} e^{\gamma x} - \frac{V_R / Z_c - I_R}{2} e^{-\gamma x}$$

Above equations involve phasors and are in ***exponential form***. If $e^{j\omega t}$ factor is brought in, it can easily be found that the term associated with $e^{\gamma x}$ is the wave traveling in the $-x$ direction, *i.e.*, traveling from S-end to R-end, and as such is called the ***incident wave***. The other term which is associated with $e^{-\gamma x}$ is the ***reflected wave***. In both the components, the voltage suffers attenuation and decrease in phase in the direction of travel.

$$V = V_{inc} + V_{ref} \qquad I = I_{inc} + I_{ref}$$

We also see that $I_{inc} = V_{inc} / Z_C$, and $I_{ref} = -V_{ref} / Z_C$

The reflected voltage and the reflected current have opposite phase.

There are various forms of the voltage and current equations.

Using hyperbolic functions relations, one can find ***the hyperbolic form*** of the expressions.

$$V = V_R \cosh \gamma x + I_R Z_c \sinh \gamma x$$

$$I = V_R \frac{\sinh \gamma x}{Z_c} + I_R \cosh \gamma x$$

If I_R is substituted by V_R / Z_L, we have the relations for V and I in terms of V_R alone.

The impedance at any point of the line looking towards the load is

$$Z_x = \frac{V}{I} = Z_c \frac{Z_L \cosh \gamma x + Z_c \sinh \gamma x}{Z_c \cosh \gamma x + Z_L \sinh \gamma x}$$

39.4 SENDING-END VOLTAGE, CURRENT, AND IMPEDANCE

The ***S-end voltage, current, and the impedance*** are obtained simply by replacing x by l in above equations. We write the equations for V_S and I_S in matrix form.

$$\begin{bmatrix} V_S \\ I_S \end{bmatrix} = \begin{bmatrix} \cosh \gamma l & Z_c \sinh \gamma l \\ \dfrac{\sinh \gamma l}{Z_c} & \cosh \gamma l \end{bmatrix} \begin{bmatrix} V_R \\ I_R \end{bmatrix} \quad \text{or} \quad \begin{bmatrix} V_S \\ I_S \end{bmatrix} = \begin{bmatrix} A & B \\ C & D \end{bmatrix} \begin{bmatrix} V_R \\ I_R \end{bmatrix}$$

or

$$V_S = AV_R + BI_R \qquad A = |A| \angle \alpha \qquad B = |B| \angle \beta$$

$$I_S = CV_R + DI_R \qquad C = |C| \angle \xi \qquad D = |D| \angle \Delta$$

Typical values are : $A = 0.95 \angle 2°$, $\beta = 80°$, $\xi = 92°$. The magnitudes of B and C increase almost linearly with line length.

The ***A-B-C-D*** constants of the line are

$$\begin{bmatrix} A & B \\ C & D \end{bmatrix} = \begin{bmatrix} \cosh \gamma l & Z_c \sinh \gamma l \\ \dfrac{\sinh \gamma l}{Z_c} & \cosh \gamma l \end{bmatrix};$$

all the parameters, in general, are complex.

We see that $AD - BC = 1$, which implies that the line is bilateral, or reciprocal, and there are three independent constants. We also see that $A = D$, therefore, the line is symmetrical and the number of independent constants is only two.

The ***voltage regulation*** of a line at a particular power factor is the rise in the receiving-end voltage when the full load at the rated voltage on it is thrown off, the sending-end voltage held constant. This is expressed as per cent of rated voltage.

Per cent voltage regulation = (rise in the receiving-end voltage/ rated voltage) × 100

$$= \frac{\left|V_{R,\text{no load}}\right| - V_{\text{rated}}}{V_{\text{rated}}} \times 100 \; ; \qquad \text{(magnitudes only)}$$

we see that $V_{R,\text{no load}} = \dfrac{V_S}{A}$. Also, $I_{S,\text{ no load}} = C\,V_{R,\text{no load}} = \dfrac{CV_S}{A}$. This is the line ***charging current***. For the power lines, the magnitude of A is less than 1; longer the line, smaller the magnitude of A; and therefore, $V_{R,\text{ no load}}$ magnitude is always greater than V_S magnitude. This is the ***Ferranti effect***. If we consider the line to be lossless, which is quite true for high voltage lines, we find $A = \cos \beta l = \cos \dfrac{2\pi}{\lambda} l = \cos (60° \times l')$, where λ is the wavelength = 6000 Km for a 50- Hz line, and l' is the line length in Mm.

For lines of lengths upto about $\lambda/60$, we may not be able to observe this effect as A magnitude remains quite close to 1.

39.5 NO LOAD LOSS IN THE LINE

The complex power loss in an open ended line is seen to be $|V_S|^2 \left(\dfrac{\tanh \gamma l}{Z_c}\right)^*$.

If we neglect the shunt conductance and the line voltage drop due to the no load current, the power loss is given by

$$\int_{x=0}^{x=l} (|V|\omega Cx)^2 R\,dx = \frac{1}{3}(|V|\omega Cl)^2 lR = \frac{1}{3} I^2 R',$$

(*c* is the line capacitance per unit length)

where $I = |V|C\omega l$, is the line charging current, and R′ is the line's total series resistance.

39.6 EQUIVALENT CIRCUITS

The symmetrical T-section, and π-section equivalent circuits of the long line can be found. In the diagram given, Z_A, Z_B and Z_C denote the impedances of the elements.

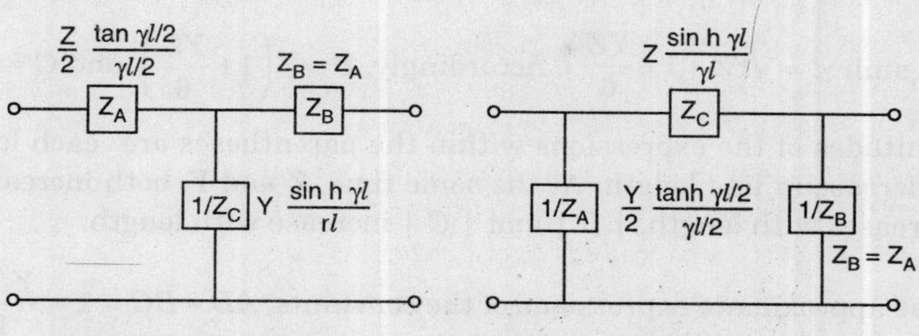

FIGURE 39.2. The T-section, and π-section equivalent circuits.
(*Z* and *Y* are respectively the total series impedance and total shunt admittance of the line)

T-section equivalent circuit:

Arms : $$Z_A = Z_B = \frac{B}{1+A} = Z_c \tanh \approx \frac{\gamma l}{2} = \frac{Z}{2}\frac{\tanh\frac{\gamma l}{2}}{\frac{\gamma l}{2}}$$

Pillar : $$Z_C = \frac{Z_A}{A-1} = Z_c/\sinh \gamma l, \text{ the corresponding admittance is } \frac{\sinh \gamma l}{Z_c} = Y\frac{\sinh \gamma l}{\gamma l}$$

π-section equivalent circuit:

Beam : $$Z_C = B = Z_c \sinh \gamma l = Z\frac{\sinh \gamma l}{\gamma l}$$

Pillars : $$Z_A = Z_B = \frac{Z_C}{A-1} = Z_c/\left(\tanh\frac{\gamma l}{2}\right), \text{ the corresponding admittance is } \frac{Y}{2}\frac{\tanh\frac{\gamma l}{2}}{\frac{\gamma l}{2}}$$

The argument γl of hyperbolic functions is complex. Cosh γl and sinh γl can be found as given below.

39.7 EVALUATING HYPERBOLIC FUNCTIONS AND APPROXIMATIONS

$$\text{Cosh } \gamma l = \cosh \alpha l \cos \beta l + j \sinh \alpha l \sin \beta l$$

$$\text{Sinh } \gamma l = \sinh \alpha l \cos \beta l + j \cosh \alpha l \sin \beta l$$

As $\gamma = \sqrt{(yz)}$ is very small in magnitude; Cosh γl and sinh γl can also be evaluated approximately by taking only first two terms in the respective series expansions.

$$\cosh \gamma l \approx \left(1+\frac{YZ}{2}\right). \text{ Accordingly, } A = D \approx \left(1+\frac{YZ}{2}\right).$$

One should not form an impression that the magnitude of A is greater than 1; it is actually less than 1.

$$\sinh \gamma l \approx \sqrt{YZ}\left(1+\frac{YZ}{6}\right). \text{ Accordingly, } B \approx Z\left(1+\frac{YZ}{6}\right), \text{ and } C \approx Y\left(1+\frac{YZ}{6}\right).$$

The magnitudes of the expressions within the parentheses are each less than 1, and decrease with increase in line length. At the same time, Z and Y, both increase with length. Thus, $|A|$ decreases with length, $|B|$ and $|C|$ increase with length.

With these approximate expressions of the constants, $AD - BC = 1 - \frac{Y^2Z^2}{12}\left(1+\frac{YZ}{3}\right)$.

If the line lengths are upto about $\lambda/60$, YZ will be quite small in comparison to 1, and therefore, further approximations can be made. For a 50-Hz, 100 km long line, YZ magnitude ≈ 0.01. Therefore, $A = D \approx 1$, $B \approx Z$, and $C \approx Y$. For lines of such lengths, Y is very small and we take $C = 0$. Then $AD - BC \approx 1$. Lines of such lengths are designated ***short lines***. For 66 or less kV lines also, the shunt capacitance is very little, and thus constant $C \approx 0$, irrespective of the length, usually found.

If required, V_R can be expressed in terms of V_S, and consequently V and I are expressed in terms of V_S alone; however, x continues to be the distance from the R-end, and not from the S-end.

If $Z_L = Z_c$, then

$$V = V_R\, e^{\gamma x} = V_S\, e^{-\gamma x'};$$

x' is the distance from the S-end, $= l - x$.

$$I = I_R\, e^{\gamma x} = I_S\, e^{-\gamma x'}$$

$Z_x = Z_c$, *i.e.*, the impedance seen at any point on the line is Z_C.

Such a line is called a ***smooth line***. There is no reflected wave.

For an ***infinite line*** also, there can be no reflected wave. $Z_x = Z_c$ at all points.

$$V = V_S e^{-\gamma x'} \quad \text{and} \quad I = I_S\, e^{-\gamma x'};$$

where x' is the distance of the point on the line from the S-end.

39.8 α AND β IN TERMS OF PRIMARY CONSTANTS

The shunt conductance G is invariably neglected in power lines analysis. Then

$$\gamma^2 = (R + j\omega L)(j\omega C).$$

Also,
$$\gamma^2 = \alpha^2 - \beta^2 + j2\alpha\beta$$

Manipulating these two equations, we can find

$$\alpha = \sqrt{\frac{\omega^2 LC}{2}\left(-1+\sqrt{1+\frac{R^2}{\omega^2 L^2}}\right)},$$

and
$$\beta = \sqrt{\frac{\omega^2 LC}{2}\left(1+\sqrt{1+\frac{R^2}{\omega^2 L^2}}\right)}$$

The distance on the line over which the phase change in the voltage or the current, in the incident, or the reflected wave is 2π radians is the wavelength, λ. In terms of the instantaneous values, at a particular time instant, it is the distance over which a complete cycle of values, in the incident, or the reflected wave occurs. $\beta\lambda = 2\pi$. Therefore, $\lambda = \frac{2\pi}{\beta}$. The instantaneous values wave is a traveling wave; traveling with a velocity $v = \frac{\omega}{\beta} = \frac{\omega\lambda}{2\pi} = f\lambda$. This is the velocity with which the energy propagates on the line. For a 50-Hz line, λ is approximately 6000 km. For all practical lines v is little dependent on frequency and is approximately 3×10^5 km/s.

Also, $\sqrt{\omega^2 LC}$ is found to be nearly the same for all lines. Its value at 50 Hz is 1.0667×10^{-3} per km = 6.1 per 100 km.

We can find ***expressions for approximate values of*** α ***and*** β by considering the fact that for transmission lines, resistance to reactance ratio is < 0.1. For higher voltages line, the ratio can be much smaller even.

$$\alpha \approx \frac{1}{2}\frac{R}{R_c}, \text{ where } R_c = \sqrt{\frac{L}{C}}$$

$$\beta \approx \sqrt{\omega^2 LC}\left(1+\frac{R^2}{8\omega^2 L^2}\right)$$

39.9 LOSSLESS LINE

The performance of *hv* lines is nearly the same as that of a lossless line.

For a ***lossless line***, $\gamma = j\omega\sqrt{(LC)}$. Therefore, $\alpha = 0$, $\beta = \omega\sqrt{(LC)}$

At 50 Hz, the phase shift (β) is about 6° per 100 km. This will enable one to find the phase shift in the voltage, or current in the incident wave, and in the reflected wave. The resultant phase shift can then be found.

$$Z_c = \sqrt{\frac{L}{C}} = R_c \qquad v = \frac{1}{\sqrt{LC}}$$

$$V = \frac{V_R + I_R Z_c}{2} e^{j\beta x} + \frac{V_R - I_R Z_c}{2} e^{-j\beta x}$$

$$I = \frac{V_R/Z_c + I_R}{2} e^{j\beta x} - \frac{V_R/Z_c - I_R}{2} e^{-j\beta x}$$

$$V = V_R \cos \beta x + jI_R R_c \sin \beta x$$

$$I = jV_R \frac{\sin \beta x}{R_c} + I_R \cos \beta x$$

$$V_S = V_R \cos \beta l + jI_R R_c \sin \beta l$$

$$I_S = jV_R \frac{\sin \beta l}{R_c} + I_R \cos \beta l$$

The line charging current is $I_{ch} = jV_R \frac{\sin \beta l}{R_c}$. The total three-phase ***Charging MVA*** = $|V_R|^2 \frac{\sin \beta l}{R_c}$, $|V_R|$ being L-to-L R-end voltage in kV. It can be viewed as if there is a lumped capacitor at the R-end of capacitance $\frac{\sin \beta l}{\omega R_c}$. The charging MVA is expressed as $|V_S|^2 \frac{\tan \beta l}{R_c}$ also, which can be viewed as a lumped capacitor at the S-end of capacitance $\frac{\tan \beta l}{\omega R_c}$.

The lines, in India, may be at most $\lambda/10$ km long, *i.e.*, 600 km long. βl is therefore, of the order of $\pi/5$ radians ≈ 0.6, *i.e.*, not large. Therefore, writing βl as $\omega \sqrt{(LC)} l = \omega \sqrt{(L'C')}$, $\cos \beta l$ may be approximated as $1 - \omega^2 L'C'/2$, and $\sin \beta l$ is approximated as

$\omega \sqrt{(L'C')}$, where $L' = L \times l$, and $C' = C \times l$. We have already noted that $A = \cosh \gamma l \approx \left(1 + \frac{YZ}{2}\right)$, and $\sinh \gamma l \approx \sqrt{YZ}\left(1 + \frac{YZ}{6}\right)$. Substituting for YZ as $-\omega^2 L'C'$, we get from this relation also, $A = \cos \beta l \approx 1 - \omega^2 L'C'/2$. For $\sinh \gamma l = j \sin \beta l$, we get $j\,\omega \sqrt{(L'C')}(1 - \omega^2 L'C'/6)$. However, $\omega^2 L'C'/6$ is less than 0.06. Thus, $\sin \beta l \approx \omega \sqrt{(L'C')}$. Then $B = jR_c \sin \beta l \approx j\,\omega L' = jX = j$ line's total series reactance. One may also find that the constant $C \approx j\,\omega C' = j\,B' = j$ line's total shunt susceptance. But then with these approximate values of the constants,

$$AD - BC = 1 + \omega^4 L'^2 C'^2/4.$$

Smaller the length of the line, more valid these approximations are.

39.10 TUNED POWER LINES

$$V_S = V_R \cos \beta l + jI_R R_c \sin \beta l$$

$$I_S = jV_R \frac{\sin \beta l}{R_c} + I_R \cos \beta l$$

In the lossless line equations given above, let $\beta l = \pi$, and therefore, $\cos \beta l = -1$. We see that $V_S = -V_R$; and $I_S = -I_R$, irrespective of the loading condition. Thus, there is no 'loss' of either the voltage or the current; the magnitudes of the voltages, and the currents at the two ends are equal under all conditions. Same result is obtained with $\beta l = 2\pi$.

β being about 6° per 100 km and frequency 50 Hz, the first condition requires that the line length should be 3000 km, and with the second possibility, the line length should be 6000 km. These lengths are too large. We increase β artificially. $\beta = \omega \sqrt{(LC)}$. It can be increased by increasing the line inductance L, or line capacitance C, or both; usually C is increased. Shunt capacitors of appropriate values at suitable intervals all along the length of the line are added to artificially enhance C, and to meet the desired condition. This is tuning of the power lines, and the lines are then called the tuned lines.

39.11 SURGE IMPEDANCE LOADING (SIL)

For a lossless line terminated in the characteristic resistance R_c,

$$V = V_R\, e^{j\beta x} = V_S\, e^{-j\beta x'};\ x' \text{ is the distance from the } S\text{-end}, = l - x.$$

$$I = I_R\, e^{j\beta x} = I_S\, e^{-j\beta x'}$$

The voltage magnitude is the same at all points on the line, only the phase varies. The same is true for the current. The reactive power consumed by the series reactance equals the reactive power generated by the shunt susceptance. Such a line is called a ***flat line***.

Surge impedance loading, also known as natural loading, of a line is the power delivered at rated voltage to a resistive load of per phase resistance $R_c = \sqrt{(L/C)}$. If the rated line-to-line voltage in kV is V_L, then

$$\text{SIL} = \frac{V_L^2}{R_C}\ MW \quad \text{(three-phase power)}$$

For a line of characteristic resistance of 400 ohms, SIL = $2.5\, V_L^2$ kW. SIL is a convenient measure of line loading. The thermal limit of a line may be around 2 SIL. Lines of lengths greater than about 400 km are loaded between 0.5 SIL to SIL. Such limits are forced due to stability considerations. Shorter lines can be loaded more than SIL.

Typical loading limits of lines:

Voltage Level (kV)	132	220	400	765
SIL (MW)	50	132	515	2200
Thermal Limit (MVA)	94	239	948	4261

39.12 LINE POWERS

The power is expressed in terms of the voltages and the *A-B-C-D* constants. The per phase complex power $S = P + j\,Q$ is given by $V\,I^*$, or $S^* = P - j\,Q = V^*I$. Starting with the general voltage and current equations, $V_S = AV_R + B\,I_R$ and $I_S = C\,V_R + D\,I_R$, and using the *S-V, I* relation, the power expressions can be found as given below. We take V_R as the reference phasor, *i.e.*, $V_R = |V_R| \angle 0$, and take $V_S = |V_S| \angle \delta$

$$P_R = \frac{|V_S||V_R|}{|B|}\cos(\beta - \delta) - \frac{|A||V_R|^2}{|B|}\cos(\beta - \alpha)$$

$$Q_R = \frac{|V_S||V_R|}{|B|}\sin(\beta - \delta) - \frac{|A||V_R|^2}{|B|}\sin(\beta - \alpha)$$

$$P_S = -\frac{|V_S||V_R|}{|B|}\cos(\beta + \delta) + \frac{|D||V_S|^2}{|B|}\cos(\beta - \Delta)$$

$$Q_S = -\frac{|V_S||V_R|}{|B|}\sin(\beta + \delta) + \frac{|D||V_S|^2}{|B|}\sin(\beta - \Delta)$$

For fixed voltages, δ is the only variable in above equations.

We see that P_R is maximum when $\delta = \beta$. And P_S is maximum when $(\delta + \beta) = 180°$. The two conditions are not the same. In actual practice, the second condition cannot be achieved, due to stability reasons.

These equations are not specific to transmission lines, but can be applied to any two-port network whose port voltages and *A-B-C-D* constants are known.

If the voltages are in kV and are line-to-neutral voltages, the powers are in MW/MVAr per phase; and if the kV is line-to-line, the powers obtained are for all the three phases.

The active power loss and the reactive power loss in the line is $P_S - P_R$, and $Q_S - Q_R$ respectively. The efficiency of the line can, therefore, be found out.

39.13 POWER CIRCLES

We can see that

$$\left\{P_R + \frac{|A||V_R|^2}{|B|}\cos(\beta - \alpha)\right\}^2 + \left\{Q_R + \frac{|A||V_R|^2}{|B|}\sin(\beta - \alpha)\right\}^2 = \left\{\frac{|V_S||V_R|}{|B|}\right\}^2$$

is an equation of a circle of radius $\frac{|V_S||V_R|}{|B|}$, and the centre located at

$= \left(-\frac{|A||V_R|^2}{|B|}\cos(\beta - \alpha), -\frac{|A||V_R|^2}{|B|}\sin(\beta - \alpha)\right)$ on the *R*-end real power, and the reactive power axes. This is the receiving-end power circle equation. We can, similarly, find the sending-end power circle equation.

$$\left\{P_S - \frac{|D||V_S|^2}{|B|}\cos(\beta - \Delta)\right\}^2 + \left\{Q_S - \frac{|D||V_S|^2}{|B|}\sin(\beta - \Delta)\right\}^2 = \left\{\frac{|V_S||V_R|}{|B|}\right\}^2$$

The radius of the S-end circle is also $\frac{|V_S||V_R|}{|B|}$, and the centre is located at $\left(\frac{|D||V_S|^2}{|B|}\cos(\beta - \Delta), \frac{|D||V_S|^2}{|B|}\sin(\beta - \Delta)\right)$ on the S-end real power, and the reactive power axes. Such circles, give complete picture of the voltages and the powers. Usually, these are drawn in terms of per unit quantities.

39.14 SHUNT COMPENSATION AND SYNCHRONOUS PHASE MODIFIER RATING

A line supplying a current I at voltage V, consumes a reactive power of I^2X in its series reactance X, and generates a reactive power of $V^2\omega C'$ from its shunt susceptance $\omega C'$. For the usual loading of lines, the reactive power consumed by the line exceeds the power generated by it. As a consequence, the R-end voltage sags. On the other hand, under unloaded, or lightly loaded conditions, the generated reactive power exceeds that consumed by it. And as a consequence, the R-end voltage swells. Both the conditions are undesirable. For EHV lines, at no point on the lines (including the R-end) the voltage should be greater than the nominal plus 5%. Therefore, under light load condition, the reactive power generated by the line need be consumed/absorbed externally so that voltage limit is not exceeded. This is achieved by installing switchable shunt reactors at the R-end. These can be switched in or out as per the requirement. This is shunt compensation. Alternatively, we can employ a ***synchronous phase modifier*** which when under excited acts as an inductor. The phase modifier provides smooth control of the reactive power.

Under heavily loaded conditions, we use shunt capacitors to supply the balance reactive power needed by the line, or the synchronous phase modifier, which when overexcited acts as a capacitor. Further, at a given pair of voltages, the reactive power which can flow in the line and reach the R-end is given by Q_R. If the reactive power requirements of the load is greater than this, the balance will have to be provided locally. This can again be met by the phase modifier. The modifier loadings, under no load condition, and under loaded condition are given by

$$Q_{pm,o} = \sqrt{R^2 - x^2} - y$$

$$Q_{pm,load} = Q_L + y - \sqrt{R^2 - (x + P_L)^2}$$

where $$R = \frac{|V_S||V_R|}{|B|},\ x = \frac{|A||V_R|^2}{|B|}\cos(\beta - \alpha),\ y = \frac{|A||V_R|^2}{|B|}\sin(\beta - \alpha)$$

P_L, and Q_L are the load active, and reactive powers. Thus, for a given set of operating conditions, the SPM rating can be worked out.

39.15 TAP CHANGING TRANSFORMER TO CONTROL VOLTAGE

To improve the R-end voltage, we can effect tap changing if the transformers at the S-end, and at the R-end of the line have such a provision. At the S-end, let nominal and actual secondary voltage to primary voltage ratios be n_1 and n_s respectively; and at the R-end the nominal and actual primary voltage to secondary voltage ratios be n_2 and n_r. Let further, V_1, and V_2 be respectively the applied voltage to the transformer at the S-end, and the output voltage (load voltage) of the transformer at the R-end. Then, under the condition $n_s n_r = n_1 n_2$, we can find

$$n_s^2\left[1-\frac{PR+QX}{n_1 n_2\,|V_1 V_2|}\right] = n_1 n_2 \frac{|V_2|}{|V_1|}$$

39.16 POWERS ON SHORT LINES

If we let $A = D = 1$, and $B = Z$, (***a short line***) then C is constrained to be zero so that the relation $AD - BC = 1$ is satisfied. $Z = |Z| \angle\theta$. The line's equivalent circuit is then simply a series impedance Z. $I_S = I_R = I$. $V_S = V_R + Z\,I$. The above relations become

$$P_R = \frac{|V_S||V_R|}{|Z|}\cos(\theta-\delta) - \frac{|V_R|^2}{|Z|}\cos\theta$$

$$Q_R = \frac{|V_S||V_R|}{|Z|}\sin(\theta-\delta) - \frac{|V_R|^2}{|Z|}\sin\theta$$

$$P_S = -\frac{|V_S||V_R|}{|Z|}\cos(\theta+\delta) + \frac{|V_S|^2}{|Z|}\cos\theta$$

$$Q_S = -\frac{|V_S||V_R|}{|Z|}\sin(\theta+\delta) + \frac{|V_S|^2}{|Z|}\sin\theta$$

If, in addition, the ***line is lossless*** so that $|Z|$ becomes X, and $\theta = 90°$. The line's equivalent circuit is then simply a series reactance jX. $V_S = V_R + jX\,I$. The power relations become

$$P_R = \frac{|V_S||V_R|}{X}\sin\delta$$

$$Q_R = \frac{|V_S||V_R|}{X}\cos\delta - \frac{|V_R|^2}{X} = \frac{|V_R|^2}{X}\left[\frac{|V_S|}{|V_R|}\cos\delta - 1\right]$$

$$P_S = \frac{|V_S||V_R|}{X}\sin\delta$$

$$Q_S = -\frac{|V_S||V_R|}{X}\cos\delta + \frac{|V_S|^2}{X} = \frac{|V_S|^2}{X}\left[1-\frac{|V_R|}{|V_S|}\cos\delta\right]$$

The reactive power loss in the line is $Q_S - Q_R$.

$$Q_S - Q_R = \frac{1}{X}\left[|V_S|^2 + |V_R|^2 - 2|V_S||V_R|\cos\delta\right] = \frac{|V_x|^2}{X},$$

where V_X is the voltage drop across the line reactance (the only impedance in the line).

The expression $P = \frac{|V_S||V_R|}{X}\sin\delta$ is the most often used expression of power for a good and quick estimate of power, *e.g.*, in stability studies.

We can bring in voltages in *pu*, the SIL, and then express power in *pu* of SIL.

$$P = \frac{|V_{S,pu}||V_{R,pu}|\text{SIL}}{\sin\frac{2\pi l}{\lambda}}\sin\delta$$

$$P_{pu} = \frac{P}{\text{SIL}} = \frac{|V_{S,pu}||V_{R,pu}|}{\sin\frac{2\pi l}{\lambda}}\sin\delta = R_c\frac{|V_{S,pu}||V_{R,pu}|}{xl}\sin\delta$$

$$= \frac{|V_{S,pu}||V_{R,pu}|}{1.0667\times l'}\sin\delta,$$

where l' is the line length in Mm. As stated earlier, $\sqrt{\omega^2 LC}$ at 50 Hz is approximately 1.0667 × 10^{-3} per km, and as such, the last relation is applicable to 50-Hz lines.

39.17 SHORT AND MEDIUM LENGTH LINES

Lines of lengths upto about λ/60 are designated as ***short lines***. A short line may be represented simply by its series impedance Z and analyzed accordingly. The results on voltage regulation and efficiency would be of acceptable accuracy. If the line is analyzed as a long line, may be there is a difference of upto about 2% in various computed values. To find any quantity of interest, the usual circuit theory analysis of this equivalent circuit can be carried out. We deal with such a circuit in finding the transformer voltage regulation also.

A few of the points concerning short lines have already been stated while dealing with long lines. Some of the relations for finding the S-end, or the R-end ***voltage magnitudes***, and a few other quantities of interests are given below. All the equations are ***algebraic equations***, involving magnitude of complex quantities. Power factor angle φ is positive for lagging load, and negative for leading load. The angle δ is positive if V_S leads V_R.

$$V_S^2 = V_R^2 + I^2Z^2 + 2\,V_R IZ\cos(\theta - \varphi_R); \qquad V_S \simeq V_R + IR\cos\varphi_R + IX\sin\varphi_R$$

or $\Delta V = V_S - V_R = \frac{PR + QX}{V_R}$; QX is the dominating component is causing the voltage drop.

$$\text{Tan}\,\varphi_S = \frac{V_R\sin\varphi_R + IX}{V_R\cos\varphi_R + IR}; \quad \text{Tan}\,\delta = \frac{IX\cos\varphi_R - IR\sin\varphi_R}{V_R + IR\cos\varphi_R + IX\sin\varphi_R}$$

For zero voltage regulation, the condition on load power factor angle is

$$-\varphi_R = 90^\circ - \theta + \sin^{-1}(IZ/2V_R),$$ it should be a leading load.

If the R-end powers, and the S-end voltage V_S are given, then the R-end voltage

$$V_R^{\,2} = \frac{B_R \pm \sqrt{B_R^{\,2} - 4C_R}}{2}, \text{ where } B_R = V_S^{\,2} - 2(P_R R + Q_R X), \text{ and } C_R = (P_R^{\,2} + Q_R^{\,2})Z^2$$

If the S-end powers, and the R-end voltage V_R are given, then the S-end voltage

$$V_S^{\,2} = \frac{B_R \pm \sqrt{B_R^{\,2} - 4C_R}}{2}, \text{ where } B_S = V_R^{\,2} + 2(P_S R + Q_S X), \text{ and } C_S = (P_S^{\,2} + Q_S^{\,2})Z^2$$

Lines of lengths from about $\lambda/60$ upto about $\lambda/15$ are designated as ***medium length lines***, and lines of longer lengths are designated as ***long lines***.

A medium length line can be analyzed by representing the line either by a π-section equivalent circuit, or by a T-section equivalent circuit. These equivalent circuits are called the nominal equivalent circuits of a line. The T-section circuit is not the delta (π) transformed into a star (T), and vice versa. Also, the performance indices found from the two representations are not exactly identical. The nominal π-circuit gives more accurate result for voltage regulation, while a nominal T-circuit gives more accurate result for line efficiency.

π-section equivalent circuit:

Beam impedance Z

Pillars admittances $\dfrac{Y}{2}$ each

T-section equivalent circuit:

Arms impedances $\dfrac{Z}{2}$ each

Pillar admittance Y

The usual circuit theory analysis of the two equivalent circuits can be carried out to find any quantity of interest.

39.18 SERIES COMPENSATION

For long distance power transmission, the power carrying capacity of the lines is limited by the line's series reactance. For transmitting the necessary power, either new lines be constructed, or the series reactance of the existing line be reduced/cancelled/compensated. The latter option turns out to be more economical for long lines. In principle, it means that suitable capacitors be put in series with the line so that the net reactance between the S-end and the R-end of the line is adequately reduced. The capacitors may be placed in the middle of the line, or placed at two locations equally spaced. Series capacitors sub-stations have to be constructed. If the capacitor reactance cancels 50% of the line's series reactance, we say it is 50% compensation. As the power carrying capacity of the line gets augmented, the stability of the system is also improved. Series capacitor also improves the voltage profile of the section of the line between the capacitor and the load. As we pass the series capacitor towards the load, the line voltage undergoes a step increase in the voltage.

With series compensation, fault current level in the line rises. The capacitor has to carry whole of the line current. Under fault conditions, the voltage across the capacitor may become several times its rated voltage. Suitable protection against such over voltages has to be provided. In practice, compensation is limited to about 70% from short-circuit current considerations.

The line inductance and the series capacitance constitute a series resonant circuit, whose resonance frequency is below the operating frequency f, say f_1. If $f - f_1$ is close to the one of the several natural torsional frequencies of the shaft system, then on the occurrence of an electrical disturbance, considerable damage to the shafts can occur. The phenomenon is known as sub-synchronous resonance. Broken shafts, due to SSR, have been reported in the literature.

39.19 ESSENCE OF LINES

Aspect	*Long line* ($l > \lambda/15$)	*Medium length line* ($\lambda/60 < l < \lambda/15$)	*Short line* (l upto $\lambda/60$)
Parameter type	*Distributed*	*Lumped*	*Lumped*
Equivalent circuit(s)	π-section- Beam $Z\frac{\sinh \gamma l}{\gamma l}$ Pillars $\frac{Y}{2}\frac{\tanh \gamma l/2}{\gamma l/2}$ each *T*-section- Arms $\frac{Z}{2}\frac{\tanh \gamma l/2}{\gamma l/2}$ each Pillar $Y\frac{\sinh \gamma l}{\gamma l}$	π-section- Beam Z Pillars $\frac{Y}{2}$ each *T*-section- Arms $\frac{Z}{2}$ each Pillar Y	Simply a series impedance Z
$Z_c = (z/y)$, $\gamma = \alpha + j\beta = \sqrt{(yz)}$, $\lambda = 2\pi/\beta$, $v = \lambda f = 3 \times 10^5$ km/s, for $f = 50$Hz, $\lambda = 6000$ km, $\beta \approx 6°$ per 100 km			
	$V_S = AV_R + B\,I_R$	$I_S = C\,V_R + D\,I_R$	
$A = D$	$\cosh \gamma l$ $\approx (1 + ½YZ)$	π-section $(1 + ½YZ)$ *T*-section $(1 + ½YZ)$	1
B	$Z_c \sinh \gamma l$ $\approx Z(1 + 1/6YZ)$	π-section Z *T*-section $Z(1 + 1/4YZ)$	Z

C	$\dfrac{\sinh \gamma l}{Z_c}$	$\approx Y(1 + 1/6YZ)$	π-section $Y(1 + 1/4YZ)$ T-section Y	0
$V_{R,\text{ no load}}$	V_S/A		π-section $V_S/(1 + YZ/2)$ T-section $V_S/(1 + YZ/2)$	V_S
$I_{s,\text{ no load}}$	CV_S/A		π-section $\dfrac{V_S Y}{2}\left(1 + \dfrac{1}{YZ/2}\right)$ T-section $V_S Y/(1 + YZ/2)$	0

39.20 MEASUREMENT OF A-B-C-D CONSTANTS

These constants can be calculated from the line's open-circuit impedance, and short-circuit impedance measurements. The impedances should be completely measured, *i.e.*, its magnitude as well as the angle be found out. The complex impedance measurement is facilitated by using a wattmeter in conjunction with voltmeter and ammeter. It also may be kept in mind that the open-circuit impedances have negative angles.

Following relations can be used. These are easily derivable.

$$A = \sqrt{\frac{Z_{SO}}{Z_{RO} - Z_{RS}}} \qquad B = AZ_{RS}$$

$$C = \frac{1}{A(Z_{RO} - Z_{RS})} \qquad D = C\,Z_{RO}$$

The symbols are self-explanatory.

39.21 TRANSIENT PERFORMANCE

Having looked a little at the steady-state performance of the lines, we next consider briefly the ***transient performance***. We will consider steep fronted voltage wave of constant magnitude impressed on the line. This knowledge enables us to understand the line's behavior when lightning strikes it, or the line is switched on to the source.

For *S-S* performance, the governing differential equation was in terms of phasors; now it is in terms of instantaneous values. Line's series resistance and shunt conductance are assumed negligible. x is the distance of the point of interest on the line from the source.

$$\frac{\partial^2 v}{\partial x^2} = LC\,\frac{\partial^2 v}{\partial t^2} \qquad \frac{\partial^2 i}{\partial x^2} = LC\,\frac{\partial^2 i}{\partial t^2}$$

The solutions are

$$v = f_1(\sqrt{(LC)}\ x - t) + f_2(\sqrt{(LC)}\ x + t)$$

$$i = \frac{1}{\sqrt{L/C}}\ [f_1(\sqrt{(LC)}\ x - t) - f_2(\sqrt{(LC)}\ x + t)]$$

These solutions are the wave solutions. The velocity of travel of the wave is $\frac{1}{\sqrt{LC}}$, which in turn is $\frac{3 \times 10^5}{\sqrt{\mu_r \varepsilon_r}}$ km per second. The impedance offered to the wave by the line is the line's characteristic resistance, $R_c = \sqrt{\frac{L}{C}}$. We write :

$$v = v' + v'', \quad \text{and} \quad i = i' + i'' = \frac{1}{R_C}[v' - v'']$$

v', i' are the incident voltage and current waves; and v'', i'' are the reflected voltage and current waves.

Whenever a wave meets a discontinuity on the line, it gets partly reflected and partly transmitted into the element at the discontinuity. Accordingly, we have reflection coefficient for the voltage and the current; and if the discontinuity element is a distributed parameter element, we have transmission coefficient as well. The reflection coefficient for the current is negative of that for the voltage, which is v''/v'.

The key point is that at the discontinuity, the solution should satisfy the line equations as given above and also the discontinuity element v-i relation.

39.22 LUMPED PARAMETER ELEMENT TERMINATION

(*a*) Termination in resistor R_L:

$$\frac{v'}{R_c} - \frac{v''}{R_c} = \frac{v' + v''}{R_L}$$

The reflection coefficient $$k = \frac{v''}{v'} = \frac{R_L - R_c}{R_L + R_c}$$

(*b*) Termination in inductor L:

$$v' = V$$

$$v' + v'' = L\frac{d}{dt}\left(\frac{v'}{R_c} - \frac{v''}{R_c}\right)$$

$$v'' = V[-1 + 2e^{(-R_c/L)t}]$$

$$v_L = v' + v'' = 2Ve^{(-R_c/L)t}$$

$$i_L = \frac{2V}{R_c}\left(1 - e^{-\frac{R_c}{L}t}\right)$$

(*c*) Termination in capacitor C:

$$\frac{v'}{R_c} - \frac{v''}{R_c} = C\frac{d}{dt}(v' + v'')$$

$$v'' = V(1 - 2e^{-t/R_cC})$$

$$v_C = v' + v'' = 2V(1 - e^{-t/R_cC})$$

$$i_C = \frac{2V}{R_c}e^{-t/R_cC}$$

All the above relations are at the termination point.

39.23 DISTRIBUTED PARAMETER ELEMENT TERMINATION

If a line is connected to another line whose characteristic resistance is different from its own, then it is a discontinuity. That another line may be a cable, a line, or a pair of parallel lines of different or the same characteristic resistances. In such cases, the wave is partly reflected and partly transmitted over the continuation line(*s*). At the junction point, the transmitted voltage wave is $v' + v''$, and the transmitted current equals $i' + i''$.

If there are more than one lines emanating from the junction point, we find the parallel sum, R_p, of the characteristic resistances R_2, R_3 etc., of these lines to find the reflection coefficient at the junction. For ease, now we drop the suffix c for denoting 'characteristic'.

The reflection coefficient for the voltage, $k = \dfrac{R_p - R_1}{R_p + R_1}$ and

The transmission coefficient for the voltage, $t_v = \dfrac{2R_p}{R_p + R_1} = 1 + k$

The transmission coefficient for the current, $t_i = \dfrac{2R_1}{R_p + R_1} = 1 - k$

On each of the emanating lines, the relation $v = i \times$ the characteristic resistance is obeyed.

It is assumed that there are no reflected waves reaching the junction from the far ends of the emanating lines.

39.24 BEWLEY LATTICE DIAGRAM

Bewly lattice diagram is a pictorial way of showing the repeated reflections at the two ends of a line and finding the value of the wave at any instant and at any point on the line. Such a diagram can be drawn for the voltage, or for the current. The transmission line of length l is shown as a horizontal line, and the left end is labeled as end 1, and the other end is labeled as end 2. At the two ends we write the respective reflection coefficients also. The time axis is taken vertically downwards. Time is measured in terms of T, the time of travel of the wave from one end of the line to the other end. T is of the order of μs. We may draw faint horizontal lines showing time T, $2T$, ...The wave originates at end 1, reaches end 2, gets reflected from end 2, this reflected wave reaches end 1, and this in turn gets reflected from end 1, reaches end 2 and so on. The incident value on getting reflected at an end gets multiplied by the reflection coefficient at that end.

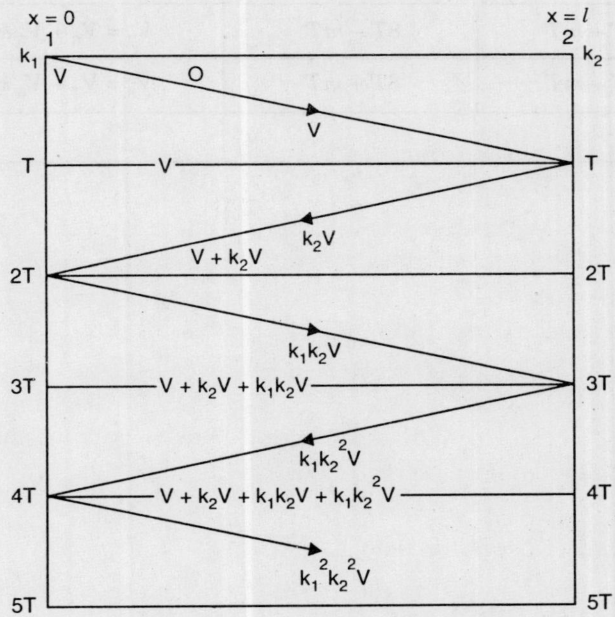

FIGURE 39.3. Bewley lattice diagram.

On examining the diagram, one can relate the time interval and the resultant value of the wave at a point of interest on the line. Let the point of interest be distant d from end 1. Let $\frac{d}{l}$ be denoted by m. The two reflection coefficients are k_1 and k_2. Wave values for some periods are as follows :

S. No.	*Period*		*Resultant wave value*
	From	*To*	
0	0	mT	$V_0 = 0$ Let $V_{-0}' = V/k_1$
1	mT	$2T - mT$	$V_1 = V_0 + V_{-0}'/k_1 \quad = V_0 + V_0'$
2	$2T - mT$	$2T + mT$	$V_2 = V_1 + V_0' k_2 \quad = V_1 + V_1'$
3	$2T + mT$	$4T - mT$	$V_3 = V_2 + V_1'k_1 \quad = V_2 + V_2'$
4	$4T - mT$	$4T + mT$	$V_4 = V_3 + V_2' k_2 \quad = V_3 + V_3'$
5	$4T + mT$	$6T - mT$	$V_5 = V_4 + V_3'k_1 \quad = V_4 + V_4'$
6	$6T - mT$	$6T + mT$	$V_6 = V_5 + V_4'k_2 \quad = V_5 + V_5'$
7	$6T + mT$	$8T - mT$	$V_7 = V_6 + V_5'k_1 \quad = V_6 + V_6'$
8	$8T - mT$	$8T + mT$	$V_8 = V_7 + V_6'k_2 \quad = V_7 + V_7'$
and so on			

Chapter 40 SYMMETRICAL FAULTS, CURRENT LIMITING REACTORS AND CIRCUIT BREAKERS

40.1 GENERAL

A symmetrical fault is a balanced three-phase fault. It may be a direct fault, *i.e.*, a dead short-circuit between the three phases at the fault point; or it may involve a balanced fault impedance, *i.e.*, the fault introduces an impedance z_f in each phase between the fault point and the short-circuit point. A symmetrical fault can be analyzed on per phase basis; for transient, as well as for the steady state conditions. When analyzed this way, the transient state solution, however, shall not include the offset component.

40.2 SHORT-CIRCUIT AT THE TERMINALS OF AN UNLOADED ALTERNATOR

The synchronous reactance x_s of a synchronous machine comprises the armature leakage reactance x_l, and the fictitious reactance x_{ar} which accounts for the armature reaction. Neglecting the resistance, the fault current under steady state condition is E_f/x_s, where E_f is the excitation emf. In steady state, the armature reaction is fully effective. We could also say that the fault current is E_r/x_l, the armature reaction having been accounted for by taking the resultant air gap emf E_r instead of E_f. E_r is much less than E_f for the sustained fault condition. The equivalent circuit for this condition is shown in Fig. 40.1.

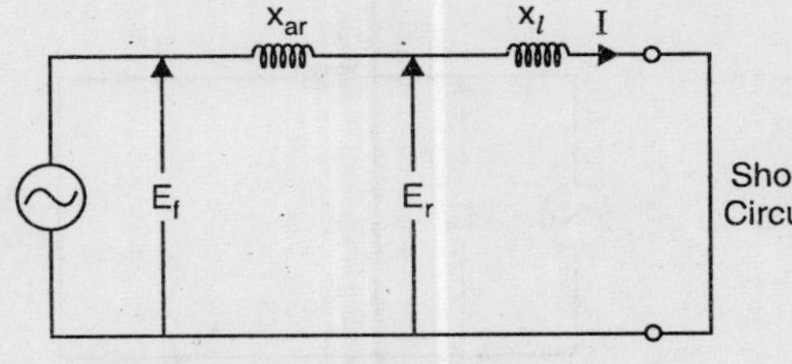

FIGURE 40.1. The voltages E_f and E_r in a synchronous generator. The steady State SC current is found using this equivalent circuit.

On the occurrence of the short circuit at the armature terminals, the armature current starts flowing and a flux ϕ_a is fired in the air gap. The fault current in the armature lags the emf by very nearly 90°, and hence has demagnetizing action on the field and the air gap flux. But the flux linkages of the field winding cannot change all of a sudden (constant flux linkages theorem), and as such, additional field current is set up instantaneously to keep the flux linkages of the field winding constant. This increases the field winding leakage flux as well, and as such, the air gap flux which is the difference of the field flux (which is constant) and its leakage flux, is decreased and hence the air gap emf which was earlier E_f, now becomes E_r'(somewhat less than E_f). Therefore, the armature fault current just when the short-circuit occurs is E_r'/x_l, and is less than E_f/x_l. $E_r'/x_l = E_f/(x_l\,E_f/E_r') = E_f/x_d'$; where $x_d' = x_l\,(E_f/E_r')$, and hence somewhat greater than x_l. The field circuit has relatively long time constant, and the transient current, given by E_f/x_d' decays only slowly. It may last about 20 cycles. The equivalent circuit for this condition is shown in Fig. 40.2

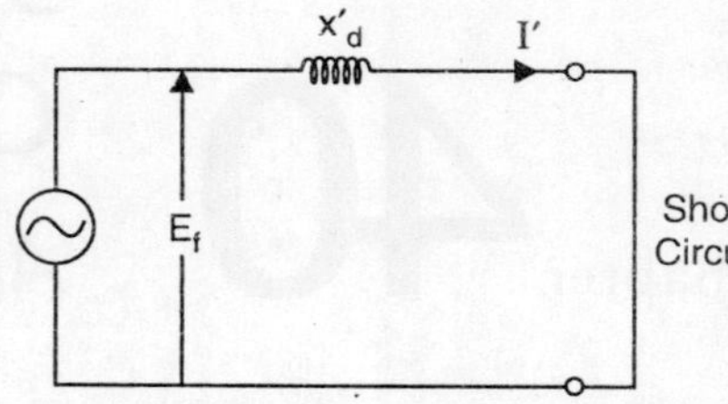

FIGURE 40.2. The equivalent circuit to find transient sc fault current.

The currents induced in the damper windings/ rotor iron by the armature reaction also assist in maintaining the fluxes of the closed circuits constant; and consequently the increase in the field winding current and therefore the increase in the field leakage flux, are slightly less in comparison to the previous situation. Consequently, the air gap flux and the air gap armature emf are only a little reduced. Thus, the fault current is only very slightly less than E_f/x_l. We say, it is E_f/x_d''; x_d'' being only slightly greater than x_l. $x_l < x_d'' < x_d'$.

The period when the damper winding currents are also flowing is the subtransient period. The time constant of the damper winding is quite short and the damper currents decay quickly, and hence the large decaying armature current (= E_f/x_d''), called the subtransient current, decays quickly to the transient current (= E_f/x_d'), which in turn is also decaying, but at relatively a slow rate. The subtransient current lasts 2 to 4 cycles. The equivalent circuit for this condition can be drawn as under.

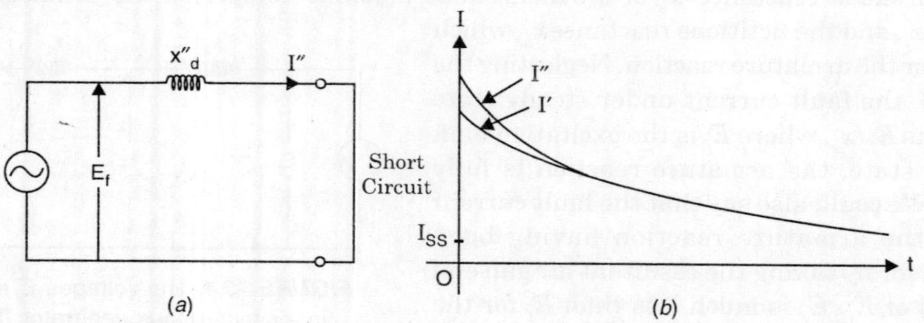

FIGURE 40.3. (*a*) The equivalent circuit to find subtransient *SC* fault current.
(*b*) The variation of the rms value of the symmetrical *SC* current with time.

In the above explanation, the dc component in the armature circuit fault current has not been accounted for. The presence of dc component makes the fault current waveshape asymmetrical. The short circuit is impressed at different points on the voltage wave of the three phases. As such, the dc content is different in each phase current. In the first cycle of the fault current, the dc content can be as high as the peak value of the sub transient symmetrical ac current. One can form a mental picture of the *sc* current by looking at the resulting current in a series *R-L* circuit when sinusoidal voltage $V_m \sin(\omega t + \alpha)$ is applied to it at $t = 0$.

$$i(t) = \frac{V_m}{Z}\left[\sin(\omega t + \alpha - \phi) - e^{-\frac{R}{L}t}\sin(\alpha - \varphi)\right]; \text{ where } Z = \sqrt{R^2 + L^2\omega^2}, \text{ and } \tan\varphi = \frac{L\omega}{R}$$

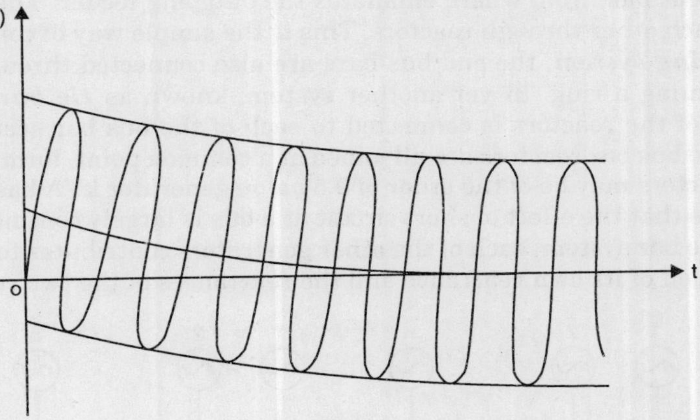

FIGURE 40.4. The current wave shape in a series R-L circuit for $0^\circ < (\alpha - \varphi) < 90^\circ$.

40.3 SHORT CIRCUITS ON POWER SYSTEM

We draw the per phase impedance or the reactance diagram of the power system. As the resistances are usually negligible, the reactance diagram suffices. Reactances included in the reactance diagram should be appropriate reactances, *e.g.*, if subtransient currents are to be determined, then the synchronous generators, synchronous motors, and the induction motors be represented by their respective subtransient reactances and the voltages behind their respective subtransient reactances. For determining the interrupting rating of the breakers, the generators are represented by their subtransient states, the synchronous motors are represented by their transient states, and the induction motors are neglected. Having obtained the appropriate reactance diagram, we look into the network at the fault point and obtain its ***Thevenin's equivalent circuit***. If prior to the occurrence of the fault, the system is not loaded, the Thevenin voltage is 1 *pu* itself. Determining the Thevenin reactance may need several series, parallel, or star-delta reductions of the network. Since, only reactances are present, it is as simple as solving a dc network. Once Thevenin's circuit is obtained, fault is simulated by closing the Thevenin's circuit directly, or through the fault impedance z_f, if any. The fault current, therefore, is determined. Again, the dc component is not included.

40.4 CURRENT LIMITING REACTORS

When the reactances of the generators are not high enough, current limiting reactors are used to limit the fault current to safe values. Such a practice is followed for older generators, whose reactances are small. Depending on the way the reactors are connected, the reactors to some extent, affect the voltages and currents under normal conditions also; *e.g.*, the generator and feeder reactors (a reactor inserted between the bus bar and the generator/feeder) cause voltage drop under normal operation as well.

In the ***bus bar system of reactors***, each of the generators of the plant is connected to its own independent bus, from where emanates the outgoing feeder. The adjacent buses are connected with each other through reactors. This is the simple way of connecting the bus bar reactors. In the ***ring system***, the end bus bars are also connected through a reactor, all the reactors thus forming a ring. In yet another system, known as ***tie bar system*** (star), one terminal of each of the reactors is connected to each of the bus bar sections, and the other terminals of these bus bar reactors are all joined at a common point, forming a star point. The rating of such reactors may be of the order of 0.5 *pu* on generator kVA base. The advantage of bus bar reactors is that the effect of short circuit at a bus is largely confined to that particular bus only. In the tie bar system, each of the other generators contributes to fault MVA through a series combination of its own reactance and the reactances of the two reactors.

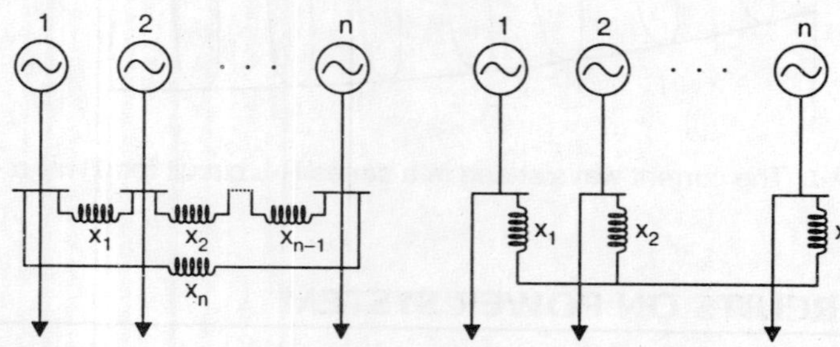

FIGURE 40.5. The ring system and the tie bar (star) system of bus bar reactors.

Even with the presence of the current limiting reactors, the approach to find the fault current due to the fault at any point on the system remains the same as given above. For example, the Thevenin reactance in the tie bar system of reactors is the parallel sum of x_g and $\dfrac{x_g + nx_r}{(n-1)}$, where n is the number of generators, or the reactors, or the bus bar sections; x_g and x_r are the reactances of the generator, and of the reactor respectively.

Usually, one works in the *pu* system.

The ***short circuit capacity*** or the short circuit strength of a bus in *pu* is simply the reciprocal of the Thevenin reactance in *pu* as found looking into the system at that bus.

40.5 CIRCUIT BREAKERS

The functions of a *CB* are that it carries the normal current without any undue heating, it opens quickly in the event of an over current or short circuit without any damage to itself, and permits its manual opening and closing as well. (The second objective is met in conjunction with protective relays.) A circuit breaker, essentially, is a mechanical device; and electrical aspects come into picture when the *CB* poles break or make currents. When a current is broken, most of the stored magnetic field energy in all the inductances is to be dissipated in the gap between the separating contacts of its poles. Lot of heat is produced and an arc is formed, which is a mixture of electrons and the positive ions of the insulating medium in the *CB*. With a view to keep the damage to *CB* and to the system due to the heavy current lowest possible, and also to keep the system running stably, it is desirable that the arc is quenched as quickly as possible.

In ac systems, the arc quenches naturally at the following current zero, it restrikes and again quenches at the next current zero; till it is quenched finally. Whether the arc restrikes or not depends on, ***which builds up faster***— the ***restriking voltage***, or the ***dielectric strength*** of the medium. We try to achieve fast rate of dielectric strength recovery, by employing one or more of the basic principles like lengthening of the arc, cooling the arc, taking away the electrons (as in an SF_6 breaker), or air blasting the arc.

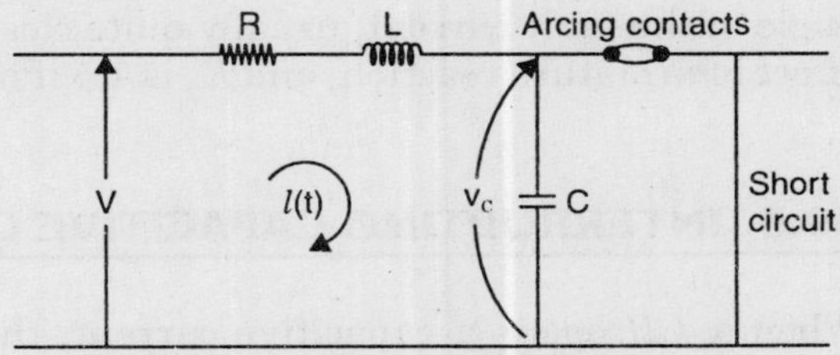

FIGURE 40.6. The circuit for finding the restriking voltage.

The electrical equivalent of the system at the *CB* contacts can be drawn as shown below.

Here, L and R are the system inductance and resistance upto the location of the *CB*; and C is the capacitance of the *CB* bushing. We count time from the instant of current zero. At that instant let the system instantaneous voltage (on per phase basis) be v. This voltage is treated constant for finding the high frequency restriking transient.

The voltage that appears across the separating contacts is the same as is across the capacitor C. The governing equation and the initial conditions which apply to the solution are :

$$v = iR + L\frac{di}{dt} + \frac{1}{C}\int i\,dt, \text{ (resistance } R \text{ is subsequently neglected)}$$

At $t = 0$, $i = 0$, and $L\dfrac{di}{dt} = v$, the final solution becomes $i = \dfrac{v}{R_o}\sin \omega_n t$

where $\omega_n = \dfrac{1}{\sqrt{LC}}$, and $R_o = \sqrt{\dfrac{L}{C}}$.

If R is considered, and is $> 2R_o$, the response becomes aperiodic.

For the short duration for which the restriking transient exists, v has been assumed to remain constant.

The voltage across the capacitor C, or across the *CB* contacts is $\mathbf{v(1 - \cos \omega_n t)}$. This expresses the restriking voltage, v_{rs}.

We see that in a time equal to π/ω_n, the voltage across the breaker contacts reaches $2v$. Thus, the average rate of rise of restriking voltage (**RRRV**) is $\frac{2v}{\pi\sqrt{LC}} = 4vf_n$. The *RRRV* is expressed in kilo volt per micro second. The frequency of the restriking transient is several kilo Hertz.

The maximum value of the *RRRV* is $v\,\omega_n = 2\pi f_n\, v$. Thus, $(RRRV)_{max} = \frac{\pi}{2}(RRRV)_{av}$.

The voltage v, which is called the **active recovery voltage**, is the instantaneous value of the system voltage that appears across the breaker contacts at the current zero; and is related to the system phase to neutral voltage V, as $v = K_1K_2\, 2 \sin \varphi\, V$. φ is the power factor angle of the fault current; usually quite close to 90°, the factor $K_1(<1)$ takes into account the effect of armature reaction, and K_2 is 1.5 if it is an ungrounded fault, else it equals 1.

40.6 INTERRUPTING CAPACITIVE CURRENT, AND INDUCTIVE CURRENTS

When a *CB* opens a capacitive current, the voltage across the contacts becomes nearly double the system peak voltage at the following system voltage peak, and the gap again breaks down. Such successive gap voltage build up at the system voltage peaks, and consequent breakdown of the gap is on account of the presence of the termination capacitance C_L, which gets charged to voltages equal to V, $-3V$, $5V$ and so on at the successive breakdowns. Thus, power frequency high voltages occur at the terminal capacitor C_L, which actually may be an open ended line of the power system.

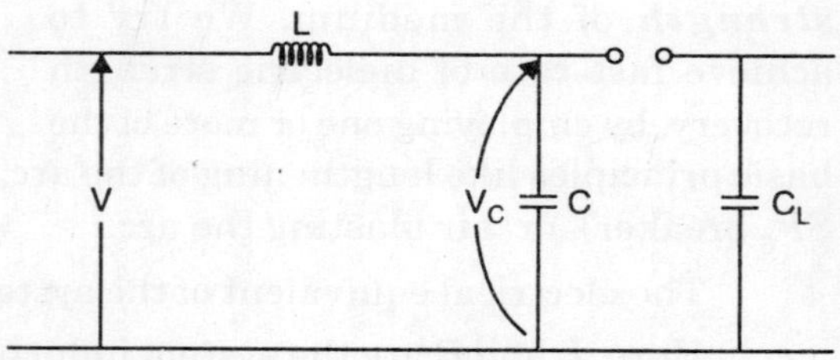

FIGURE 40.7. The circuit when interrupting capacitive current.

Similarly, if an inductive current flowing in a high value inductor L_L is broken by the breaker before the natural current zero on account of its too fast an action (*e.g.*, an air blast *CB*), a sudden high voltage appears across the capacitor C', and hence across the contacts, which again breaks the gap and reignites the arc. The voltage across the contacts is v_C', governed by the energy relation $\frac{1}{2} L_L\, i^2 = \frac{1}{2} C'\, v_C'^2$, where i is the current in the inductor at the instant the current is broken. Such current interruption before the natural zero is known as ***current chopping***. Current chopping, and hence the high voltages, occur repeatedly till natural zero is reached.

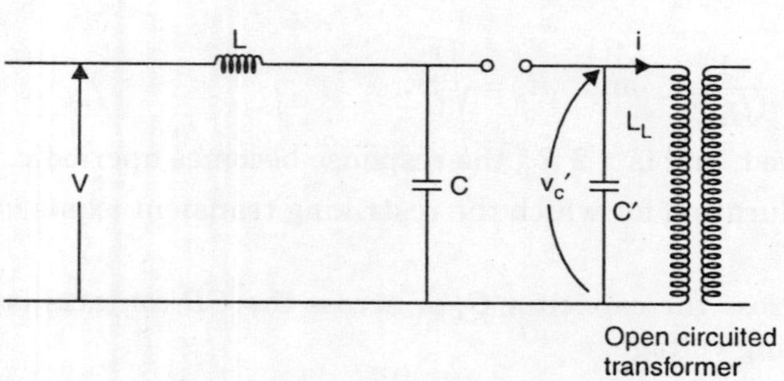

FIGURE 40.8. The circuit when interrupting inductive current.

Both the adverse conditions can be prevented by what is known as, ***'resistance switching'***, wherein a resistance of suitable value is automatically inserted across the contacts so that some or the major part of the current flows through this resistance, thereby energy in the arc is reduced and ionization conditions may not be adequate for the subsequent strike. The circuit effectively becomes that of L, C, and R in parallel.

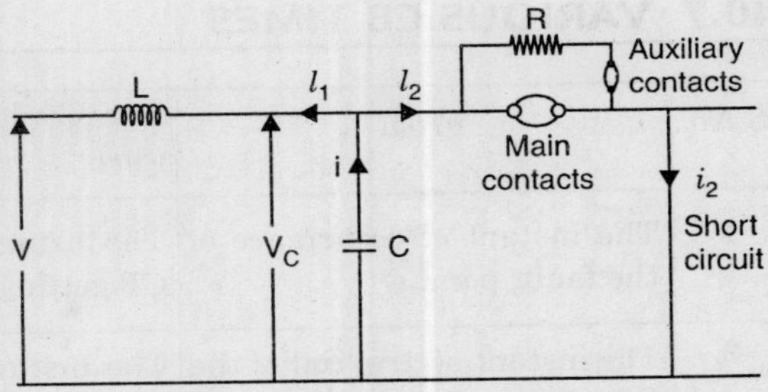

FIGURE 40.9. The resistance switching.

The governing equations are :

$$i = i_1 + i_2, \quad -C\frac{dv}{dt} = \int \frac{v_C - v}{L}\,dt + \frac{v_C}{R}, \text{ or } C\frac{d^2 v_C}{dt^2} + \frac{1}{R}\frac{dv_C}{dt} + \frac{v_C - v}{L} = 0$$

The restriking voltage is given by

$$v_{rs} = v\left[1 - e^{-at}\left(\cos\sqrt{b}\,t + \frac{a}{b}\sin\sqrt{b}\,t\right)\right], \text{ where } a = \frac{1}{2RC}, \text{ and } b = \frac{1}{LC} - \frac{1}{4R^2C^2}.$$

The natural frequency of oscillation is

$$f_n = \frac{1}{2\pi}\sqrt{\frac{1}{LC} - \frac{1}{4R^2C^2}}.$$

The resistance for non oscillatory transient is

$$R = \frac{1}{2}\sqrt{\frac{L}{C}}.$$

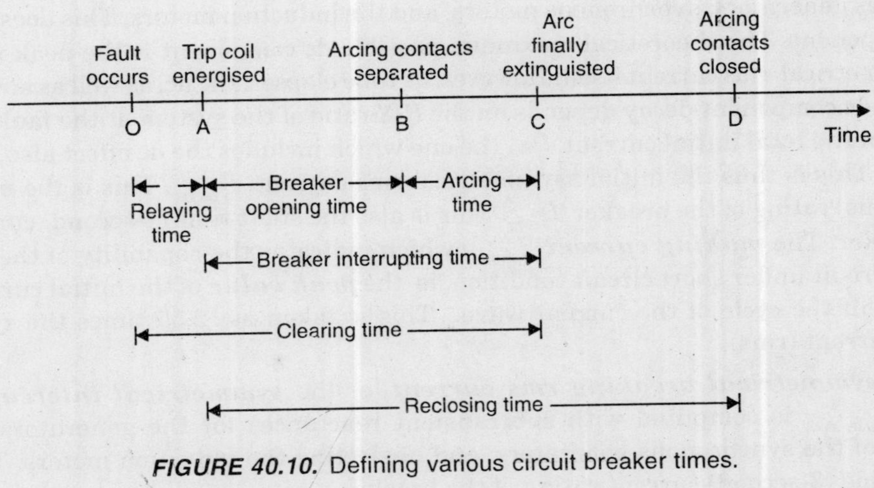

FIGURE 40.10. Defining various circuit breaker times.

40.7 VARIOUS CB TIMES

S.No.	From	To	Nomenclature	Approximate time
1	The instant of occurrence of the fault, point *O*	The instant of trip coil of the *CB* getting energized, point *A*	Relaying time, or the tripping delay	≈ Half a cycle
2	The instant of trip coil of the CB getting energized, point *A*	The instant of the parting contacts of the *CB* completely separated, point *B*	Breaker opening time	≈ 2 cycles
3	The instant of the parting contacts of the *CB* completely separated, point *B*	The instant of the arc finally extinguished, point *C*	Arcing time	≈ 1 cycle
4	The instant of trip coil of the *CB* getting energized, point A	The instant of the arc finally extinguished, point *C*	Breaker interrupting time	≈ 3 cycles
5	The instant of occurrence of the fault, point *O*	The instant of the arc finally extinguished, point *C*	Clearing time	≈ 5 cycles
6	The instant of trip coil of the *CB* getting energized, point *A*	The instant of closing of contacts, point *D*	Reclosing time	

40.8 VARIOUS CB CURRENTS

The term initial refers to the very first cycle of the fault current. The ***initial symmetrical rms current*** $I_{i,sym}$, denoted by I'' also, is the one found with subtransient reactances for the synchronous generators, synchronous motors, and the induction motors. This does not include the dc component. The theoretical maximum possible dc component is the peak value of the initial symmetrical rms current itself. However, as time elapses, the dc, as well as ac components decay. The dc component decay depends on the *R/X* ratio of the system at the fault point. The rms value of the total initial current, *i.e.*, the one which includes the dc effect also, is taken as $1.6 \times I_{i,sym}$. This is thus the initial asymmetrical rms current, $I_{i,asym}$. This is the ***momentary current*** (rms) rating of the breaker, I_{mom}. This is also the short time (1-second) current rating of the breaker. The ***making current***, I_{mak}, which indicates the capability of the breaker to make the circuit under short circuit condition, is the ***peak value*** of the initial current, which occurs at half the cycle of the current wave. This is taken as 2.55 times the symmetrical breaking current (rms).

The ***symmetrical breaking rms current***, or the ***symmetrical interrupting rms current***, $I_{bre,\ sym}$ is computed with subtransient reactances for the generators, transient reactances of the synchronous generators, and neglecting the induction motors. This is also the short time (3-second) current rating of the breaker.

The ***asymmetrical breaking***, or the interrupting current, $I_{bre,asym}$ is obtained by multiplying the symmetrical interrupting, or the breaking current by a suitable factor, as given below, which is related to the speed of the *CB*.

Breaker	Multiplying factor
2-cycle breaker	1.4
3-cycle breaker	1.2
5-cycle breaker	1.1
8-cycle, or slower breaker	1.0

Such ratings have to be specified, as very large electromagnetic forces are caused on the parts of the breaker when carrying such large currents; and the breaker should be able to withstand these forces.

The ***normal current***, I_n (obviously symmetrical) is the rms ac current which the breaker is able to carry continuously without exceeding the prescribed limits of the temperatures of any of its parts.

The ***short time current*** rating is the rms ac current which the breaker can carry safely for 3 seconds, if the ratio of the symmetrical breaking current to normal current is less than 40, or for 1 second, otherwise.

These currents are generally expressed in kilo amperes.

We express the breaker capacities in MVA also. For a 3-phase breaker, if V is the line to line rated service voltage in kV, and I the rated symmetrical breaking current in *kA*, then the capacity corresponding to the current I is $\sqrt{3}\ VI$ MVA.

In addition to the breaking current, momentary current, and making current ratings, a *CB* is to conform to the ***operating duty*** as *B*-3-*MB*-3-*MB*. This means that the *CB* breaks, after a 3-minute interval it is able to make, followed immediately by break, and again a 3-minute interval, and make and break operation successfully.

Chapter 41

SYMMETRICAL COMPONENTS AND UNSYMMETRICAL FAULTS

41.1 GENERAL

A balanced three-phase system can always be analyzed as a single-phase circuit. An unbalanced system of known impedances can be analyzed directly by applying the usual methods of circuit analysis treating the three phase circuits as three circuits, and solving the equations simultaneously. But it would be more convenient if one could use the balanced system analysis approach.

Also, an unbalanced three-phase system cannot be solved as a three-phase circuit if a synchronous machine is present in the system, because we are confronted with the difficulty of appropriate impedance. The synchronous impedance can be used only if the generator operation is balanced.

It is with these views that we resort to the method of symmetrical components to analyze an unbalanced three-phase system.

According to the method of symmetrical components, a set of n-phase unbalanced n phasors can be resolved into $(n - 1)$ sets of balanced n-phase phasors, and one set of all-in-phase n phasors. Confining ourselves to three-phase systems only, a set of unbalanced three-phase phasors can be resolved into two sets of balanced three-phase phasors, and a set of three all-in-phase phasors. We call them components of the original system. The set which has the same phase sequence as the given system has, is called the set of positive sequence components, and the one which has the sequence opposite to that of the given, is called the set of negative sequence components. The all-in-phase set is called the set of zero sequence components.

41.2 PHASE SEQUENCE

The anticlockwise direction of rotation of the phasors is taken as the positive direction of rotation. The three phases are denoted by letters a, b, and c. If in the rotation, a is followed by

b, and b in turn is followed by c, the phase sequence is a-b-c. This phase sequence could also be said to be b-c-a, or c-a-b. This is the positive phase sequence. In the negative phase sequence, a is followed by c, and c in turn is followed by b. The sequence, is thus, a-c-b, or c-b-a, or b-a-c. This is called the negative phase sequence, because if the phasors rotate in the negative (clockwise) direction, then the sequence is a-b-c itself.

In the analysis, the phase a is usually taken as the reference phase. The positive sequence components are denoted by the additional suffix 1, the negative sequence components are denoted by the additional suffix 2, and the zero sequence components by the additional suffix 0. All the equations given below involve complex variables; and as such, the complex, or the ***phasor algebra is to be used in manipulating them***. The results obtained are with phase a as the reference phase.

41.3 RELATION BETWEEN SYMMETRICAL COMPONENTS AND PHASE VARIABLES

We can find the components of the voltages, or the currents, or the fluxes, or any other three-phase quantity of interest. The approach remains the same. Consider, *e.g.*, the voltages-$\mathbf{V}_a$, $\mathbf{V}_b$, and $\mathbf{V}_c$ (these may be line to neutral voltages).

$$\mathbf{V_a = V_{ao} + V_{a1} + V_{a2}}, \quad \mathbf{V_b = V_{b0} + V_{b1} + V_{b2}}, \quad \text{and} \quad \mathbf{V_c = V_{co} + V_{c1} + V_{c2}}$$

$\mathbf{V}_{a0}$, $\mathbf{V}_{b0}$, and $\mathbf{V}_{c0}$ constitute the zero sequence components, $\mathbf{V}_{a1}$, $\mathbf{V}_{b1}$, and $\mathbf{V}_{c1}$ constitute the positive sequence components, and $\mathbf{V}_{a2}$, $\mathbf{V}_{b2}$, and $\mathbf{V}_{c2}$ constitute the negative sequence components of the original phase voltages $\mathbf{V}_a$, $\mathbf{V}_b$, and $\mathbf{V}_c$. $\mathbf{V}_{a0}$, $\mathbf{V}_{b0}$, and $\mathbf{V}_{c0}$ are all in phase, $\mathbf{V}_{a1}$, $\mathbf{V}_{b1}$, and $\mathbf{V}_{c1}$ have a mutual phase difference of 120°, and $\mathbf{V}_{a2}$, $\mathbf{V}_{b2}$, and $\mathbf{V}_{c2}$ also have a mutual phase difference of 120°. We let the phase operator

$$\mathbf{a} = 1\angle 120°, \qquad \mathbf{a}^2 = 1\angle 240° = 1\angle -120°, \qquad \text{and} \quad \mathbf{a}^3 = 1\angle 0°.$$

We see that $\mathbf{a} = -0.5 + j\,0.867$, $\quad \mathbf{a}^2 = -0.5 - j\,0.867$, and $\mathbf{a + a^2 + a^3} = 0$.

Several other relations amongst a's can be found.

We can write $\mathbf{V}_{c0} = \mathbf{V}_{b0} = \mathbf{V}_{a0}$; $\mathbf{V}_{c1} = \boldsymbol{a}\,\mathbf{V}_{a1}$, $\mathbf{V}_{b1} = \mathbf{a}^2\mathbf{V}_{a1}$, and $\mathbf{V}_{c2} = \mathbf{a}^2\,\mathbf{V}_{a2}$, $\mathbf{V}_{b2} = \mathbf{a}\,\mathbf{V}_{a2}$.

We express these relations in the matrix form.

$$\begin{bmatrix} \mathbf{V_a} \\ \mathbf{V_b} \\ \mathbf{V_c} \end{bmatrix} = \begin{bmatrix} 1 & 1 & 1 \\ 1 & \mathbf{a}^2 & \mathbf{a} \\ 1 & \mathbf{a} & \mathbf{a}^2 \end{bmatrix} \begin{bmatrix} \mathbf{V_{a0}} \\ \mathbf{V_{a1}} \\ \mathbf{V_{a2}} \end{bmatrix}$$

The transformation matrix $\begin{bmatrix} 1 & 1 & 1 \\ 1 & \mathbf{a}^2 & \mathbf{a} \\ 1 & \mathbf{a} & \mathbf{a}^2 \end{bmatrix}$ is denoted by $\mathbf{A}$.

The inverse relations are

$$\begin{bmatrix} \mathbf{V_{a0}} \\ \mathbf{V_{a1}} \\ \mathbf{V_{a2}} \end{bmatrix} = \frac{1}{3} \begin{bmatrix} 1 & 1 & 1 \\ 1 & \mathbf{a} & \mathbf{a}^2 \\ 1 & \mathbf{a}^2 & \mathbf{a} \end{bmatrix} \begin{bmatrix} \mathbf{V_a} \\ \mathbf{V_b} \\ \mathbf{V_c} \end{bmatrix}. \text{ Thus, we note that } \mathbf{A}^{-1} = \frac{1}{3} \begin{bmatrix} 1 & 1 & 1 \\ 1 & \mathbf{a} & \mathbf{a}^2 \\ 1 & \mathbf{a}^2 & \mathbf{a} \end{bmatrix}$$

Identical relations can be written for the currents.

$$\mathbf{I}_{ph} = \mathbf{A}\,\mathbf{I}_{sy}, \quad \text{and} \quad \mathbf{I}_{sy} = \mathbf{A}^{-1}\mathbf{I}_{ph}$$

It is noted that the line voltages do not have any zero sequence component, as $\mathbf{V}_{ab} + \mathbf{V}_{bc} + \mathbf{V}_{ca}$ is always zero. The phase voltages may have. In the absence of a connection between the source neutral and the load neutral, the sum of the three line currents I_a, I_b, and I_c is also zero; as such the line currents in the absence of a neutral connection, also do not have any zero sequence component. If a neutral connection is there, the zero sequence component may be present. In that case, $\mathbf{I}_n = (\mathbf{I}_a + \mathbf{I}_b + \mathbf{I}_c) = 3\,\mathbf{I}_0$. The delta currents may, or may not have a zero sequence component. Obviously, for a balanced system, the zero sequence component as well as the negative sequence component are both zero.

It is stated that the transformation is not power invariant, as we find that $\mathbf{V}_a\mathbf{I}_a^* + \mathbf{V}_b\mathbf{I}_b^* + \mathbf{V}_c\mathbf{I}_c^* \neq \mathbf{V}_{a0}\mathbf{I}_{a0}^* + \mathbf{V}_{a1}\mathbf{I}_{a1}^* + \mathbf{V}_{a2}\mathbf{I}_{a2}^*$, but is equal to $3(\mathbf{V}_{a0}\mathbf{I}_{a0}^* + \mathbf{V}_{a1}\mathbf{I}_{a1}^* + \mathbf{V}_{a2}\mathbf{I}_{a2}^*)$.

41.4 SEQUENCE IMPEDANCES

The impedance offered by a three-phase circuit, device, or an apparatus when only zero sequence currents flow, is called the ***zero sequence impedance***, denoted by $\mathbf{Z}_0$. Similarly, we can define the positive sequence impedance $\mathbf{Z}_1$, and the negative sequence impedance $\mathbf{Z}_2$. The impedance we normally speak of is understood to be the positive sequence impedance $\mathbf{Z}_1$.

For transformers, all the three impedances are identical (the zero sequence impedance is slightly different from others). For transmission lines, the positive sequence reactance X_1 = the negative sequence reactance X_2, and the zero sequence reactance X_o is 2 to 3.5 times X_1. For synchronous machines, $X_1 \approx 1.25\ pu$, $X_2 \approx 0.20\ pu$, and X_o also $\approx 0.20\ pu$, on the base of machine rating.

41.5 THREE UNEQUAL IMPEDANCES

Let the current flowing from a towards a', through the impedance $\mathbf{Z}_a$ be $\mathbf{I}_a$, and the voltage across its terminals aa' be $\mathbf{V}_{aa'}$, the current flowing through the impedance $\mathbf{Z}_b$ be $\mathbf{I}_b$, and the voltage across its terminals $b\,b'$ be $\mathbf{V}_{bb'}$, and similarly for the third impedance $\mathbf{Z}_c$, the current is $\mathbf{I}_c$ and the voltage $\mathbf{V}_{cc'}$. We thus have voltages $\mathbf{V}_{aa'}$, $\mathbf{V}_{bb'}$, and $\mathbf{V}_{cc'}$; and the currents $\mathbf{I}_a$, $\mathbf{I}_b$, and $\mathbf{I}_c$. We can find the components of the voltages, and those of the currents using the transformation relations. Assuming ***no mutual couplings*** between $\mathbf{Z}_a$, $\mathbf{Z}_b$, and $\mathbf{Z}_c$, the results are :

$$\begin{bmatrix} \mathbf{V}_{aa'0} \\ \mathbf{V}_{aa'1} \\ \mathbf{V}_{aa'2} \end{bmatrix} = \frac{1}{3} \begin{bmatrix} (\mathbf{Z}_a + \mathbf{Z}_b + \mathbf{Z}_c) & (\mathbf{Z}_a + \mathbf{a}^2\mathbf{Z}_b + \mathbf{a}\mathbf{Z}_c) & (\mathbf{Z}_a + \mathbf{a}\mathbf{Z}_b + \mathbf{a}^2\mathbf{Z}_c) \\ (\mathbf{Z}_a + \mathbf{a}\mathbf{Z}_b + \mathbf{a}^2\mathbf{Z}_c) & (\mathbf{Z}_a + \mathbf{Z}_b + \mathbf{Z}_c) & (\mathbf{Z}_a + \mathbf{a}^2\mathbf{Z}_b + \mathbf{a}\mathbf{Z}_c) \\ (\mathbf{Z}_a + \mathbf{a}^2\mathbf{Z}_b + \mathbf{a}\mathbf{Z}_c) & (\mathbf{Z}_a + \mathbf{a}\mathbf{Z}_b + \mathbf{a}^2\mathbf{Z}_c) & (\mathbf{Z}_a + \mathbf{Z}_b + \mathbf{Z}_c) \end{bmatrix} \begin{bmatrix} \mathbf{I}_{a0} \\ \mathbf{I}_{a1} \\ \mathbf{I}_{a2} \end{bmatrix}$$

If all the impedances are $\mathbf{Z}_a$ each, then we have

$$\mathbf{V}_{aa'0} = \mathbf{Z}_a\,\mathbf{I}_{a0}$$

$$V_{aa'1} = Z_a \, I_{a1}$$
$$V_{aa'2} = Z_a \, I_{a2}$$

It is thus seen that the symmetrical components of unbalanced currents flowing in a balanced *Y*-connected load, *i.e.*, in three equal impedances, produce voltage drops of like sequence only. The other component currents do not have any effect. It is an important conclusion.

41.6 SEQUENCE NETWORKS

Due to the above conclusion, we can draw a particular sequence network in which the current and the voltage are of that particular sequence alone. The other sequence components do not interfere. A network which satisfies the positive sequence quantities relations of the given three-phase system can be called a positive sequence network. This network is a network of identical impedances (positive sequence impedance) in the three phases. It is a balanced network. Like any three-phase balanced system, we can represent it by the reference voltage, *i.e.*, phase a positive sequence voltage, phase a positive sequence current, and the positive sequence impedance. This is the per phase ***positive sequence network***. The reference for reckoning the voltage is the zero potential point, as it is a balanced system.

Statements as given above, apply to the ***negative sequence network*** also, the word 'positive', replaced by the word 'negative'.

These statements apply to the ***zero sequence network*** too, the word 'positive', replaced by the word 'zero'. The differences are that it is not a balanced three-phase system, and that the reference for reckoning the voltage is the potential of the ground at the point in the system at which any particular voltage is specified.

In a balanced three-phase system, there is no need of a neutral connection for the currents to complete their paths, as the sum of the three currents at the neutral point is zero ; but in case of a zero sequence network, the three currents are to be in phase and at the neutral point their sum cannot be zero. The zero sequence currents can complete their path only if the neutral connection is provided.

41.7 ZERO SEQUENCE DIAGRAMS OF THREE-PHASE TRANSFORMERS FOR VARIOUS CONNECTIONS

In a transformer, currents cannot flow in the windings of one circuit alone (except for the small magnetizing current in the primary circuit). If currents flow in the windings on the secondary side, there has to be corresponding currents in the windings on the primary side also. It has already been pointed out that the zero sequence currents can flow only if a closed path is available to them. Thus, in a three-phase transformer, the path for the zero sequence currents to flow should simultaneously be available in the primary winding, as well as in the secondary winding. Zero sequence currents can flow in a three-phase transformer if

(*i*) it is a *Y*-*Y* connected transformer, both the *Y* neutrals are grounded, and the external circuits provide the paths on the two sides.

(*ii*) it is a *Y*-Δ connected transformer, *Y* neutral is grounded, and the external circuit on the *Y* side provides the path.

(*iii*) it is a Δ-Δ connected transformer.

The grounding of the neutrals may be direct grounding, or grounding through some impedance.

There cannot be any zero sequence current in the external circuit connected to the Δ-connected winding of the transformer. The external circuit connected to the *Y*-connected transformer winding with grounded neutral, can allow zero sequence current only if the other end (source, or the load) neutral also is grounded. Given below are various connections of three-phase transformers, and their corresponding zero sequence diagrams. $\mathbf{Z}_o$ is the zero sequence impedance of the transformer.

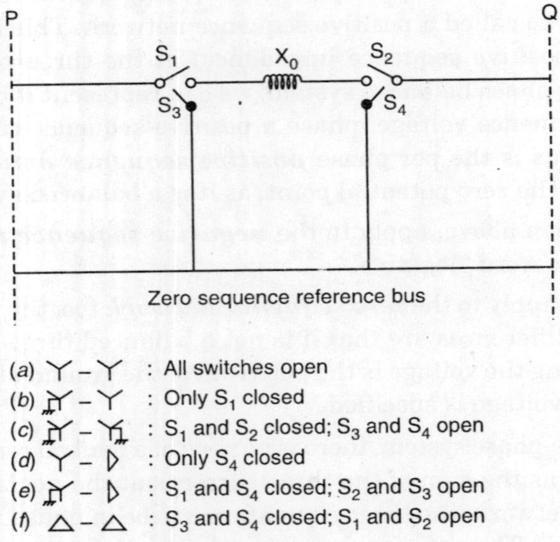

FIGURE 41.1. Various three-phase transformer connections and the corresponding zero sequence diagrams.

41.8 BASIC V-I COMPONENTS RELATIONS AND THE SEQUENCE DIAGRAMS FOR AN UNLOADED ALTERNATOR

The pertinent equations in the matrix form are

$$\begin{bmatrix} \mathbf{V}_{a0} \\ \mathbf{V}_{a1} \\ \mathbf{V}_{a2} \end{bmatrix} = \begin{bmatrix} 0 \\ \mathbf{E}_1 \\ 0 \end{bmatrix} - \begin{bmatrix} \mathbf{Z}_o & 0 & 0 \\ 0 & \mathbf{Z}_1 & 0 \\ 0 & 0 & \mathbf{Z}_2 \end{bmatrix} \begin{bmatrix} \mathbf{I}_0 \\ \mathbf{I}_{a1} \\ \mathbf{I}_{a2} \end{bmatrix},$$

In the expanded form these equations are

$$\mathbf{V}_{a0} = -\mathbf{Z}_0 \mathbf{I}_{a0}$$

$$\mathbf{V}_{a1} = \mathbf{E}_1 - \mathbf{Z}_1\mathbf{I}_{a1}$$
$$\mathbf{V}_{a2} = -\mathbf{Z}_2\,\mathbf{I}_{a2}$$

The generator generates balanced voltages only. $\mathbf{E}_1$ is the line to neutral, phase a normal excitation generated emf, *i.e.*, $\mathbf{E}_1 = \mathbf{E}_a$. There can be no such emf as $\mathbf{E}_0$, or $\mathbf{E}_2$. $\mathbf{Z}_1$, and $\mathbf{Z}_2$ are the generator sequence impedances. $\mathbf{Z}_0$ is the sum of the zero sequence impedance of the generator and $3\mathbf{z}_n$, *i.e.*, $\mathbf{Z}_o = \mathbf{Z}_{og} + 3\mathbf{z}_n$. As the resistance/reactance ratio is generally quite small, the impedances can be replaced by the corresponding reactances (with j attached).

If the subtransient fault current is sought, the emf $\mathbf{E}_1$ is the emf behind the subtransient reactance, and the impedances/reactances are the subtransient impedances/reactances. Similar statements apply, if the desired current is the transient fault current, or the sustained fault current. For an unloaded generator, each of these three voltages is $\mathbf{E}_a$ itself.

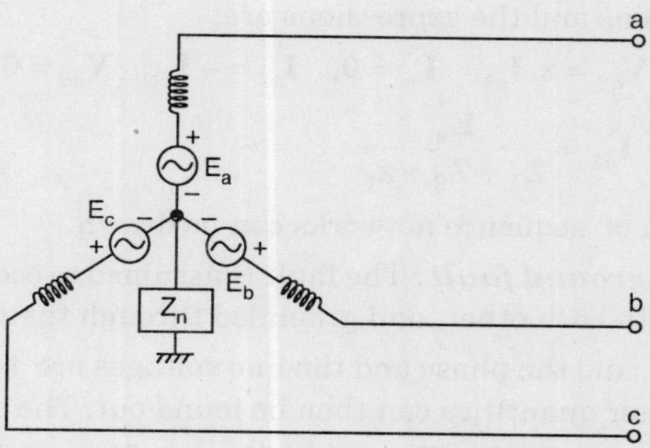

FIGURE 41.2. A normally excited and unloaded alternator.

41.9 UNSYMMETRICAL FAULTS

With this background of symmetrical components, we are in a position to analyze the unsymmetrical faults at the terminals of an unloaded alternator.

In all the fault analysis, the objective is to find the fault current, and the terminal voltages under fault conditions. For the type of fault to be analyzed, we first identify the terminal conditions imposed by the fault. These conditions are used in the relations $\mathbf{I}_{sy} = \mathbf{A}^{-1}\mathbf{I}_{ph}$ and/or $\mathbf{V}_{sy} = \mathbf{A}^{-1}\mathbf{V}_{ph}$ to find the interrelations amongst the components. The interrelations found can be used in the basic *v-i* components relations as given above to finally find an expression for $\mathbf{I}_{a1}$. Once $\mathbf{I}_{a1}$ is found, other components, and then the phase quantities can be determined.

Depending upon the relations found between the component currents, and between the component voltages, we can draw the interconnected sequence networks.

(*i*) ***Single line-to-ground fault*** : The fault is assumed to occur on phase a. The phase a gets faulted to ground through the fault impedance $\mathbf{z}_f$. The fault current $\mathbf{I}_a$ is to be determined. The terminal conditions are:

$$\mathbf{V}_a = \mathbf{z}_f\,\mathbf{I}_a, \qquad \mathbf{I}_b = 0, \quad \text{and} \quad \mathbf{I}_c = 0$$

The resulting relations and the expressions are:

$$\mathbf{I}_{a0} = \mathbf{I}_{a1} = \mathbf{I}_{a2}$$

$$\mathbf{I}_{a1} = \frac{\mathbf{E_a}}{\mathbf{Z}_o + \mathbf{Z}_1 + \mathbf{Z}_2 + 3\mathbf{z_f}}, \quad \mathbf{I}_a = 3Ia = 1.$$

The interconnection of sequence networks can be drawn.

(*ii*) ***Line-to-line fault*** : The fault is assumed to occur between phases *b* and *c*, through the fault impedance $\mathbf{z}_f$.

The fault current I_b, and the phase and the line voltages are to be determined. Only $\mathbf{I}_{a1}$ would be evaluated, as other quantities can then be found out. The terminal conditions are:

$$\mathbf{I}_a = 0, \quad \mathbf{I}_c = -\mathbf{I}_b, \quad \text{and} \quad \mathbf{V}_c = \mathbf{V}_b - \mathbf{z}_f\mathbf{I}_b$$

The resulting relations and the expressions are:

$$\mathbf{V}_{a1} - \mathbf{V}_{a2} = \mathbf{z}_f\mathbf{I}_{a1}, \quad \mathbf{I}_{ao} = 0, \quad \mathbf{I}_{a2} = -\mathbf{I}_{a1}, \quad \mathbf{V}_{a0} = 0$$

$$\mathbf{I}_{a1} = \frac{\mathbf{E_a}}{\mathbf{Z}_1 + \mathbf{Z}_2 + \mathbf{z_f}}$$

The interconnection of sequence networks can be drawn.

(*iii*) ***Line-to-line-to-ground fault*** **:** The fault is assumed to occur on phases *b* and *c*. The two phases get shorted with each other, and grounded through the fault impedance $\mathbf{z}_f$.

The fault current I_b, and the phase and the line voltages are to be determined. Only $\mathbf{I}_{a1}$ would be evaluated, as other quantities can then be found out. The terminal conditions are:

$$\mathbf{I}_a = 0, \mathbf{V}_c = \mathbf{V}_b, \quad \text{and} \quad \mathbf{V}_c \text{ or } \mathbf{V}_b = \mathbf{z}_f(\mathbf{I}b + \mathbf{I}_c)$$

Applying the terminal conditions, we get

$$\mathbf{I}_{a0} = \frac{\mathbf{V}_b}{3\mathbf{z}_f}, \quad \text{and} \quad \mathbf{V}_{a2} = \mathbf{V}_{a1}$$

Using the basic **v-i** components relations, we get

$$\mathbf{I}_{a2} = -\frac{\mathbf{E}_a - \mathbf{Z}_1\mathbf{I}_{a1}}{\mathbf{Z}_2}, \quad \text{and} \quad \mathbf{I}_{a0} = -\frac{\mathbf{E}_a - \mathbf{Z}_1\mathbf{I}_{a1}}{\mathbf{Z}_0 + 3\mathbf{z}_f}$$

$$\mathbf{I}_{a1} = \frac{\mathbf{E}_a}{\mathbf{Z}_1 + \mathbf{Z}_p},$$

where $\mathbf{Z}_p$ is the parallel sum of Z_2 and $\mathbf{Z}_0'$; $\mathbf{Z}_0'$ being equal to $\mathbf{Z}_0 + 3\mathbf{z_f}$.

The interconnection of sequence networks can be drawn.

41.10 SYMMETRICAL THREE-PHASE FAULT

All the three terminals of the generator get shorted with each other through a three-phase fault impedance, comprising an impedance of z_f in each phase. The terminals may or may not further get grounded. The neutral may or may not be grounded. The fault being symmetrical, there is no need to bring in the sequence components.

$$\mathbf{I}_a = \frac{\mathbf{E_a}}{\mathbf{Z_1 + z_f}}, \quad \mathbf{V}_a = \mathbf{aV}_b = \mathbf{a^2\, V}_c \,.\, \mathbf{V}_n \text{ would be zero.}$$

Except the positive sequence component of the current, all the components of the voltages, and of the currents are zero.

41.11 FAULT AT THE TERMINALS OF A LOADED GENERATOR

If prior to the occurrence of the fault, the generator is supplying a current I_a at terminal voltage V_{ta}, the excitation emf E_1 in the basic *v-i* component relations need to be replaced by the voltage behind the subtransient reactance, if subtransient current is to be found out; or by the voltage behind the transient reactance, if transient current is required.

The voltage behind the subtransient reactance, $\mathbf{E_a}'' = \mathbf{V_{ta}} + \mathbf{j\, I_a X_1}''$

The voltage behind the transient reactance, $\mathbf{E_a}' = \mathbf{V_{ta}} + \mathbf{j\, I_a X_1}'$

For finding the sustained fault current, no change is required.

41.12 FAULT AT A POINT IN THE POWER SYSTEM

For finding the current flowing into the fault in a power system, we first need to draw the three sequence networks of the pre-fault power system, preserving the identity of the fault point. Next the Thevenin's equivalent networks of these sequence networks, looking into the network at the fault point, are obtained. Finally, these Thevenin's equivalent networks are appropriately inter-connected for the type of fault being studied, and then the positive sequence component of the current into the fault is obtained in the usual way. Other quantities of interest can then be found out.

This approach is quite general. It can be applied to simple cases as well; like—a single unloaded alternator, a loaded alternator, or several generators operating in parallel.

Chapter 42 NEUTRAL GROUNDING

42.1 GENERAL

Connecting the star points of the transformers, generators, and other star-connected equipment in the power systems to the general mass of the earth is called grounding the neutrals. This connection to the earth may be a direct connection, or a connection through some impedance. In either case we say that the ***neutral is grounded***. If the neutral is not connected thus, we say it is an isolated or ***ungrounded neutral***.

The neutral connection does not come into picture under balanced conditions and at fundamental frequency. However, if there is an unbalanced operation involving earth, either due to a fault, or due to the unbalanced loading, the zero sequence components of the currents flow through the neutral connection. Such a current flow in the system can be detected and protective action taken.

In the absence of the neutral grounding, the small balanced capacitive currents in the lines which flow in the normal conditions become unbalanced under fault or abnormal conditions and flow into the faulted line at the earth fault point. Due to a phenomenon called arcing ground, the system experiences high frequency voltage oscillations which cause it to acquire high voltage. The phenomenon is due to the cyclic charging and discharging of the line capacitances during the earth fault. Also, the normal power frequency voltages of the healthy phases w.r.t. earth (or the ground) rise to $\sqrt{3}$ times the normal phase value.

The primary purpose of neutral grounding is to allow sufficient earth fault current to flow so as to operate the protective relays and isolate the faulty line, and to prevent arcing ground by neutralizing the capacitive current at the earth fault point.

42.2 GROUNDING TYPES

(*a*) Solid grounding

(*b*) Resistance grounding

(*c*) Reactance grounding

(*d*) Resonant grounding

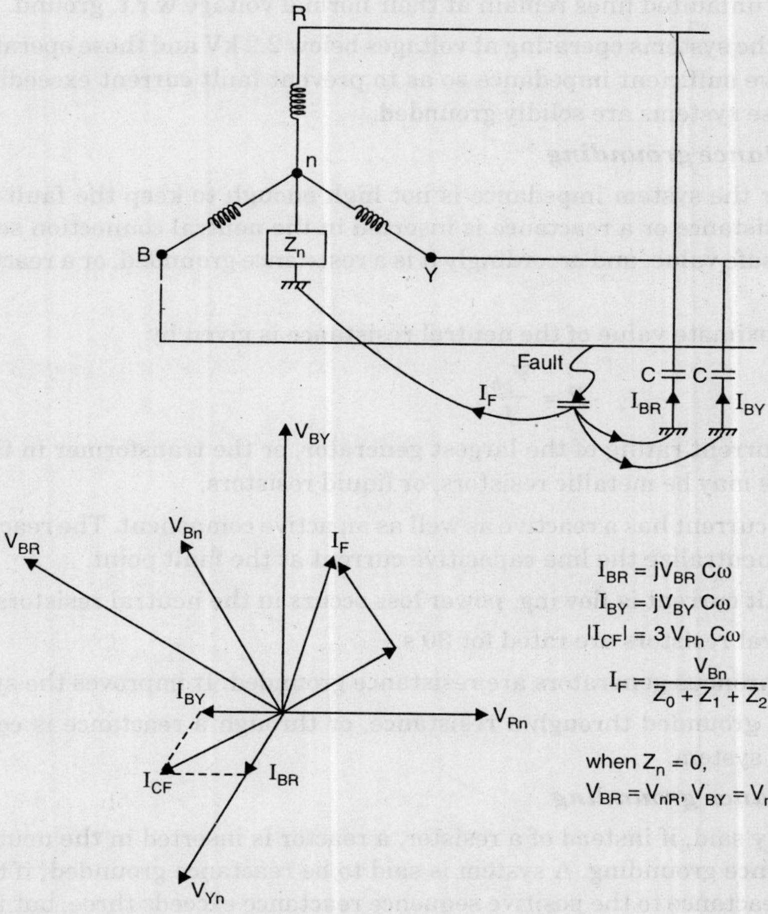

FIGURE 42.1. The system neutral grounded through an impedance, an earth fault on line B, and the corresponding phasor diagram.

(*a*) *Solid grounding*

A direct metallic connection is made between the neutral point and the earth. It is also known as direct or effective grounding.

A system, or a portion of the system is said to be effectively grounded when for all points or for the specified portion of the system, the ratio of the zero sequence reactance to the positive sequence reactance is not greater than three, and the ratio of the zero sequence resistance to the positive sequence reactance is not greater than one for any condition of operation and for any amount of generator capacity.

The earth fault current is predominantly inductive and flows through the fault point and the neutral connection. Depending on the various sequence impedances (which are mainly reactances) upto the fault point, the fault current is usually large enough so as to neutralize completely the capacitive current at the fault point. Thus, the possibility of arcing ground is eliminated. The unfaulted lines remain at their normal voltage w.r.t. ground.

Usually, the systems operating at voltages below 2.2 kV and those operating at voltages above 33 kV have sufficient impedance so as to prevent fault current exceeding safe values, and as such these systems are solidly grounded.

(*b*) ***Resistance grounding***

Whenever the system impedance is not high enough to keep the fault current within safe value; a resistance or a reactance is inserted in the neutral connection so as to limit the fault current to safe value, and accordingly it is a resistance grounded, or a reactance grounded system.

The approximate value of the neutral resistance is given by

$$R = \frac{V_{ph}}{I},$$

where I is the current rating of the largest generator, or the transformer in the system. The neutral resistors may be metallic resistors, or liquid resistors.

The fault current has a reactive as well as an active component. The reactive component can be made to neutralize the line capacitive current at the fault point.

When fault current is flowing, power loss occurs in the neutral resistors.

The neutral resistors are rated for 30 s.

The synchronous generators are resistance grounded. It improves the system stability.

A system grounded through a resistance, or through a reactance is costlier than the direct grounded system.

(*c*) ***Reactance grounding***

As already said, if instead of a resistor, a reactor is inserted in the neutral connection, then it is reactance grounding. A system is said to be reactance grounded, if the ratio of the zero sequence reactance to the positive sequence reactance exceeds three, but is still less than that required in resonant grounding. The earth fault current is predominantly reactive as in the case of direct grounding.

During fault current flow, there is negligible power loss in the reactor as compared to that in the resistor in the resistance grounding. The neutral reactor acts as an open circuit to the voltage surges, and a positive reflection occurs, thereby leading to build up of high voltage. The synchronous motors and the synchronous capacitors are reactance grounded.

(*d*) ***Resonant grounding***

In all the above three grounding types, sufficient earth fault current is permitted so as to cause the operation of the protective gear and consequently isolate the faulty line. If we wish that only the arcing ground be eliminated and even the faulted line continues to operate, we should reduce the earth fault current in the neutral connection to such a low value that the relays do not detect it. Thus, the reactance in the neutral connection has to be increased accordingly. The inductance resonates with the line capacitance, and therefore, it is called resonant grounding. We also call it arc suppression coil grounding, or the fault neutralizer grounding, or the Peterson coil grounding. Since, the capacitive current in the fault is 3 V_{ph}/X_C, and the fault current through the neutral inductor is V_{ph}/X_L, the inductance of the neutral inductor is given by

$$L = \frac{1}{3\omega^2 C}$$, where C is the capacitance of each of the lines to ground.

The current rating of the coil is 3 V_{ph}/X_C.

Usually, taps are provided on the inductor so as to achieve resonance for varying network conditions.

In resonant grounding, the voltages of the healthy phases w.r.t. ground is $\sqrt{3}$ times the normal phase voltage. From this point of view, the system is as bad as not grounded.

This type of grounding can be used in transmission systems where it is known that the earth faults are mostly of transient nature and the faulty line need not be isolated.

Chapter 43 POWER SYSTEM STABILITY

43.1 GENERAL

Stability of a power system is that attribute of the system due to which, on the occurrence of a ***disturbance***, it develops counter forces within it so as to overcome the destabilizing effect of the disturbance, and it is able to continue operating normally, or in equilibrium. Essentially, the stability problem is the problem of keeping the rotors of all the synchronous machines in the system running at synchronous speeds, even in the face of disturbances. The disturbance may be a change in the load, a change in the network configuration, a fault, or a change in the control parameters of the automatic voltage regulator of the generator. Further, the change may be too sudden, *i.e.*, a step change, or it may be a gradual change. In stability studies, the aim is to find the impact of the disturbance on the rotor dynamics. The ***system model*** to study rotor dynamics leads to ***nonlinear differential and algebraic equations***.

When the changes are gradual and small, the stability referred to is the ***steady-state stability***. The system model can be simplified for such studies, and the resulting governing equations are linear equations.

An extension of steady-state stability is known as the ***dynamic stability***. The dynamic stability is concerned with the effect of continually occurring small and slow disturbances, taking into account the action of automatic control devices in the system. It may not lead to instability, but can cause large amplitude oscillations in the rotor speed. Such a dynamic state may last several seconds.

When the changes are sudden and large, the stability referred to is the ***transient state stability***. During the initial (***first swing*** of the rotor) transient stability period (about 1s), the effects of the control actions like the input valve opening, or closing; or the AVR action, are not present. When the effects of control devices are also included in modeling the system dynamics, and the periods of study are relatively longer; the stability referred to is the ***multiswing*** transient stability.

We are mostly concerned with first swing transient stability.

43.2 SWING EQUATION

We derive here the machine rotor dynamics equation considering it in generating mode. In steady-state operation, the rotor is running at synchronous speed. Neglecting rotational losses, the torque T_m provided by the prime mover is balanced by the electromagnetic torque, T_e. Whenever, due to any reason, T_m and T_e differ, the rotor undergoes acceleration/deceleration, and the rotor speed deviates from the synchronous speed.

$$J\frac{d\omega_r}{dt} = T_m - T_e,$$

J is the moment of inertia of the rotating masses, and ω_r is the rotor angular speed. Damping has been neglected.

It is more usual to express the above equation in terms of powers. Multiplying throughout by ω_r,

$$J\,\omega_r\frac{d\omega_r}{dt} = P_m - P_e$$

The rotor speed does not deviate too far from the synchronous speed, ω_s. Therefore,

$$J\,\omega_s\frac{d\omega_r}{dt} = P_m - P_e$$

The quantity, $J\,\omega_s$ is denoted by M, which is called the angular momentum of the rotor. It is a constant for a given machine.

$$M\frac{d\omega_r}{dt} = P_m - P_e$$

Let θ_r be the angle which the rotor makes in the direction of rotation with a fixed axis in space, and θ_s be the angle which a synchronously rotating axis makes with that fixed axis. θ_s is given by $\omega_s t + \theta_{so}$. Further, let δ be the angle by which the rotor axis is ahead of the synchronously rotating axis. Then $\theta_r = \theta_s + \delta = \omega_s t + \theta_{so} + \delta$.

The rotor angular speed, $\omega_r = \dfrac{d\theta_r}{dt} = \omega_s + \dfrac{d\delta}{dt}$

The rotor relative angular speed, $\omega_{rr} = \omega_r - \omega_s = \dfrac{d\delta}{dt}$

The rotor angular acceleration, $\alpha_r = \dfrac{d\omega_r}{dt} = \dfrac{d^2\delta}{dt^2}$

Therefore, the acceleration equation becomes

$$M\frac{d^2\delta}{dt^2} = P_m - P_e$$

In all the above relations, the angles are in mechanical radians.

If the powers are in MW, M should be in mega joule seconds per mechanical radian. If the angle δ is in electrical radians, value of M is in MJs/electrical radian. Like wise, if the angle δ is in electrical degrees, value of M is in MJs/electrical degree.

Above is one of the various forms of the ***swing equation***.

We bring in a constant of the machine, called the ***inertia constant***; denoted by H.

It is the stored kinetic energy in MJ in the rotating masses at the synchronous speed per MVA of the machine rating. Clearly, the unit of H is second.

$$H = \frac{\frac{1}{2} J\omega_s^2}{S_{\text{machine}} \times 10^6},$$

where S_{machine} is the machine rating in MVA. It is stressed that ω_s in this relation is in mechanical radians per second.

$$H = \frac{\frac{1}{2} M\omega_s}{S_{\text{machine}}}, \quad \text{where } M \text{ is in } MJ\text{s per mechanical radian}$$

$$M = \frac{2\, S_{\text{machine}}\, H}{\omega_s}.$$

The speed of the rotor in electrical degrees per second is $360f$, f being the frequency of the induced voltage in the armature.

Therefore, M in MJs per electrical degree $= \dfrac{S_{\text{machine}}\, H}{180f}$

The swing equation, therefore, becomes

$$\frac{2\, S_{\text{machine}}\, H}{\omega_s} \frac{d^2\delta}{dt^2} = P_m - P_e;$$

The angle of ω_s, and the angle δ are in mechanical radians; however, because one angle is in the numerator, and the other in the denominator, these can be in any identical unit. Powers are in MW.

The other form of the equation is

$$\frac{S_{\text{machine}}\, H}{180f} \frac{d^2\delta}{dt^2} = P_m - P_e; \quad \text{the angle } \delta \text{ in electrical degrees, powers in MW.}$$

If the powers are expressed in per unit of S_{machine}, then

$$\frac{2H}{\omega_s} \frac{d^2\delta}{dt^2} = P_{m,pu} - P_{e,pu};$$

The angle in ω_s, and the angle δ; both have to be in the same unit.

$$\frac{H}{180f} \frac{d^2\delta}{dt^2} = P_{m,pu} - P_{e,pu}; \quad \delta \text{ in electrical degrees.}$$

These were the various forms of the swing equation.

For the first swing transient stability studies, P_m is treated as constant. The electrical power delivered by the generator during the transient period is given by $\dfrac{E_1'E_2'}{X} \sin\delta$, where E_1'

is the generator voltage behind the transient reactance, E_2' is the voltage of the infinite bus into which the generator feeds power, or is the voltage behind the transient reactance of the synchronous motor if the load is a synchronous motor, X is the transfer reactance between the two voltages, and δ is the angle by which E_1' leads E_2'. X includes the transient reactances of the machines. The two voltages are also assumed to remain constant during the transient period. P_e is, therefore, $P_{max} \sin \delta$, where $P_{max} = E_1' E_2'/X$, a constant.

The angle δ in the swing equation, and the one in the P_e equation may differ by a constant. The $\frac{d}{dt}$, or $\frac{d^2}{dt^2}$ of both the angles are identical. We are only interested in knowing how the change in the machine angle δ varies with time during the transient period and not in the absolute value of δ.

$$\frac{2H}{\omega_s}\frac{d^2\delta}{dt^2} = P_{m,pu} - P_{max,pu} \sin \delta; \text{ or } M'\frac{d^2\delta}{dt^2} = P_{m,pu} - P_{max,pu} \sin \delta, \text{ where } M' = \frac{2H}{\omega_s}.$$

If we solve this non-linear equation for δ as a function of time t and plot the resulting solution, the plot is the ***swing curve*** of the machine. Obviously, in steady state operation, RHS equals zero and δ stays constant.

This equation is generally solved using numerical methods. However, if the aim is only to determine whether the machine will remain stable or not on the occurrence of a disturbance, there is no need to solve the equation; rather we apply the criterion, called the equal area criterion.

43.3 EQUIVALENT MACHINE

When several machines are being considered in a particular problem, their powers and inertia constants are to be expressed on a common MVA base, S_{common}. Else we work in terms of actual values.

$$P_{pu,\,common} = P_{pu,\,machine}\,(S_{machine} / S_{common})$$
$$H_{pu,\,common} = H_{pu,\,machine}\,(S_{machine} / S_{common})$$

It is possible to reduce a number of electrically close machines to a single equivalent machine. Then we have fewer swing equations to solve. The machines are electrically close, if they are not separated by high reactance circuits. The rotors of such machines swing together on the occurrence of a disturbance, *i.e.*, their swinging is in phase with each other. This is called coherency.

For such machines,

$$H_{eqt} = H_1 + H_2 + \ldots + H_n,$$
$$P_{m,\,eqt} = P_{m1} + P_{m2} + \ldots + P_{mn}$$
$$P_{e,\,eqt} = P_{e1} + P_{e2} + \ldots + P_{en}$$
$$\frac{d^2\delta_1}{dt^2} = \frac{d^2\delta_2}{dt^2} = \ldots = \frac{d^2\delta_n}{dt^2}$$

If two machines are separated by a significant reactance, and these do not swing together, we have

$$\frac{2H_{12}}{\omega_s}\frac{d^2\delta_{12}}{dt^2} = P_{m,pu,12} - P_{e,pu,12}$$

$$\delta_{12} = \delta_1 - \delta_2$$

$$H_{12} = H_1H_2/(H_1+H_2)$$

$$P_{m,pu,12} = (P_{m,pu,1}H_2 - P_{m,pu,2}H_1)/(H_1+H_2)$$

$$P_{e,pu,12} = (P_{e,pu,1}H_2 - P_{e,pu,2}H_1)/(H_1+H_2)$$

43.4 SYNCHRONIZING POWER COEFFICIENT

We see that the electrical power output curve is a sine curve. For a particular power output, the angle δ has to be the one which is less than 90º, say δ_o, though the angle $\delta = 180º - \delta_o$ also results in the same power. The former point is the stable point of operation, whereas the latter one is not; as even the slightest disturbance at that point causes instability. This is described by the synchronizing power coefficient, C_{sp}. In the region from $\delta = 0º$ to $\delta = 90º$, this coefficient is positive, meaning that on disturbance, the machine tends to bring the rotor back to synchronous speed; and in the region from $\delta = 90º$ to $\delta = 180º$, the synchronizing power is negative, meaning that on disturbance the rotor angle goes on increasing and the machine falls. The C_{sp} is actually $\frac{dP_e}{d\delta} = P_{max}\cos\delta$. Thus if the machine is heavily loaded and therefore δ is large, the synchronizing power coefficient is small.

The frequency of natural undamped oscillations of the rotor is

$$f_n = \frac{1}{2\pi}\sqrt{\frac{\omega_s C_{sp}}{2H}} \text{ Hz,}$$ and the time period of the oscillations is $T = 1/f_n$

This frequency is of the order of 1 Hz.

43.5 EQUAL AREA CRITERION OF TRANSIENT STABILITY

On the occurrence of a disturbance, the machine switches over to another power angle curve, generally with reduced P_{max}. The mechanical power input being constant, the rotor undergoes acceleration. During this course, δ and P_e increase. If the conditions are such that δ stops increasing any further, *i.e.*, the rotor speed is again synchronous; we conclude that the stability will be maintained. Thus, $\frac{d\delta}{dt}$ becoming zero is taken as the indication that the stability is not lost. When this fact is shown on the power angle curve, we identify two areas; one of which is the accelerating area, and the other is the decelerating area. When the available decelerating area is more or equal to the accelerating area, stability is maintained.

$$\int_{\delta=\delta_o}^{\delta=\delta_x} (P_m - r_1 P_{\max} \sin\delta)\, d\delta \le \int_{\delta=\delta_x}^{\delta=\delta_{\max}} (r_1 P_{\max} \sin\delta - P_m)\, d\delta,$$

where $r_1 P_{\max}$ is the maximum power on the new curve, $P_m = P_{\max} \sin\delta_o = r_1 P_{\max} \sin\delta_x = r_1 P_{\max} \sin\delta_{\max}$. $\delta_{\max} > 90^\circ$.

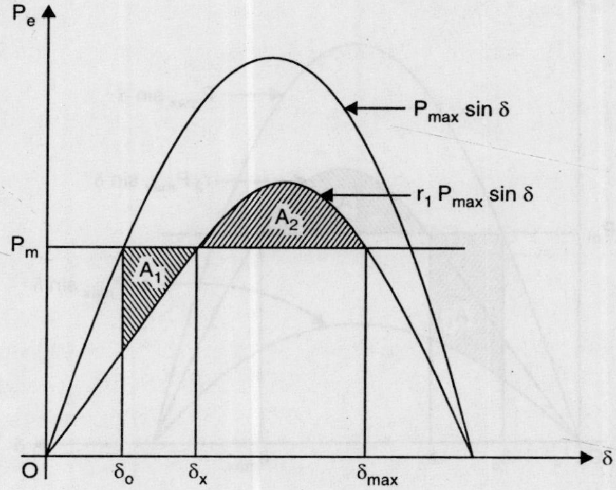

FIGURE 43.1. Accelerating area and decelerating area, the machine acting as a generator.

43.6 FAULT CLEARING AND CRITICAL CLEARING ANGLE

Depending on the nature of the disturbance, r_1 may be very small, it can even be zero (a bolted three-phase fault). Usually, corrective action occurs immediately, the fault is cleared at angle $\delta = \delta_c$, and the operating characteristic shifts to yet another power angle curve, having maximum power $r_2 P_{\max}$. $r_2 > r_1$; and both are < 1. We thus have three P – δ curves.

The accelerating area, $$A_{acc} = \int_{\delta=\delta_o}^{\delta=\delta_c} (P_m - r_1 P_{\max} \sin\delta)\, d\delta$$

The available decelerating area, $$A_{dec} = \int_{\delta=\delta_c}^{\delta=\delta_{\max}} (r_2 P_{\max} \sin\delta - P_m)\, d\delta$$

For stability, again, A_{dec} should be greater than or equal to $A_{acc.}$

By equating the two areas, one can derive an expression for δ_{cr}, so that stability is just maintained. If the fault clearing is delayed beyond δ_{cr}, the machine will lose stability.

$$\cos\delta_{cr} = \frac{(P_m / P_{\max})(\delta_{\max} - \delta_o) + r_2 \cos\delta_{\max} - r_1 \cos\delta_o}{r_2 - r_1}$$

$$= \frac{\sin\delta_o\,(\delta_{max} - \delta_o) + r_2\cos\delta_{max} - r_1\cos\delta_o}{r_2 - r_1}$$

The angles in $(\delta_{max} - \delta_o)$ are to be in radians.

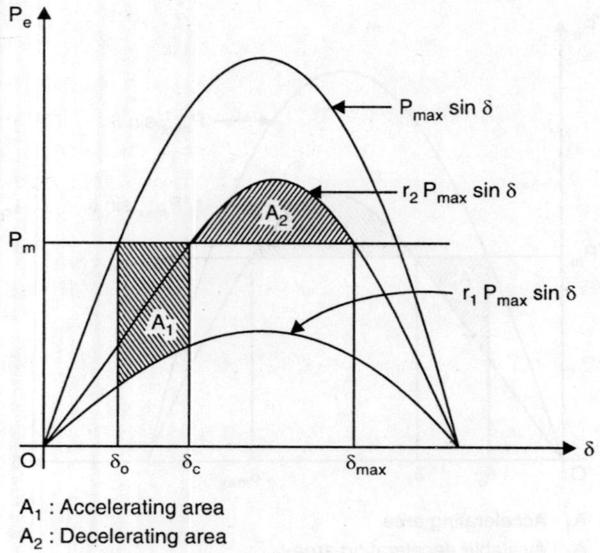

FIGURE 43.2. Accelerating area and decelerating area. Fault is cleared at δ_c.

We note that $P_m = P_{max} \sin\delta_o = r_2 P_{max} \sin\delta_{max}$. It has been assumed that $P_m < r_2 P_{max}$. δ_{max} is greater than $\pi/2$.

$$\delta_{max} = \pi - \sin^{-1}\frac{P_m}{r_2 P_{max}} = \pi - \sin^{-1}\frac{\sin\delta_o}{r_2}$$

As a special case, let $r_1 = 0$, and $r_2 = 1$. Then

$$\cos\delta_{cr} = (\pi - 2\,\delta_o)\sin\delta_o - \cos\delta_o.$$

For this special case, the accelerating power is constant; and therefore, the relative speed increases linearly till fault clearing occurs.

If the fault is sustained, then $r_2 = r_1 = r$. There is nothing like fault clearing in this case. The critical relation between r and δ_o can be found by applying the equal-area criterion and is given by the following equation.

$$\sin\delta_o = r\,\frac{\cos\delta_0 - \cos\delta_{max}}{\delta_{max} - \delta_0}, \text{ or } \quad r = \frac{(\delta_{max} - \delta_0)\sin\delta_0}{\cos\delta_0 - \cos\delta_{max}}, \quad \text{with } \delta_{max} = \pi - \sin^{-1}\frac{\sin\delta_0}{r}$$

In the case of a ***synchronous motor***, the stability criterion becomes that the available accelerating area should be more or equal to the decelerating area.

The ways to ***enhance the transient stability*** can be arrived at by noting that high value of P_{max} is advantageous. Lowering P_m (if we could) after the fault has occurred, keeping r_1 and r_2 high can also be implemented.

43.7 SOLVING SWING EQUATION

We take the swing equation as $M \dfrac{d^2\delta}{dt^2} = P_a$, where P_a is the accelerating power.

Usually, digital solutions of the swing equation are obtained. For hand calculations, we can apply the step-by-step method. In this method, the period from the instant the fault occurred, upto about one second is considered. The period is divided into a large number of short intervals ($\Delta t = 0.05$ s). During an interval, the angular velocity is assumed to remain constant at a value found at the middle of the interval. And from the mid point of an interval to the mid point of the following interval, the accelerating power is assumed to remain constant at a value computed at the start of that following interval.

The change in the rotor angle, $\Delta\delta_j$, undergone during *j*th interval is given by the product of the rotor relative velocity, $\omega_{rr'j}$ during the *j*th interval and Δt.

At the beginning of an interval *j*, the change in the angular velocity is $P_{a,j}\,\Delta t/M$, $P_{a,j}$ is the accelerating power at the beginning of *j*th interval. $P_{a,j}$ can be computed as the angle δ at the end of $(j-1)$th interval, *i.e.*, at the beginning of the *j*th interval would have been determined.

$$\Delta\delta_j = \Delta t\ \omega_{rr'j}$$

$$\Delta\delta_{j+1} = \Delta t\ \omega_{rr'j+1}$$

$$\Delta\delta_{j+1} = \Delta\delta_j + \Delta t\,(\omega_{rr'j+1} - \omega_{rr'j}) = \Delta\delta_j + \Delta t\, P_{a,j+1}\,\Delta t/M = \Delta\delta_j + \frac{P_{a,j+1}}{M}(\Delta t)^2$$

$$\delta_{j+1} = \delta_j + \Delta\delta_{j+1}.\ \delta_j \text{ is the rotor angle at the end of } j\text{th interval.}$$

If, at the beginning of an interval, a discontinuity in the accelerating power occurs we take the average of the two accelerating powers.

In the above relations, $\Delta\delta$ is in electrical degrees, Δt is in seconds, P_a is in MW, and M is in MJs per electrical degree.

We could have used the equation $\dfrac{2H}{\omega_s}\dfrac{d^2\delta}{dt^2} = P_{a,pu}$

The relation would be $\Delta\delta_{j+1} = \Delta\delta_j + \dfrac{\omega_s}{2H} P_{a,pu,j+1}(\Delta t)^2$; where ω_s is in electrical degrees per second. We can also find the relation

$$\Delta\delta_{j+1} = \Delta\delta_j + \frac{180f}{H} P_{a,pu,j+1}(\Delta t)^2 = \Delta\delta_j + kP_{a,pu,j+1},\ \text{where } k = \frac{180f}{H}(\Delta t)^2.$$

Chapter 44

INCIDENCE AND IMMITTANCE MATRICES

44.1 TOPOLOGICAL CONSIDERATIONS

The structure obtained after we replace each element of a network by a line segment, is the graph of the network. In the graph, the nature of the network elements is lost, but the connectivity is preserved. It has the same number of elements and nodes as the original network has. By assigning directions to the various elements, we get the oriented or the directed graph. Let the graph have e number of elements or edges, and $(n + 1)$ number of nodes.

A tree of a graph is one of its subgraphs which has all the $(n + 1)$ nodes, and some such edges that there exists a path from every node to every other node, and there are no closed paths. A graph may have many trees. The elements of the tree are called the twigs or the branches. There are $b = (n + 1 - 1) = n$ twigs. The graph less the tree, called the cotree, is the complement of the tree. The cotree has $(e - b)$ edges, *i.e.*, $l = (e - n)$ edges. An edge of a cotree is called a link. A tree and its cotree taken together constitute the original graph.

44.2 INCIDENCE MATRICES

There are several incidence matrices of the graph which convey the incidence of its edges on the nodes, loops, cutsets etc. While forming these matrices, we number the tree branches first, and then we number the other edges of the graph. Further, in the matrices, the edges are taken downwards; and the nodes, loops, cutsets etc., rightwards. In all these matrices entries are 0, 1, or – 1.

- ***Element-node incidence matrix*** $\hat{A}$

It is obtained from the graph. It is an $e \times (n + 1)$ matrix.

A_{ij} is 1 if edge i is incident to node j and directed away from it, and is – 1 if the edge is directed towards it. If the edge i is not incident to node j, the entry is zero.

As every edge is incident to two nodes, there are exactly two non-zero entries in any row; 1, – 1, and as such the sum of all the entries in a row is zero.

• ***Bus incidence matrix A***

It is obtained from $\hat{A}$ itself simply by deleting a column corresponding to the reference node or the reference bus. It is an $e \times n$ matrix. A is usually partitioned as $\begin{bmatrix} A_b \\ A_l \end{bmatrix}$; the suffix b is for the branches and l for links. A_b is the $b \times n$ submatrix and A_l is $l \times n$ submatrix.

• ***Branch path incidence matrix K***

It is obtained from the tree. It is a $b \times n$ matrix.

K_{ij} is 1 if in the tree, edge i is in the path from node j to the reference node and is in the direction of traverse, and is – 1 if the edge is in the direction opposite to the direction of traverse. If the edge i is not in the said path, the corresponding entry is zero.

• ***Basic loop or circuit incidence matrix C***

A basic or the fundamental loop is obtained by placing a link across its end nodes on the tree. A basic loop thus has exactly one link. The loop is assigned the same orientation as that of defining link. First loop is obtained from first link; second loop is obtained from second link and so on. There would be as many basic loops as the number of links, *i.e.*, l. C is $e \times l$ matrix.

C_{ij} is 1 if edge i is in the loop j and is oriented in the orientation of the loop, and is – 1 if the edge is oriented opposite to the orientation of this loop. If the edge i is not in the said loop, the corresponding entry is zero.

C is usually partitioned as $\begin{bmatrix} C_b \\ U_l \end{bmatrix}$, where C_b is the $b \times l$ submatrix, and U_l is $l \times l$ unity or identity submatrix.

• ***Basic cutset incidence matrix B***

A basic cutset of a graph is that minimal set of its edges which when removed from it leaves it cut or separated in two parts. A basic cutset has exactly one twig. Thus, every twig defines a cutset. A cutset is oriented in the same direction as the direction of the defining twig. First cutset is obtained from first twig; second cutset is obtained from second twig and so on. There would be as many cutsets as the number of twigs, *i.e.*, b. B is an $e \times b$ or $e \times n$ matrix.

B_{ij} is 1 if edge i is in the cutset j and is oriented in the orientation of the cutset, and is – 1 if the edge is oriented opposite to the orientation of this cutset. If the edge i is not in the said cutset, the corresponding entry is zero. B is usually partitioned as $\begin{bmatrix} U_b \\ B_l \end{bmatrix}$, where U_b is $b \times b$ unity or identity submatrix, and B_l is $l \times b$ submatrix.

44.3 SOME RELATIONS

Let $\bar{i}$ and $\bar{v}$ the element currents and the element voltages vectors respectively. These can be expressed in the partitioned forms.

$$\bar{i} = \begin{bmatrix} \bar{i}_b \\ \bar{i}_l \end{bmatrix} \quad \bar{v} = \begin{bmatrix} \bar{v}_b \\ \bar{v}_l \end{bmatrix}$$

The following relations can be developed and verified.

$A^t\,\bar{i} = 0,\qquad B^t\,\bar{i} = 0\qquad$ KCL

$\bar{i}_b = -\,[A_b{}^t]^{-1}A_l{}^t\,\bar{i}_l,\qquad \bar{i}_b = -\,B_l{}^t\,\bar{i}_l\qquad$ It shows that $B_l = A_l A_b{}^{-1}$

Also, $A_b\,K^t = U,\qquad$ therefore, $\qquad B_l = A_l\,K^t$

$C^t\bar{v} = 0,\qquad$ KVL

44.4 PRIMITIVE NETWORK

A primitive network is, in fact, not a network in the real sense of the word, but it is simply the set of unconnected elements of the network. The collective elements' $v - i$ relations are depicted through the so called ***primitive network impedance or admittance matrix***.

The network may be a dc, 1-phase ac, or per phase equivalent of a balanced 3-phase ac network. The parameters and the variables, in general, are therefore complex quantities. As such, complex algebra rules are to be applied. Further, these are all steady state quantities.

The sources associated with an element pq can all be reduced to either a voltage source e_{pq}, or a current source j_{pq}. We can then have either the impedance form or the admittance form of the element $v - i$ relation. Marking the polarities of the sources appropriately, we have $v - i$ relation for the element pq as

$e_{pq} + v_{pq} = z_{pq,\,pq}\, i_{pq}$, where $z_{pq,\,pq}$ is the self impedance of the element pq, or

$j_{pq} + i_{pq} = y_{pq,\,pq}\, v_{pq}$, where $y_{pq,pq}$ is the self admittance of the element pq.

It is seen that $\quad j_{pq} = -\,e_{pq}\, y_{pq,\,pq}$, or $\; e_{pq} = -\,j_{pq}\, z_{pq,\,pq}$

For two elements pq and rs, we may express the v-i relation in the impedance matrix form as

$$\begin{bmatrix} e_{pq} \\ e_{rs} \end{bmatrix} + \begin{bmatrix} v_{pq} \\ v_{rs} \end{bmatrix} = \begin{bmatrix} z_{pq,pq} & z_{pq,rs} \\ z_{rs,pq} & z_{rs,rs} \end{bmatrix} \begin{bmatrix} i_{pq} \\ i_{rs} \end{bmatrix}.$$

The admittance form can be similarly easily obtained.

For a set of n unconnected elements, we may write in the compact matrix form as

$$\bar{e} + \bar{v} = [z]\,\bar{i} \quad \text{or} \quad \bar{j} + \bar{i} = [y]\bar{v}$$

$\bar{e}$, $\bar{v}$, $\bar{j}$ and $\bar{i}$ are $n \times 1$ column vectors, and $[z]$ and $[y]$ are $n \times n$ primitive network impedance and admittance matrices respectively.

If there is no mutual coupling between a pair of elements pq and rs, then $z_{pq,rs} = 0, y_{pq,rs} = 0$.

If there are no mutual couplings between any pair of elements of the complete set of the elements, [z] and [y] are diagonal matrices.

44.5 NETWORK MATRICES

Given a network and the excitation, we are required to find the response. The network is described by its associated admittance or impedance matrices; and the excitation is a vector of injected bus currents, the bus voltages, or loop voltages, etc.

As already noted, an e-edge, (n+1)-node network has n fundamental or basic cutsets and l = (e-n) basic loops. In the bus frame of reference, we have bus admittance matrix and the bus impedance matrix. The network equation is $\bar{I}_{BUS} = Y_{BUS}\ \bar{E}_{BUS}$ or $\bar{E}_{BUS} = Z_{BUS}\ \bar{I}_{BUS}$. $\bar{E}_{BUS}$ is the n-element vector of n bus voltages, voltages being reckoned w.r.t. the reference node, $\bar{I}_{BUS}$ is the n-element vector of n injected currents or the source currents at the buses, Z_{BUS} is the $n \times n$ bus impedance matrix, and Y_{Bus} is $n \times n$ bus admittance matrix. The $\bar{I}_{BUS} = Y_{BUS}\bar{E}_{BUS}$ is the set of equations obtained on applying the nodal method of circuit analysis. $\bar{I}_{BUS}$ is given, and we find $\bar{E}_{BUS}$. If there are no mutual couplings, one can find Y_{BUS} by inspection. Y_{jj} is the sum of the admittances connected at bus j, and Y_{ij} is the negative of the admittance of the element connected between buses i and j. If no element is connected between buses i and j, Y_{ij} is zero. In general, we find Y_{BUS} from the relation $Y_{BUS} = A^t[y]A$. $\bar{E}_{BUS} = Z_{BUS}\bar{I}_{BUS}$ is the inverse set of $\bar{I}_{BUS} = Y_{BUS}\bar{E}_{BUS}$. Z_{BUS} is Y_{BUS}^{-1}. However, we can find Z_{BUS} directly also by applying the so called building algorithm.

The elements of Y_{BUS} are short circuit bus admittances and it is a sparse (thinly populated) matrix, while the elements of Z_{BUS} are open circuit bus impedances and it is a full matrix.

Equations as above can be written in the loop-, and the branch frame of reference also; for instance we have $\bar{E}_{LOOP} = Z_{LOOP}\bar{I}_{LOOP}$. $\bar{E}_{LOOP}$ and $\bar{I}_{LOOP}$ are l-element column vectors of loop source voltages and fictitious loop currents respectively, and Z_{LOOP} is $l \times l$ loop impedance matrix. This set of equations follows from the loop method of circuit analysis. Again, if there are no mutual couplings, Z_{jj} is the sum of all the impedances in loop j, and if the loop currents I_i and I_j flow in the same direction in the element common to the two loops, then Z_{ij} is the impedance of this common element, else negative of the common element impedance. If there is no common element between loop i and loop j, Z_{ij} is zero. In general, we find Z_{LOOP} from the relation $Z_{LOOP} = C^t[z]C$. $\bar{I}_{LOOP} = Y_{LOOP}\bar{E}_{LOOP}$ is the inverse set of $\bar{E}_{LOOP} = Z_{LOOP}\ \bar{I}_{LOOP}$. Y_{LOOP} is Z_{LOOP}^{-1}. In the branch frame of reference, we have $Y_{BR} = B^t[y]B$. $Z_{BR} = Y_{BR}^{-1}$.

By applying KCL and KVL to the primitive network, we can find the following simple relations also: $\bar{I}_{BUS} = A^t\ \bar{j}$, $\bar{E}_{LOOP} = C^t\bar{e}$, $\bar{I}_{BR} = B^t\ \bar{j}$, $\bar{v} = A\bar{E}_{BUS}$, $\bar{e}$ and $\bar{v} = B\bar{E}_{BR}$.

Some other interrelations amongst the various admittance and impedance matrices are as follows :

$$Y_{BR} = KY_{\text{BUS}}K^t,\ Z_{BR} = A_b Z_{\text{BUS}} A_b^{\ t},\ Y_{\text{BUS}} = A_b^{\ t} Y_{\text{BR}} A_b, \text{ and } Z_{\text{BUS}} = K^t Z_{BR} K.$$

44.6 Z_{BUS} BUILDING RELATIONS

Bus impedance matrix Z_{BUS} can be formed by building the given network from zero level. We usually begin with the reference bus. We go on adding elements one by one till the given network is obtained. In the process, we are simultaneously updating the current Z_{BUS}. At a particular stage, we may be adding an element between an already existing bus and a new bus, or an element across two already existing buses. Former is addition of a branch, and the latter, addition of a link. On adding a branch, a new bus is created and the Z_{BUS} dimension increases by one as a new row and a new column corresponding to the new bus are added. On adding a link, no new bus is created; the elements of Z_{BUS}, however, get changed.

The elements of the current network obtained may have mutual couplings or there may be no coupling.

Given below are the relations which can be used in building Z_{BUS}. At the stage being considered, m buses have appeared in the network. The new element appears across buses p and q. p belongs to the already included buses in the partial m-bus network. If the new element is a branch, q is a new bus; and if the new element is a link, q also belongs to the already included buses.

The formulas given below are basically derived from the primitive network equation for the partial network and the new element pq, and the relations which exist between the bus voltages and an injected current at bus i.

$\bar{j} + \bar{i} = [y]\bar{v}$. As $\bar{j}$ does not exist, we have $\bar{i} = [y]\bar{v}$. $\begin{bmatrix} i_{pq} \\ i_{\rho\sigma} \end{bmatrix} = \begin{bmatrix} y_{pq,pq} & \bar{y}_{pq,\rho\sigma} \\ \bar{y}_{\rho\sigma,pq} & y_{\rho\sigma,\rho\sigma} \end{bmatrix} \begin{bmatrix} v_{pq} \\ \bar{v}_{\rho\sigma} \end{bmatrix}$

$i_{pq} = y_{pq,pq}\, v_{pq} + \bar{y}_{pq,\rho\sigma}\, \bar{v}_{\rho\sigma}$. As i_{pq} is zero, we have $E_p - E_q = -\dfrac{\bar{y}_{pq,\rho\sigma}\,(\bar{E}_\rho - \bar{E}_\sigma)}{y_{pq,pq}}$.

On injecting a current I_i at bus i, i varying from 1 to m, the voltages acquired by the various buses are related to Z_{BUS} elements as $E_k = Z_{ki}\, I_i$; $k = 1, 2, \ldots, m, q$. For all these cases i_{pq} is zero. When we inject I_q at bus q, then the relation is $E_k = Z_{kq}\, I_q$, and i_{pq} for this lone case is $-I_q$. Incorporating these key points, we get the relations for modifying Z_{BUS} when a branch is added. For developing the formulas when a link is added, temporarily a bus l is introduced between buses p and q in a manner that the branch addition formulas can be employed. The bus l is, in the end, eliminated.

(A) ***Addition of a branch***:

(*a*) No mutual couplings:

(*i*) p is the reference bus:

$$Z_{qi} = 0, \qquad i = 1, 2, \ldots, m;\ i \neq q$$

$$Z_{qq} = z_{pq,pq}$$

(*ii*) *p* is not the reference bus:

$$Z_{qi} = Z_{pi}, \qquad i = 1, 2, \ldots, m; i \neq q$$

$$Z_{qq} = Z_{pq} + z_{pq,\,pq}$$

(*b*) Mutual couplings present :

(*i*) *p* is the reference bus :

$$Z_{qi} = \frac{\bar{y}_{pq,\,\rho\sigma}\,(\bar{Z}_{\rho i} - \bar{Z}_{\sigma i})}{y_{pq,\,pq}}, \qquad i = 1, 2, \ldots, m; i \neq q$$

$$Z_{qq} = \frac{1 + \bar{y}_{pq,\,\rho\sigma}\,(\bar{Z}_{\rho q} - \bar{Z}_{\sigma q})}{y_{pq,\,pq}}$$

(*ii*) *p* is not the reference bus :

$$Z_{qi} = Z_{pi} + \frac{\bar{y}_{pq,\,\rho\sigma}\,(\bar{Z}_{\rho i} - \bar{Z}_{\sigma i})}{y_{pq,\,pq}}, \qquad i = 1, 2, \ldots, m; i \neq q$$

$$Z_{qq} = Z_{pq} + \frac{1 + \bar{y}_{pq,\,\rho\sigma}\,(\bar{Z}_{\rho q} - \bar{Z}_{\sigma q})}{y_{pq,\,pq}}$$

The index $\rho\sigma$ varies so as to span all the elements already present in the *m*-bus partial network. $\bar{y}_{pq,\,\rho\sigma}$ is the row vector of the primitive network admittance matrix [*y*] of all the elements so far assembled including the incoming element *pq*. [*y*] is found by taking inverse of [*z*]. Thus, when we bring in a new element, we find [*z*] and [*y*] afresh. These comments apply to (*B*) also.

(*B*) ***Addition of a link***:

Here, temporarily, an additional new bus *l* (it is not any of the buses of the network) is introduced between buses *p* and *q*.

(*a*) No mutual couplings

(*i*) *p* is the reference bus :

$$Z_{li} = -Z_{qi}, \qquad i = 1, 2, \ldots, m; i \neq l$$

$$Z_{ll} = -Z_{ql} + z_{pq,\,pq}$$

(*ii*) *p* is not the reference bus :

$$Z_{li} = Z_{pi} - Z_{qi}, \qquad i = 1, 2, \ldots, m; i \neq l$$

$$Z_{ll} = Z_{pl} - Z_{ql} + z_{pq,\,pq}$$

(*b*) Mutual couplings present

(*i*) *p* is the reference bus :

$$Z_{li} = -Z_{qi} + \frac{\bar{y}_{pq,\,\rho\sigma}\,(\bar{Z}_{\rho i} - \bar{Z}_{\sigma i})}{y_{pq,\,pq}}, \qquad i = 1, 2, \ldots, m; i \neq l$$

$$Z_{ll} = -Z_{ql} + \frac{1 + \bar{y}_{pq,\,\rho\sigma}\,(\bar{Z}_{\rho l} - \bar{Z}_{\sigma l})}{y_{pq,\,pq}}$$

(ii) p is not the reference bus

$$Z_{li} = Z_{pi} - Z_{qi} + \frac{\bar{y}_{pq,\rho\sigma}(\bar{Z}_{\rho i} - \bar{Z}_{\sigma i})}{y_{pq,pq}}, \quad i = 1, 2, \ldots, m;\ i \neq l$$

$$Z_{ll} = Z_{pl} - Z_{ql} + \frac{1 + \bar{y}_{pq,\rho\sigma}(\bar{Z}_{\rho l} - \bar{Z}_{\sigma l})}{y_{pq,pq}}$$

The temporarily introduced bus l is eliminated to obtain the new Z_{BUS} which includes the effect of the link pq, using the relation

$$Z_{ij,\,new} = Z_{ij,\,original} - \frac{Z_{il}Z_{lj}}{Z_{ll}}, \quad i, j = 1, 2, \ldots, m$$

44.7 MODIFYING Z_{BUS} FOR CHANGE OF REFERENCE BUS

Let the network have $(n + 1)$ buses. Then the given Z_{BUS} is an $n \times n$ matrix with bus r as the reference bus. We wish to find the new Z_{BUS} with bus r brought in and bus k as the reference bus.

We see that

$$\bar{E}^t_{BUS} = [E_1 \ \ E_2 \ \ . \ \ . \ \ . \ \ E_k \ \ . \ \ . \ \ . \ \ E_n]$$

and $$\bar{E}^t_{BUS,\,new} = [E_1 \ \ E_2 \ \ . \ \ . \ \ . \ \ E_r \ \ . \ \ . \ \ . \ \ E_n]$$

Similarly,

$$\bar{I}^t_{BUS} = [I_1 \ \ I_2 \ \ . \ \ . \ \ . \ \ I_k \ \ . \ \ . \ \ . \ \ I_n]$$

and $$\bar{I}^t_{BUS,\,new} = [I_1 \ \ I_2 \ \ . \ \ . \ \ . \ \ I_r \ \ . \ \ . \ \ . \ \ I_n]$$

By applying KVL to the bus voltages, one can find

$$\bar{E}^{New}_{BUS} = \begin{bmatrix} E_1 \\ E_2 \\ \cdot \\ E_r \\ \cdot \\ \cdot \\ E_n \end{bmatrix} = \begin{bmatrix} 1 & & -1 & & \\ & 1 & -1 & & \\ & & \cdot & & \\ & & \cdot & & \\ & & \cdot & 1 & \\ & & -1 & & 1 \end{bmatrix} \begin{bmatrix} E_1 \\ E_2 \\ \cdot \\ E_k \\ \cdot \\ \cdot \\ E_n \end{bmatrix} = T\,\bar{E}^{Given}_{BUS}$$

By noting that in the new reference network $I_k = -(I_1 + I_2 + \ldots + I_r + \ldots I_n)$, we have

$$\bar{I}^{Given}_{BUS} = \begin{bmatrix} I_1 \\ I_2 \\ \cdot \\ I_k \\ \cdot \\ \cdot \\ I_n \end{bmatrix} = \begin{bmatrix} 1 & & & & & \\ & 1 & & & & \\ & & & & & \\ -1 & -1 & \cdot & \cdot & \cdot & -1 \\ & & & & & \\ & & & & 1 & \\ & & & & & 1 \end{bmatrix} \begin{bmatrix} I_1 \\ I_2 \\ \cdot \\ I_r \\ \cdot \\ \cdot \\ I_n \end{bmatrix} = T^t\,\bar{I}^{New}_{BUS}$$

We then have

$$\bar{E}_{BUS}^{New} = T\,\bar{E}_{BUS}^{Given} = T\,Z_{BUS}^{Given}\,\bar{I}_{BUS}^{Given} = T\,Z_{BUS}^{Given}\,T^t\,\bar{I}_{BUS}^{New} = Z_{BUS}^{New}\,\bar{I}_{BUS}^{New}$$

Thus, $$Z_{BUS}^{New} = T Z_{BUS}^{Given} T^t$$

This new Z_{BUS} can be obtained by following the procedure given below.

1. Subtract the elements of the row k (the new ref bus) from each of the other rows and negate the sign of the elements of the row k in the given Z_{BUS}.
2. Subtract the elements of column k from each of the other columns and negate the sign of the elements of the column k in the new matrix obtained in step 1.

This is the new Z_{BUS}, with bus k as the new reference bus and the bus r (the original reference bus) occupying the position of bus k in the original matrix.

Chapter 45 THREE PHASE SHORT CIRCUIT CALCULATIONS

45.1 THREE-PHASE ELEMENTS

For a three-phase element pq, the number of quantities increase by a factor of three. Denoting the phases by a, b, and c, the v-i relations in the impedance and the admittance form for a three-phase element are as under:

$$\begin{bmatrix} e_{pq}^{a} \\ e_{pq}^{b} \\ e_{pq}^{c} \end{bmatrix} + \begin{bmatrix} v_{pq}^{a} \\ v_{pq}^{b} \\ v_{pq}^{c} \end{bmatrix} = \begin{bmatrix} z_{pq}^{aa} & z_{pq}^{ab} & z_{pq}^{ac} \\ z_{pq}^{ba} & z_{pq}^{bb} & z_{pq}^{bc} \\ z_{pq}^{ca} & z_{pq}^{cb} & z_{pq}^{cc} \end{bmatrix} \begin{bmatrix} i_{pq}^{a} \\ i_{pq}^{b} \\ i_{pq}^{c} \end{bmatrix}, \quad \begin{bmatrix} j_{pq}^{a} \\ j_{pq}^{b} \\ j_{pq}^{c} \end{bmatrix} + \begin{bmatrix} i_{pq}^{a} \\ i_{pq}^{b} \\ i_{pq}^{c} \end{bmatrix} = \begin{bmatrix} y_{pq}^{aa} & y_{pq}^{ab} & y_{pq}^{ac} \\ y_{pq}^{ba} & y_{pq}^{bb} & y_{pq}^{bc} \\ y_{pq}^{ca} & y_{pq}^{cb} & y_{pq}^{cc} \end{bmatrix} \begin{bmatrix} v_{pq}^{a} \\ v_{pq}^{b} \\ v_{pq}^{c} \end{bmatrix}$$

In compact form these are written as

$$e_{pq}^{abc} + v_{pq}^{abc} = z_{pq}^{abc}\, i_{pq}^{abc} \qquad j_{pq}^{abc} + i_{pq}^{abc} = y_{pq}^{abc}\, v_{pq}^{abc}$$

As in the case of single-phase elements, here too, we have

$$e_{pq}^{abc} = -z_{pq}^{abc}\, j_{pq}^{abc} \qquad j_{pq}^{abc} = -y_{pq}^{abc}\, e_{pq}^{abc}$$

For a three-phase network, the primitive network equations are written as

$$\bar{e}^{abc} + \bar{v}^{abc} = [z^{abc}]\, \bar{i}^{abc} \qquad \bar{j}^{abc} + \bar{i}^{abc} = [\bar{y}^{abc}]\, \bar{v}^{abc}$$

For the element pq, if there are no mutual couplings, z_{pq}^{abc} and y_{pq}^{abc} are simply diagonal matrices.

If the element pq is a ***balanced stationary element***, z_{pq}^{abc} and y_{pq}^{abc} assume symmetrical form.

$$\begin{bmatrix} z_{pq}^{s} & z_{pq}^{m} & z_{pq}^{m} \\ z_{pq}^{m} & z_{pq}^{s} & z_{pq}^{m} \\ z_{pq}^{m} & z_{pq}^{m} & z_{pq}^{s} \end{bmatrix} \qquad \begin{bmatrix} y_{pq}^{s} & y_{pq}^{m} & y_{pq}^{m} \\ y_{pq}^{m} & y_{pq}^{s} & y_{pq}^{m} \\ y_{pq}^{m} & y_{pq}^{m} & y_{pq}^{s} \end{bmatrix}$$

If the element pq is a ***balanced rotating element***, z_{pq}^{abc} and y_{pq}^{abc} assume the following form.

$$\begin{bmatrix} z_{pq}^{s} & z_{pq}^{m1} & z_{pq}^{m2} \\ z_{pq}^{m2} & z_{pq}^{s} & z_{pq}^{m1} \\ z_{pq}^{m1} & z_{pq}^{m2} & z_{pq}^{s} \end{bmatrix} \qquad \begin{bmatrix} y_{pq}^{s} & y_{pq}^{m1} & y_{pq}^{m2} \\ y_{pq}^{m2} & y_{pq}^{s} & y_{pq}^{m1} \\ y_{pq}^{m1} & y_{pq}^{m2} & y_{pq}^{s} \end{bmatrix}$$

For ***balanced excitation***, the balanced element or the balanced network can be analyzed on per phase basis. We can find that for the element pq

$$e_{pq}^{a} + v_{pq}^{a} = (z_{pq}^{s} + a^{2} z_{pq}^{m1} + a z_{pq}^{m2})\, i_{pq}^{a} \qquad j_{pq}^{a} + i_{pq}^{a} = (y_{pq}^{s} + a^{2} y_{pq}^{m1} + a y_{pq}^{m2})\, v_{pq}^{a}$$

where $a = 1\angle 120° = -0.5 + j\,0.867 \quad a^2 = 1\angle 240° = -0.5 - j0.867$

We see that $a^* = a^2$ and $a^{2*} = a$.

For a stationary element, the equations get simplified to

$$e_{pq}^{a} + v_{pq}^{a} = (z_{pq}^{s} - z_{pq}^{m})\, i_{pq}^{a} \qquad j_{pq}^{a} + i_{pq}^{a} = (y_{pq}^{s} - y_{pq}^{m})\, v_{pq}^{a}$$

The values of the variables for the phases b and c can be found by using relations such as

$$v_{pq}^{abc} = \begin{bmatrix} 1 \\ a^2 \\ a \end{bmatrix} v_{pq}^{a}.$$

45.2 TRANSFORMATION

If the excitation is unbalanced, the response too would be unbalanced. To ease the analysis, we resort to some kind of transformation. With suitable transformation, the system is analyzed on per transformed variable basis. The transformation should be power invariant. If the transformation relation is expressed as

$v_{pq}^{abc} = T v_{pq}^{ijk}$, where T is the power invariant transformation matrix, we find that T is a unitary matrix, *i.e.*, $T^{-1} = T^{*t}$. This is a convenient property. In the new $i\ j\ k$ variables domain, we can find that $z_{pq}^{ijk} = T^{-1}\, z_{pq}^{abc}\, T = T^{*t}\, z_{pq}^{abc}\, T$.

A commonly used transformation matrix is the ***symmetrical components transformation matrix T_s***.

$$T_s = \frac{1}{\sqrt{3}}\begin{bmatrix} 1 & 1 & 1 \\ 1 & a^2 & a \\ 1 & a & a^2 \end{bmatrix};\ T_s^{t} = T_s;\ T_s^{*t} = \frac{1}{\sqrt{3}}\begin{bmatrix} 1 & 1 & 1 \\ 1 & a & a^2 \\ 1 & a^2 & a \end{bmatrix};\ T_s^{-1} = \frac{1}{\sqrt{3}}\begin{bmatrix} 1 & 1 & 1 \\ 1 & a & a^2 \\ 1 & a^2 & a \end{bmatrix} = T_s^{*t}$$

The transformed components $i\ j\ k$ are respectively 0 1 2. The transformed impedance of the element pq is given by

$$z_{pq}^{012} = T_s^{*t} z_{pq}^{abc} T_s = \begin{bmatrix} z_{pq}^s + z_{pq}^{m1} + z_{pq}^{m2} & 0 & 0 \\ 0 & z_{pq}^s + a^2 z_{pq}^{m1} + a z_{pq}^{m2} & 0 \\ 0 & 0 & z_{pq}^s + a z_{pq}^{m1} + a^2 z_{pq}^{m2} \end{bmatrix}$$

We observe that z_{pq}^{abc} gets ***diagonalized.***

$$z_{pq}^0 = z_{pq}^s + z_{pq}^{m1} + z_{pq}^{m2} \quad z_{pq}^1 = z_{pq}^s + a^2 z_{pq}^{m1} + a z_{pq}^{m2} \quad z_{pq}^2 = z_{pq}^s + a z_{pq}^{m1} + a^2 z_{pq}^{m2}$$

For a ***stationary element***, the above sequence impedances get simplified to

$$z_{pq}^0 = z_{pq}^s + 2z_{pq}^m \quad z_{pq}^1 = z_{pq}^s - z_{pq}^m \quad z_{pq}^2 = z_{pq}^s - z_{pq}^m$$

45.3 GENERAL SHORT CIRCUIT EQUATIONS

In short circuit calculations, the resistances and the shunt elements are neglected, the machines are represented by their internal voltages behind the appropriate reactance, and all transformers are assumed to be at their nominal taps.

The bus impedance matrix Z_{BUS} is formed which takes into effect the machine reactances and the transmission network reactances. The fault occurs at bus p through a fault element which has equal impedance z_F in its each phase and an impedance z_g between the star point and the ground for faults involving ground. It is represented by its admittance matrix Z_F^{abc}, or Y_F^{abc}. Z_F^{abc} can be obtained by applying $\bar{E} = [Z]\bar{I}$ to the fault element. Current sources are applied successively to the three phase terminals *a, b, c* of the element. A reference or ground point is required to apply the current source. Similarly, Y_F^{abc} is obtained by applying $\bar{I} = [Y]\bar{E}$. For faults which do not involve ground, we use Y_F^{abc}. $I_{p,F}^{abc}$ is the fault current at the faulted bus. No currents flow into other buses. For the fault condition, one can write

$\bar{E}_{\text{BUS},F}^{abc} = \bar{E}_{\text{BUS},0}^{abc} - Z_{\text{BUS}}^{abc} \bar{I}_{\text{BUS},F}^{abc}$. Solving this equation, we get

$$I_{p,F}^{abc} = (Z_F^{abc} + Z_{pp}^{abc})^{-1} E_{p,0}^{abc}$$

$$E_{p,F}^{abc} = Z_F^{abc} (Z_F^{abc} + Z_{pp}^{abc})^{-1} E_{p,0}^{abc}$$

$$E_{i,F}^{abc} = E_{i,0}^{abc} - Z_{ip}^{abc} (Z_F^{abc} + Z_{pp}^{abc})^{-1} E_{p,0}^{abc} \; ; i = 1, 2, \ldots, n; \; i \neq p$$

When the fault element is represented as admittance matrix Y_F^{abc}, we find the following relations.

$$E_{p,F}^{abc} = (U + Z_{pp}^{abc} Y_F^{abc})^{-1} E_{p,0}^{abc}$$

$$I_{p,F}^{abc} = Y_F^{abc} (U + Z_{pp}^{abc} Y_F^{abc})^{-1} E_{p,0}^{abc}$$

$$E_{i,F}^{abc} = E_{i,0}^{abc} - Z_{ip}^{abc} \, Y_F^{abc} (U + Z_{pp}^{abc} Y_F^{abc})^{-1} E_{p,0}^{abc} \; ; i = 1, 2, \ldots, n; \; i \neq p$$

The element currents are found from the relation

$$i_{ij,F}^{abc} = \bar{y}_{ij,\rho\sigma}^{abc} (\bar{E}_{\rho,F}^{abc} - \bar{E}_{\sigma,F}^{abc})$$

For the balanced three-phase networks, all the above equations can be expressed in terms of 0 1 2 variables and parameters. We simply replace *abc* by 012 in above equations.

$$I_{p,F}^{012} = (Z_F^{012} + Z_{pp}^{012})^{-1} E_{p,0}^{012}$$

$$E_{p,F}^{012} = Z_F^{012} (Z_F^{012} + Z_{pp}^{012})^{-1} E_{p,0}^{012}$$

$$E_{i,F}^{012} = E_{i,0}^{012} - Z_{ip}^{012} (Z_F^{012} + Z_{pp}^{012})^{-1} E_{p,0}^{012} \; ; i = 1, 2, \ldots, n; i \neq p$$

$$E_{p,F}^{012} = (U + Z_{pp}^{012} Y_F^{012})^{-1} E_{p,0}^{012}$$

$$I_{p,F}^{012} = Y_F^{012} \, (U + Z_{pp}^{012} Y_F^{012})^{-1} E_{p,0}^{012}$$

$$E_{i,F}^{012} = E_{i,0}^{012} - Z_{ip}^{012} Y_F^{012} \, (U + Z_{pp}^{012} Y_F^{012})^{-1} E_{p,0}^{012}; \; i = 1, 2, \ldots, n; i \neq p$$

$$i_{ij,F}^{012} = \overline{y}_{ij,\rho\sigma}^{012} \, (\overline{E}_{\rho,F}^{012} - \overline{E}_{\sigma,F}^{012})$$

45.4 VARIOUS TYPES OF FAULTS

A fault on the power system may be symmetrical, or unsymmetrical. A three-phase to ground fault or a three-phase fault is a symmetrical or a balanced fault. This fault can be analyzed by using the per phase equivalent circuit. There is no need of resorting to 012 quantities.

$$I_{p,F}^{a} = \frac{E_{p,0}}{Z_F + Z_{pp}^{1}}, \quad E_{p,F}^{a} = z_F I_{p,F}^{a}, \quad E_{i,F}^{a} = E_{i,0} - Z_{ip}^{1} I_{p,F}^{a}$$

The phase *b* and the phase *c* values during fault can be obtained from above equations by giving appropriate phase rotation.

It is common to take pre-fault voltages of all the buses equal, and equal to 1 per unit.

The unsymmetrical faults are line to ground fault, line to line fault, and line to line to ground fault. For analyzing the unsymmetrical faults we take recourse to 012 equations. The second fault does not involve ground, as such we are constrained to use equations involving Y_F^{012}.

For the other two faults which involve ground, we do have Z_F^{abc}, but a few of the entries in it are infinite. As such, Z_F^{012} is not defined in these cases also. We can have Y_F^{abc} for these faults. No entry is infinite in Y_F^{abc}, and therefore Y_F^{012} exists. Y_F^{012} for these faults are as under:

Fault	Y_F^{012}
Line to ground	$\dfrac{1}{3(z_F + z_g)} \begin{bmatrix} 1 & 1 & 1 \\ 1 & 1 & 1 \\ 1 & 1 & 1 \end{bmatrix}$
Line to line	$\dfrac{1}{2z_F} \begin{bmatrix} 0 & 0 & 0 \\ 0 & 1 & -1 \\ 0 & -1 & 1 \end{bmatrix}$

Line to line to ground $\dfrac{1}{3(z_F^2+2z_Fz_g)}\begin{bmatrix} 2z_F & -z_F & -z_F \\ -z_F & (2z_F+3z_g) & -(z_F+3z_g) \\ -z_F & -(z_F+3z_g) & (2z_F+3z_g)\end{bmatrix}$

In all the above fault analysis cases we find $I_{p,F}^{012}$, $E_{p,F}^{012}$, and $E_{i,F}^{012}$ by applying the general short circuit equations involving Y_F^{012}. Then applying the transformation, we can find $I_{p,F}^{abc}$, $E_{p,F}^{abc}$, and $E_{i,F}^{abc}$ for example, $I_{p,F}^{abc} = T_s I_{p,F}^{012}$.

The relations for ***single line to ground fault*** follow.

Let $DZ = z_{pp}^0 + 2z_{pp}^1 + 3z_f$. This is used, for ease, in the relations which follow. Further, the negative sequence impedance and the positive sequence impedance are assumed to be equal, *e.g.*, $z_{ip}{}^2 = z_{ip}{}^1$, $z_{pp}{}^2 = z_{pp}{}^1$.

Symmetrical components / ***Three phase components***

$$I_{p,F}^{012} = \frac{\sqrt{3}E_{p,0}}{DZ}\begin{bmatrix}1\\1\\1\end{bmatrix} \qquad I_{p,F}^{abc} = \frac{3E_{p,0}}{DZ}\begin{bmatrix}1\\0\\0\end{bmatrix}$$

$$E_{p,F}^{012} = \frac{\sqrt{3}E_{p,0}}{DZ}\begin{bmatrix}-z_{pp}^0\\ z_{pp}^0+z_{pp}^1+3z_F\\ -z_{pp}^1\end{bmatrix} \qquad E_{p,F}^{abc} = E_{p,0}\begin{bmatrix}\dfrac{3z_F}{DZ}\\ a^2\dfrac{z_{pp}^0-z_{pp}^1}{DZ}\\ a-\dfrac{z_{pp}^0-z_{pp}^1}{DZ}\end{bmatrix}$$

$$E_{i,F}^{012} = E_{i,0}\begin{bmatrix}0\\ \sqrt{3}\\ 0\end{bmatrix} - \frac{\sqrt{3}E_{p,0}}{DZ}\begin{bmatrix}z_{ip}^0\\ z_{ip}^1\\ z_{ip}^1\end{bmatrix},\quad i \neq p \qquad E_{i,F}^{abc} = E_{i,0}\begin{bmatrix}1\\a^2\\a\end{bmatrix} - E_{p,0}\begin{bmatrix}\dfrac{z_{ip}^0+2z_{ip}^1}{DZ}\\ \dfrac{z_{ip}^0-z_{ip}^1}{DZ}\\ \dfrac{z_{ip}^0-z_{ip}^1}{DZ}\end{bmatrix},\quad i \neq p$$

Along similar lines, we analyze ***line to line fault***.

Let $D = 1 + z_{pp}{}^1 y_F$, where $y_F = 1/z_F$. Then

$$I_{p,F}^{012} = \frac{\sqrt{3}E_{p,0}\,y_F}{2D}\begin{bmatrix}0\\1\\-1\end{bmatrix} \qquad I_{p,F}^{abc} = \frac{E_{p,0}\,y_F}{2D}\begin{bmatrix}0\\ a(a-1)\\ -a(a-1)\end{bmatrix}$$

$$E_{p,F}^{012} = \frac{\sqrt{3}E_{p,0}}{D}\begin{bmatrix}0\\ 1+z_{pp}^1\,y_F/2\\ z_{pp}^1\,y_F/2\end{bmatrix} \qquad E_{p,F}^{abc} = \frac{E_{p,0}}{D}\begin{bmatrix}1+z_{pp}^1\,y_F\\ a^2-z_{pp}^1\,y_F/2\\ a-z_{pp}^1\,y_F/2\end{bmatrix}$$

$$E_{i,F}^{012} = E_{i,0}\begin{bmatrix}0\\ \sqrt{3}\\ 0\end{bmatrix} - \frac{\sqrt{3}E_{p,0}\,y_F}{2D}\begin{bmatrix}0\\ z_{ip}^1\\ -z_{ip}^1\end{bmatrix} \qquad E_{i,F}^{abc} = E_{i,0}\begin{bmatrix}1\\a^2\\a\end{bmatrix} - \frac{E_{p,0}\,y_F}{2D}\begin{bmatrix}0\\ a(a-1)z_{ip}^1\\ -a(a-1)z_{ip}^1\end{bmatrix}$$

Line to line to ground fault can be similarly examined.

Chapter **46** LOAD FLOW ANALYSIS

46.1 GENERAL

Load flow analysis in power systems is carried out to find out the system bus voltages. The system network parameters and the injections at the buses are specified. In our normal nodal method of circuit analysis, the circuit parameters and the current sources at various nodes are specified, nodal equations are written and we can straight way solve these linear equations for node voltages. But in power systems, instead of the currents, the powers at the buses are given and the nodal equations then become non-linear equations in voltages.

Consider any four-node circuit. Taking one of the nodes as reference node, we can write

$$I_1 = Y_{11}E_1 + Y_{12}E_2 + Y_{13}E_3$$
$$I_2 = Y_{21}E_1 + Y_{22}E_2 + Y_{23}E_3$$
$$I_3 = Y_{31}E_1 + Y_{32}E_2 + Y_{33}E_3$$

If instead of the currents, we are given *EI* products, then the equations become

$$S_1 = E_1I_1 = Y_{11}E_1^{\,2} + Y_{12}E_1E_2 + Y_{13}\,E_1\,E_3$$
$$S_2 = E_2I_2 = Y_{21}E_1E_2 + Y_{22}\,E_2^{\,2} + Y_{23}\,E_2\,E_3$$
$$S_3 = E_3I_3 = Y_{31}E_1E_3 + Y_{32}E_2\,E_3 + Y_{33}\,E_3^{\,2}$$

Now these equations are non-linear equations in voltages. We use some suitable numerical or iterative method to solve these non-linear equations.

It may be noted that all the quantities involved are complex quantities and as such, these should be manipulated accordingly. ***No bold facing is being used for denoting complex quantities.***

For solving equations such as above, iterative techniques are employed. The iterative methods begin with suitable assumed or known initial solution, and successively improve the solution to acceptable accuracy. Whether the solution will converge quickly, slowly, or will not converge depends on the initial solution, the solution technique used, and the system characteristics.

46.2 BUSES

At the power system buses, generators, loads, VAR compensators etc. are connected. The generations, P_G and Q_G are considered control variables, and the demands P_D and Q_D are the called the disturbance variables. Q_G may be provided by the generators or/and by the VAR compensators or capacitors which may be of fixed or variable values. For a particular period, P_D and Q_D are known from the load forecast and are assumed to stay constant during that period. The net real power injection at any bus i is $P_i = P_{G\,i} - P_{D\,i}$ and the net reactive power injection is $Q_i = Q_{G\,i} - Q_{D\,i}$. The bus voltages magnitudes $|E_i|$ and the angles δ_i are called the state variables. We therefore, note that each bus is associated with four quantities–P_i, Q_i, $|E_i|$, and δ_i. For load flow analysis, we classify the buses in three broad categories.

1. *PQ*, or load, or type 1 buses. At these buses, the net active power injection P_i and the net reactive power injection Q_i are specified. A pure load bus has no generator connected to it. Usually, at *PQ* buses, P_i and Q_i are both negative. A bus having only a fixed capacitor connected to it is also a *PQ* bus with $P = 0$, and Q positive. At the *PQ* buses, we are required to find the bus voltage magnitude $|E_i|$ and the angle δ_i, or the real and the imaginary parts e_i and f_i respectively of the bus voltage E_i.
2. *PE*, or generator, or voltage controlled, or type 2 buses. At these buses, P_i and $|E_i|$ are specified, and the *LFA* provides Q_i and δ_i. A controllable reactive power source must be connected to such buses.
3. $E\delta$, or slack, or swing, or reference, or type 3 bus. At this bus voltage is completely specified, *i.e.*, $|E|$ and δ are both specified and we find P and Q. It has a large generator connected to it.

We cannot specify P and Q arbitrarily at all the buses, as then we may get absurd solutions for voltages, or solution may not converge. ΣP_i must be > 0, and ΣQ_i can be positive, zero, or negative.

About 85% of the buses are *PQ* buses, and 15% are *PE* buses. It is a common practice to label the slack bus as bus 1, *PE* buses are numbered next, and then we number the *PQ* buses. Further, the slack bus voltage is usually taken $1\angle 0°$ *pu*. In flat start of the computations, the initial solution of voltages for all the *PQ* buses is taken $1\angle 0°$ *pu*, and the angles of voltages for all the *PE* buses are taken 0°.

46.3 POWER EQUATIONS

Including the ground bus, there are $(n + 1)$ buses. In forming Y_{BUS}, the ground bus is taken as the reference bus. Y_{BUS} is a sparse and symmetrical $n \times n$ matrix.

We have $$\bar{I}_{BUS} = Y_{BUS}\,\bar{E}_{BUS}$$

One bus is the slack bus whose voltage is known. Thus, we are to solve for $(n - 1)$ bus voltages.

For any bus i, we can write,

$$I_i = \sum_{k=1}^{k=n} Y_{ik}\,E_k$$

The complex power S_i is $P_i + jQ_i = E_i I_i^*$, and its conjugate is $S_i^* = P_i - j\,Q_i = E_i^* I_i$.

$$S_i^* = E_i^* \sum_{k=1}^{k=n} Y_{ik} E_k$$

$$P_i = \text{Real of } E_i^* \sum_{k=1}^{k=n} Y_{ik} E_k$$

and
$$Q_i = -\text{ Imag of } E_i^* \sum_{k=1}^{k=n} Y_{ik} E_k$$

The voltages and the elements Y_{ik} of the bus admittance matrix Y_{BUS} are expressed in polar form as

$$E_i = |E_i| \angle \delta_i, \quad Y_{ik} = |Y_{ik}| \angle -\theta_{ik}.$$

The angle δ_i is with respect to the voltage of the slack bus voltage. θ_{ii} may be around 80°, and θ_{ik} around –100°.

$$P_i = \sum_{k=1}^{k=n} |E_i E_k Y_{ik}| \cos(\delta_i - \delta_k + \theta_{ik}) = \sum_{k=1}^{k=n} |E_i E_k Y_{ik}| \cos(\delta_{ik} + \theta_{ik})\ ; \text{ where } \delta_{ik} = \delta_i - \delta_k$$

$$Q_i = \sum_{k=1}^{k=n} |E_i E_k Y_{ik}| \sin(\delta_i - \delta_k + \theta_{ik}) = \sum_{k=1}^{k=n} |E_i E_k Y_{ik}| \sin(\delta_{ik} + \theta_{ik})$$

The above two equations are the static power flow equations in angle form. These equations can be expressed in conductance and susceptance form. First, we let

$$Y_{ik} = G_{ik} - jB_{ik}$$

$$Y_{ik} = |Y_{ik}| \angle -\theta_{ik} = |Y_{ik}| (\cos\theta_{ik} - j\sin\theta_{ik})$$

Thus,
$$G_{ik} = |Y_{ik}| \cos\theta_{ik}$$

$$B_{ik} = |Y_{ik}| \sin\theta_{ik}$$

With the notation we have followed, G_{ii}, B_{ii} are both positive ; and G_{ik}, B_{ik} are both negative.

$$P_i = \sum_{k=1}^{k=n} |E_i E_k| (G_{ik} \cos\delta_{ik} - B_{ik} \sin\delta_{ik})$$

$$Q_i = \sum_{k=1}^{k=n} |E_i E_k| (G_{ik} \sin\delta_{ik} + B_{ik} \cos\delta_{ik})$$

Further, the static power flow equations can also be expressed in terms of e_i and f_i, the real and imaginary parts of the voltage E_i, $i = 1, 2, \ldots, n$.

We let
$$E_i = e_i + jf_i$$

$$P_i = \sum_{k=1}^{k=n} \{e_i (e_k G_{ik} + f_k B_{ik}) + f_i (f_k G_{ik} - e_k B_{ik})\}$$

$$Q_i = \sum_{k=1}^{k=n} \{f_i\,(e_k\,G_{ik} + f_k\,B_{ik}) - e_i\,(f_k\,G_{ik} - e_k\,B_{ik})\}$$

46.4 GAUSS-SEIDEL METHOD

The current at bus i is seen to be

$$I_i = \sum_{k=1}^{k=n} Y_{ik}\,E_k = Y_{ii}\,E_i + \sum_{\substack{k=1\\k\neq i}}^{k=n} Y_{ik}\,E_k. \text{ It leads to}$$

$$E_i = \frac{1}{y_{ii}}\left(I_i - \sum_{\substack{k=1\\k\neq i}}^{k=n} Y_{ik}E_k\right) = \frac{1}{Y_{ii}}\left(\frac{P_i - jQ_i}{E_i^*} - \sum_{\substack{k=1\\k\neq i}}^{k=n} Y_{ik}E_k\right)$$

$$= \frac{T_i}{E_i^*} - \sum_{\substack{k=1\\k\neq i}}^{k=n} A_{ik}\,E_k;\ i = 2, 3, \ldots, n.$$

We have set $\dfrac{P_i - jQ_i}{Y_{ii}} = T_i$, and $\dfrac{Y_{ik}}{Y_{ii}} = A_{ik}$.

The equation $E_i = \dfrac{T_i}{E_i^*} - \sum_{\substack{k=1\\k\neq i}}^{k=n} A_{ik}\,E_k$ is in terms of the given quantities and the voltages.

The iterative process can be started.

Assume first that there are ***no PE buses.***

One begins with the assumed initial solution of the voltages, $E_i^{\,0}$.

As per the Gauss iterative method, we obtain $E_i^{\,1}$ from above equation for E_i by using $E_i^{\,0}$. At the end of r iterations or $(r + 1)$ computation cycles, we have

$$E_i^{\,r+1} = \frac{T_i}{E_i^{\,r*}} - \sum_{\substack{k=1\\k\neq i}}^{k=n} A_{ik}\,E^r{}_k \quad i = 2, 3, \ldots, n.$$

In the *G–S* iterative method, the new voltage estimate obtained is utilized in the subsequent computations in that iteration. Thus, we have

$$E_i^{\,r+1} = \frac{T_i}{E_i^{\,r*}} - \sum_{k=1}^{k=i-1} A_{ik}\,E_k^{\,r+1} - \sum_{k=i+1}^{k=n} A_{ik}\,E_k^{\,r} \quad i = 2, 3, \ldots, n.$$

The computations are continued till the $\left|E_i^{\,r+1} - E_i^{\,r}\right|$ for each of the buses falls to a certain acceptable very small value. Then further computations are stopped and the voltage solution is said to have been reached.

It is reminded that we are dealing with complex quantities all through.

To accelerate the convergence, we sometimes apply ***acceleration factor*** to the correction. Basically, we enhance the voltage estimate obtained a little for further computations.

$$E_{i,\,acc}^{r+1} = E_{i,\,acc}^{r} + \alpha(E_i^{r+1} - E_{i,\,acc}^{r})$$

The value of α depends on the problem and the initial solution assumed; an inappropriate value may even cause the solution to diverge. Generally $\alpha = 1.6$ is found workable. α is not assigned a value greater than 2. If working in terms of components, independent acceleration factors are applied to the real part and the imaginary part of the voltage.

PE buses can also be incorporated in the G–S method of LFA.

For the PE buses, voltage magnitudes are already specified, and the voltage angles are to be determined. To begin with, we assume that all these angles to be zero. Further, for these buses Q is not specified, and to be able to use the voltage equation to compute the new estimate E_i for the current *PE* bus, we need the value of Q_i for this bus. The value of Q_i is found from the Q_i equation using the latest estimates of the voltages. This computed Q_i is checked against $Q_{i,\,min}$ and $Q_{i,\,max}$ limits of the bus i. If the limits are not violated, E_i^{r+1} is found using this computed Q_i and the latest estimates of the voltages in the E_i equation. Then we set,

$$E_i^{r+1} = |\,E_{i,\,specified}\,| \angle E_i^{r+1}.$$

If either of the two limits is violated, then this Q_i is replaced by the violated Q limit value and this bus is treated as a *PQ* bus in this computation and we continue with the computed E_i^{r+1} rather than $|\,E_{i,\,specified}\,| \angle E_i^{r+1}$. If in any subsequent iteration, Q for a particular originally *PE* bus comes within limits, its status of *PE* bus is restored and we proceed as explained above. Such approach applies to all the *PE* buses as we meet them to compute their voltages.

46.5 NEWTON- RAPHSON METHOD

The $N - R$ method is based on the fact that if a good approximate solution of the given function $f(x) = K$ is known, it can be improved further by applying corrections successively; and the correction at any step is given by

$$\Delta x^r = \frac{K - f(x)|_{x^r}}{f'(x)|_{x^r}} = \frac{\Delta f^r}{f'(x)|_{x^r}} \quad \text{and} \quad \text{then } x^{r+1} = x^r + \Delta x^r$$

The above relation follows when we expand $f(x^r + \Delta x^r)$ by Taylor's series and truncate the series after the first derivative term, and equate it to K.

In the above relations, we can begin with a solution x^o and go on improving the solution successively. It is assumed that the derivative of the function exists at point $x = x^r$, and also that the derivative is not too small.

This is the case of a single variable function, it can be easily extended to multivariable functions. For instance, if we have two equations in two variables

$$f_1(x_1, x_2) = K_1 \quad \text{and} \quad f_2(x_1, x_2) = K_2 \quad \text{then}$$

$$\begin{bmatrix} \dfrac{\partial f_1}{\partial x_1} & \dfrac{\partial f_1}{\partial x_2} \\ \dfrac{\partial f_2}{\partial x_1} & \dfrac{\partial f_2}{\partial x_2} \end{bmatrix}_r \begin{bmatrix} \Delta x_1 \\ \Delta x_2 \end{bmatrix}_r = \begin{bmatrix} \Delta f_1 \\ \Delta f_2 \end{bmatrix}_r, \text{ where } \begin{bmatrix} \Delta f_1 \\ \Delta f_2 \end{bmatrix}_r = \begin{bmatrix} K_1 - f_1(x_1, x_2)_r \\ K_2 - f_2(x_1, x_2)_r \end{bmatrix}.$$

and

$$\begin{bmatrix} \Delta x_1 \\ \Delta x_2 \end{bmatrix}_r = \begin{bmatrix} \dfrac{\partial f_1}{\partial x_1} & \dfrac{\partial f_1}{\partial x_2} \\ \dfrac{\partial f_2}{\partial x_1} & \dfrac{\partial f_2}{\partial x_2} \end{bmatrix}_r^{-1} \begin{bmatrix} \Delta f_1 \\ \Delta f_2 \end{bmatrix}_r, \quad \begin{bmatrix} \Delta x_1 \\ \Delta x_2 \end{bmatrix}_{r+1} = \begin{bmatrix} x_1 \\ x_2 \end{bmatrix}_r + \begin{bmatrix} \Delta x_1 \\ \Delta x_2 \end{bmatrix}_r$$

Similar expressions will follow for n functions in n variables.

$$\begin{bmatrix} \dfrac{\partial f_1}{\partial x_1} & \dfrac{\partial f_1}{\partial x_2} & \cdots & \dfrac{\partial f_1}{\partial x_n} \\ \dfrac{\partial f_2}{\partial x_1} & \dfrac{\partial f_2}{\partial x_2} & \cdots & \dfrac{\partial f_2}{\partial x_n} \\ \vdots & & & \vdots \\ \dfrac{\partial f_n}{\partial x_1} & \dfrac{\partial f_n}{\partial x_2} & \cdots & \dfrac{\partial f_n}{\partial x_n} \end{bmatrix}_r \begin{bmatrix} \Delta x_1 \\ \Delta x_2 \\ \vdots \\ \Delta x_n \end{bmatrix}_r = \begin{bmatrix} \Delta f_1 \\ \Delta f_2 \\ \vdots \\ \Delta f_n \end{bmatrix}_r = \begin{bmatrix} K_1 - f_1 \\ K_2 - f_2 \\ \vdots \\ K_n - f_n \end{bmatrix}_r$$

This can be expressed compactly using matrix notation as

$$\left[\frac{\partial \mathbf{f}}{\partial \mathbf{x}}\right] [\Delta \mathbf{x}] = [\Delta \mathbf{f}] = [\mathbf{K} - \mathbf{f}]$$

$\Delta \mathbf{x}$ is called the correction vector, and $\Delta \mathbf{f}$ is the vector of residuals. The square matrix of the partial derivatives of **f** w.r.t. **x** is called the Jacobian.

We apply this technique of improving the solutions to LFA. The LFA can be carried out either in the polar form, or in the rectangular form of voltages. The variables are then either the magnitudes and the angles, or the real parts and the imaginary parts of the bus voltages. The functions f are the active power and the reactive power equations for the buses. And finally, the K's are the specified active powers and the reactive powers at the buses. To find the residuals, the P's and the Q's are computed for each bus using the current estimates of voltages in every iteration.

As there are n proper buses, out of which one bus is a slack bus whose voltage is already known and remains fixed throughout, we have $n - 1$ buses, from bus 2 to bus n, whose voltages are to be determined.

If these $n - 1$ buses are all PQ buses, then we have $2(n - 1)$ variables; $n - 1$ voltage magnitudes and $n - 1$ voltage angles, or $n - 1$ real parts e and $n - 1$ imaginary parts f. The Jacobian is $2(n-1) \times 2(n-1)$ matrix of partial derivatives of P's and Q's. The partial derivatives of P and Q w.r.t. δ and $|E|$, or w.r.t. e and f are found from the expressions for P and Q, and are computed using the current estimates of voltages. Likewise we compute the real power P and the reactive power Q at the $n - 1$ buses using the current estimates of voltages. The residuals are therefore, found out by subtracting the computed powers from the corresponding specified powers. Thus, in the following set of equations, all the quantities are known in value and by solving this set of $2(n-1)$ equations, the correction vector is found. The next estimates of the voltages are then found simply by adding the corrections to the current estimates.

$$\delta_i^{r+1} = \delta_i^r + \Delta\delta_i^r \qquad |E_i^{r+1}| = |E_i^r| + \Delta|E_i^r|$$

Similar equations can be written in terms of e and f.

$$\begin{bmatrix} \frac{\partial P_2}{\partial \delta_2} & \cdots & \frac{\partial P_2}{\partial \delta_n} & \frac{\partial P_2}{\partial |E_2|} & \cdots & \frac{\partial P_2}{\partial |E_n|} \\ \vdots & & \vdots & \vdots & & \vdots \\ \frac{\partial P_n}{\partial \delta_2} & \cdots & \frac{\partial P_n}{\partial \delta_n} & \frac{\partial P_n}{\partial |E_2|} & \cdots & \frac{\partial P_n}{\partial |E_n|} \\ \frac{\partial Q_2}{\partial \delta_2} & \cdots & \frac{\partial Q_2}{\partial \delta_2} & \frac{\partial Q_2}{\partial |E_2|} & \cdots & \frac{\partial Q_2}{\partial |E_n|} \\ \vdots & & \vdots & \vdots & & \vdots \\ \frac{\partial Q_n}{\partial \delta_2} & \cdots & \frac{\partial Q_n}{\partial \delta_n} & \frac{\partial Q_n}{\partial |E_2|} & \cdots & \frac{\partial Q_n}{\partial |E_n|} \end{bmatrix}_r \begin{bmatrix} \Delta\delta_2 \\ \vdots \\ \Delta\delta_n \\ \Delta|E_2| \\ \vdots \\ \Delta|E_n| \end{bmatrix}_r = \begin{bmatrix} \Delta P_2 \\ \vdots \\ \Delta P_n \\ \Delta Q_2 \\ \vdots \\ \Delta Q_n \end{bmatrix}_r = \begin{bmatrix} P_{2.sp} - P_{2,co} \\ \vdots \\ P_{n.sp} - P_{n,co} \\ Q_{2.sp} - Q_{2,co} \\ \vdots \\ Q_{n.sp} - Q_{n,co} \end{bmatrix}_r$$

In compact form the above set of $2(n-1)$ equations is written as

$$\begin{bmatrix} \mathbf{J}_1 & \mathbf{J}_2 \\ \mathbf{J}_3 & \mathbf{J}_4 \end{bmatrix} \begin{bmatrix} \Delta\delta \\ \Delta|\mathbf{E}| \end{bmatrix} = \begin{bmatrix} \Delta\mathbf{P} \\ \Delta\mathbf{Q} \end{bmatrix}$$

The elements of $\mathbf{J}_1$, $\mathbf{J}_2$, $\mathbf{J}_3$, and $\mathbf{J}_4$ submatrices are divided in two categories—the off-diagonal elements and the diagonal elements. The expressions for these elements can be easily found from the expressions for power P, and the reactive power Q, and are given below—in rectangular as well as polar form.

46.6 RECTANGULAR FORM

$$\begin{bmatrix} \mathbf{J}_1 & \mathbf{J}_2 \\ \mathbf{J}_3 & \mathbf{J}_4 \end{bmatrix} \begin{bmatrix} \Delta\mathbf{e} \\ \Delta\mathbf{f} \end{bmatrix} = \begin{bmatrix} \Delta\mathbf{P} \\ \Delta\mathbf{Q} \end{bmatrix}$$

First we note that $\quad I_i = c_i + j\,d_i = \sum_{k=1}^{k=n} Y_{ik} E_k = Y_{ii} E_i + \sum_{\substack{k=1 \\ k \neq i}}^{k=n} Y_{ik} E_k$

$$= (G_{ii} - jB_{ii})(e_i + f_i) + \sum_{\substack{k=1 \\ k \neq i}}^{k=n} (G_{ik} - jB_{ik})(e_k + jf_k)$$

This gives $\quad c_i = e_i G_{ii} + f_i B_{ii} + \sum_{\substack{k=1 \\ k \neq i}}^{k=n} (e_k G_{ik} + f_k B_{ik})$

$$d_i = f_i G_{ii} - e_i B_{ii} + \sum_{\substack{k=1 \\ k \neq i}}^{k=n} (f_k G_{ik} - e_k B_{ik}) \quad i = 2, 3, \ldots, n$$

$\mathbf{J}_1$ –

Off-diagonal elements

$$\frac{\partial P_i}{\partial e_k} = e_i G_{ik} - f_i B_{ik} \quad k \neq i$$

Diagonal elements

$$\frac{\partial P_i}{\partial e_i} = 2e_i G_{ii} + f_i B_{ii} - f_i B_{ii} + \sum_{\substack{k=1 \\ k \neq i}}^{k=n} (e_k G_{ik} + f_k B_{ik}) = e_i G_{ii} - f_i B_{ii} + c_i$$

$\mathbf{J}_2$ –

Off-diagonal elements

$$\frac{\partial P_i}{\partial f_k} = e_i B_{ik} + f_i G_{ik} \quad k \neq i$$

Diagonal elements

$$\frac{\partial P_i}{\partial f_i} = e_i B_{ii} + 2f_i G_{ii} - e_i B_{ii} + \sum_{\substack{k=1 \\ k \neq i}}^{k=n} (f_k G_{ik} - e_k B_{ik}) = e_i B_{ii} + f_i G_{ii} + d_i$$

$\mathbf{J}_3$ –

Off-diagonal elements

$$\frac{\partial Q_i}{\partial e_k} = e_i B_{ik} + f_i G_{ik} \quad k \neq i$$

Diagonal elements

$$\frac{\partial Q_i}{\partial e_i} = f_i G_{ii} - f_i G_{ii} + 2 e_i B_{ii} - \sum_{\substack{k=1 \\ k \neq i}}^{k=n} (f_k G_{ik} - e_k B_{ik}) = e_i B_{ii} + f_i G_{ii} - d_i$$

$\mathbf{J}_4$ –

Off-diagonal elements

$$\frac{\partial Q_i}{\partial f_k} = -e_i G_{ik} + f_i B_{ik} \quad k \neq i$$

Diagonal elements

$$\frac{\partial Q_i}{\partial f_i} = e_i G_{ii} + 2f_i B_{ii} - e_i G_{ii} + \sum_{\substack{k=1 \\ k \neq i}}^{k=n} (e_k G_{ik} + f_k B_{ik}) = -e_i G_{ii} + f_i B_{ii} + c_i$$

We observe that $J_{1,\,ik} + J_{4,\,ik} = 0,$ $\quad J_{2,\,ik} - J_{3,\,ik} = 0$

and $\quad J_{1,\,ii} + J_{4,\,ii} = 2c_i$ $\quad J_{2,\,ii} - J_{3,\,ii} = 2d_i$

It is found that neglecting the off-diagonal terms in J_1, J_2, J_3, and J_4 does not cause any significant error in the solution. Then

$$\frac{\partial P_i}{\partial e_i}\Delta e_i + \frac{\partial P_i}{\partial f_i}\Delta f_i = \Delta P_i$$

and
$$\frac{\partial Q_i}{\partial e_i}\Delta e_i + \frac{\partial Q_i}{\partial f_i}\Delta f_i = \Delta Q_i \quad i = 2, 3, \ldots, n$$

The equations have got decoupled and the pair of equations for each bus can be solved independently for its correction vector.

The *PE* buses can also be included. The set of equations then is

$$\begin{matrix} PQ \text{ and } PE \text{ buses} \\ \text{Only } PQ \text{ buses} \\ \text{Only } PE \text{ buses} \end{matrix} \begin{bmatrix} J_1 & J_2 \\ J_3 & J_4 \\ J_5 & J_6 \end{bmatrix} \begin{bmatrix} \Delta e \\ \Delta f \end{bmatrix} = \begin{bmatrix} \Delta P \\ \Delta Q \\ \Delta |E|^2 \end{bmatrix}$$

The elements of J_5 and J_6 are the derivative of $|E|^2$ w.r.t. e and f. We have $|E_i|^2 = e_i^2 + f_i^2$.

$$\frac{\partial |E_i|^2}{\partial e_k} = 0, \quad \frac{\partial |E_i|^2}{\partial e_i} = 2e_i, \quad \frac{\partial |E_i|^2}{\partial e_k} = 0 \quad \text{and} \quad \frac{\partial |E_i|^2}{\partial f_i} = 2f_i$$

Thus, the Jacobian is completely determined and we can find the correction vector.

The polar form of the N–R method is , however, more often used.

46.7 POLAR FORM

J_1 –

Off-diagonal elements

$$\frac{\partial P_i}{\partial \delta_k} = |E_i E_k Y_{ik}| \sin(\delta_{ik} + \theta_{ik})$$

Diagonal elements

$$\frac{\partial P_i}{\partial \delta_i} = -\sum_{\substack{k=1 \\ k \neq i}}^{k=n} |E_i E_k Y_{ik}| \sin(\delta_{ik} + \theta_{ik}) = |E_i E_i Y_{ii}| \sin\theta_{ii} - \sum_{k=1}^{k=n} |E_i E_k Y_{ik}| \sin(\delta_{ik} + \theta_{ik})$$

$$= |E_i E_i Y_{ii}| \sin\theta_{ii} - Q_i$$

J_2 –

Off-diagonal elements

$$\frac{\partial P_i}{\partial |E_k|} = |E_i Y_{ik}| \cos(\delta_{ik} + \theta_{ik})$$

Diagonal elements

$$\frac{\partial P_i}{\partial |E_i|} = 2\,|E_i Y_{ii}|\cos\theta_{ii} + \sum_{\substack{k=1\\k\neq i}}^{k=n} |E_k Y_{ik}|\cos(\delta_{ik}+\theta_{ik})$$

J_3 –

Off-diagonal elements

$$\frac{\partial Q_i}{\partial \delta_k} = -\,|E_i E_k Y_{ik}|\cos(\delta_{ik}+\theta_{ik})$$

Diagonal elements

$$\frac{\partial Q_i}{\partial \delta_i} = \sum_{\substack{k=1\\k\neq i}}^{k=n} |E_i E_k Y_{ik}|\cos(\delta_{ik}+\theta_{ik})$$

$$= -\,|E_i E_i Y_{ik}|\cos\theta_{ii} + \sum_{k=1}^{k=n} |E_i E_k Y_{ik}|\cos(\delta_{ik}+\theta_{ik})$$

$$= -\,|E_i E_i Y_{ii}|\cos\theta_{ii} + P_i$$

J_4 –

Off-diagonal elements

$$\frac{\partial Q_i}{\partial |E_k|} = |E_i Y_{ik}|\sin(\delta_{ik}+\theta_{ik})$$

Diagonal elements

$$\frac{\partial Q_i}{\partial |E_i|} = 2\,|E_i Y_{ii}|\sin\theta_{ii} + \sum_{\substack{k=1\\k\neq i}}^{k=n} |E_k Y_{ik}|\sin(\delta_{ik}+\theta_{ik})$$

$$= \frac{1}{|E_i|}\left[\,|E_i E_i Y_{ii}|\sin\theta_{ii} + Q_i\right]$$

Knowing the Jacobian completely, the correction vector can be found.

46.8 SIMPLIFICATIONS

This full N-R method is laborious to program and the Jacobian has to be computed afresh in each iteration. Several simplifications are introduced in order to reduce the computational effort of an iteration without seriously affecting the overall computational effort and time.

Knowing that under the usual operating conditions, P is a weak function of $|E|$, and Q is a weak function of δ, we can altogether neglect J_2 and J_3.

Then ΔP and ΔQ equations get ***decoupled***, *i.e.*, $J_1 \Delta\delta = \Delta P$ and $J_4\,\Delta|E| = \Delta Q$. This is the decoupled form of N-R method.

Further, the voltage angle differences δ_{ik} are quite small, and their effects are ignored. Then

$$\frac{\partial P_i}{\partial \delta_k} = | E_i E_k | B_{ik} \qquad k \neq i$$

$$\frac{\partial P_i}{\partial \delta_i} = | E_i |^2 B_{ii} - Q_i.$$

In this simplified expression too, Q_i is considered small in comparison to $|E_i|^2 B_{ii}$. As such,

$$\frac{\partial P_i}{\partial \delta_i} = | E_i E_i | B_{ii}$$

Turning to J_4,

$$\frac{\partial Q_i}{\partial | E_k |} = | E_i | B_{ik} = \frac{1}{| E_k |} | E_i E_k | B_{ik} \qquad k \neq i$$

$$\frac{\partial Q_i}{\partial | E_i |} = \frac{1}{| E_i |} [| E_i E_i Y_{ii} | \sin \theta_{ii} + Q_i] \approx \frac{1}{| E_i |} | E_i E_i | B_{ii}.$$

Thus, we can summarize as under.

Submatrix	Off-diagonal elements	Diagonal elements
J_1	$\| E_i E_k \| B_{ik}$	$\| E_i E_i \| B_{ii}$
J_4	$\frac{1}{\| E_k \|} \| E_i E_k \| B_{ik}$	$\frac{1}{\| E_i \|} \| E_i E_i \| B_{ii}$

We shall therefore have,

$$\sum_{k=2}^{n} | E_i E_k | B_{ik} \, \Delta\delta_k = \Delta P_i \quad \text{or} \quad \sum_{k=2}^{n} | E_k | B_{ik} \, \Delta\delta_k = \frac{\Delta P_i}{| E_i |}; \qquad i = 2, 3, \ldots, n$$

$$\sum_{k=2}^{n} | E_i E_k | B_{ik} \frac{\Delta | E_k |}{| E_k |} = \Delta Q_i \quad \text{or} \quad \sum_{k=2}^{n} | E_k | B_{ik} \frac{\Delta | E_k |}{| E_k |} = \frac{\Delta Q_i}{| E_i |}; \qquad i = 2, 3, \ldots, n$$

In matrix form,

$$[\mathbf{B}'] \, [\Delta\delta] = \left[\frac{\Delta \mathbf{P}}{|\mathbf{E}|} \right] \quad \text{and} \quad [\mathbf{B}'] \left[\frac{\Delta |\mathbf{E}|}{|\mathbf{E}|} \right] = \left[\frac{\Delta \mathbf{Q}}{|\mathbf{E}|} \right]$$

Further, if we set $|E_k| = 1\,pu$ on the LHS of first scalar equation, and rearrange the two scalar equations,

$$\sum_{k=2}^{n} B_{ik} \, \Delta\delta_k = \frac{\Delta P_i}{| E_i |} \qquad i = 2, 3, \ldots, n$$

$$\sum_{k=2}^{n} B_{ik} \, \Delta | E_k | = \frac{\Delta Q_i}{| E_i |} \qquad i = 2, 3, \ldots, n$$

In matrix form,

$$[\mathbf{B}]\,[\Delta\delta] = \left[\frac{\Delta \mathbf{P}}{|\mathbf{E}|}\right] \quad \text{and} \quad [\mathbf{B}]\,[\Delta|\,\mathbf{E}\,|\,] = \left[\frac{\Delta \mathbf{Q}}{|\mathbf{E}|}\right]$$

Thus, J_1 and J_4 got simplified merely to **[B]**, an $(n-1) \times (n-1)$ constant matrix of susceptances as obtained from Y_{BUS}. These two matrix equations are the equations of ***Fast decoupled method.***

If there are m PE buses, the number of rows in the second equation ($\Delta\mathbf{Q}$ equation) becomes $(n-1-m)$.

If resistances of the lines are neglected, then the elements of **[B]** are simply the reciprocals of line reactances, with appropriate signs.

We continue simplifying further. We neglect the $\Delta\mathbf{Q}$ equation altogether, and set $|\,\mathbf{E}\,| = 1pu$ in the ΔP equation. Then we have,

$[\mathbf{B}]\,[\Delta\delta] = [\Delta\mathbf{P}]$ From this, we find the voltages as $E_i = 1\angle\delta_i$.

This is DC power flow analysis.

Once we have found out all the bus voltages of the power system applying any of the iterative methods, we can next find other quantities of interest, *e.g.*, line current, MW flow, MVAr flow, slack bus power, and system active and reactive power losses.

The current entering the line at bus i connected between buses i and k is

$$i_{ik} = (E_i - E_k)\, y_{ik} + E_i \frac{y_{ik}'}{2}$$

This includes the line charging current at bus i.

The power entering the line at bus i, which is the line flow from bus i to bus k, is

$$P_{ik} - jQ_{ik} = E_i^*\, i_{ik} = E_i^*\left[(E_i - E_k)\, y_{ik} + E_i \frac{y_{ik}'}{2}\right];$$

similarly, the line flow from bus k to bus i is

$$P_{ki} - jQ_{ki} = E_k^*\, i_{ki} = E_k^*\left[(E_k - E_i)\, y_{ki} + E_k \frac{y_{ik}'}{2}\right]; \qquad y_{ki} = y_{ik},\; y'_{ki} = y'_{ik}$$

The ***complex power loss in the line*** is the sum of the two powers

$$P_{ik,\,\text{loss}} - jQ_{ik,\,\text{loss}} = \left[|\,E_i\,|^2 + |\,E_k\,|^2\,|\,2 - E_i\,||\,E_k\,|\cos\delta_{ik}\,|\right] y_{ik} + \left[|\,E_i\,|^2 + |\,E_k\,|^2\right]\frac{y_{ik}'}{2}$$

$$= |\,E_{ik}\,|^2\, y_{ik} + \left[|\,E_i\,|^2 + |\,E_k\,|^2\, \frac{y_{ik}'}{2}\right]; \; |\,E_{ik}\,|$$

is the magnitude of the voltage of bus i w.r.t. bus k, and $\delta_{ik} = \delta_i - \delta_k$.

The ***slack bus (s) power*** is

$$S_s^* = E_s^* I_s = E_s^* \sum_{k=1}^{n} Y_{sk} E_k$$

The complex power loss in the network is

$$S^*_{\text{loss}} = \sum_{k=1}^{n} S_k^*$$

46.9 OFF-NOMINAL RATIO TRANSFORMERS

We find that transformers with off-nominal ratio, and phase shifting transformers are also used in power systems. This modifies the system Y_{BUS}.

If the per unit ratio of the transformer voltage on bus i side to the voltage on bus k side is a, where a is, in general, a complex quantity so that phase shifting transformers are also covered, and that the effect of off-nominal ratio or phase shift is represented at bus k then, to obtain the new Y_{BUS}, add $(-a+1)y$ to Y_{ik}, $(-a^*+1)y$ to Y_{ki}, and $(|a|^2 - 1)y$ to Y_{kk}, where y is the admittance of the transformer, or of the line, or of both of them in series. Thus, the ***change matrix*** ΔY_{BUS} is

	.	i	.	k	.
i	.	0	.	$(-a+1)y$	.
	.	.	.	.	.
k	.	$(-a^*+1)y$	.	$(\lvert a \rvert^2 - 1)y$	.
	.	.	.	.	.

If there are two such transformers, one at bus i, and the other at bus k, with a_i and a_k as their respective per unit transformation ratios at bus i and at bus k ; and further, that the total admittance spanning the two buses is y, then ΔY_{BUS} is

.	.	i	.	k	.
.	.	.	.	.	.
i	.	$(\lvert a_i \rvert^2 - 1)y$	.	$(-a_i^* a_k + 1)y$	.
.	.	.	.	.	.
k	.	$(-a_i a_k^* + 1)y$	.	$(\lvert a_k \rvert^2 - 1)y$	.
.	.	.	.	.	.

With the coming in of such transformers, and consequently change in Y_{BUS}, the expressions for P, Q, and the elements of the Jacobian matrix will all have the per unit transformation ratios (a's) present in them.

If at a bus having such a transformer, $|E|$ is held constant; then 'a', instead of $|E|$ becomes the state variable for that bus, and the corresponding derivative in the Jacobian matrix has to found w.r.t. 'a'.

Chapter 47

ECONOMIC LOAD DISPATCH AND UNIT COMMITMENT

47.1 GENERAL

As the loads on the power plants vary, so does their power generations. How should the plant load be distributed amongst the running units ? There is an infinite number of combinations of loads on the units to meet a given load. We wish that the loading of the various units be such that the total cost of meeting the load is minimum. Similarly, the plants in a power system should be loaded in a way that the total operating cost is minimum possible.

The operating cost of a thermal unit or a nuclear unit comprises the cost of fuel, oil, water, labour, and maintenance; the first being the major component. The costs at a certain power are expressed in currency units per hour, say, Rs/hr. It is usually taken as directly proportional to the fuel input per hour, which may be measured in quantity of fuel input per hour, like tons of coal per hour, or as heat units per hour, like kilocalories per hour, or British thermal units per hour.

Like any other device, a thermal unit too has an upper limit on its power generating capability, P_{Max}; but it has a lower limit too, P_{Min}, due to the fact that some minimum steam has to keep flowing in the boiler tubes from thermal stability point of view of the boilers.

47.2 INPUT-OUTPUT AND IC CURVES

The input (fuel)-output (power) curve is a monotonically increasing curve, and happens to be such that the fuel input per hour per MW, *i.e.*, kCal/MWhr decreases somewhat with power output upto about 85% of P_{Max}, and beyond this point, it increases with output power. This curve is called the heat rate curve. We also have another curve, called the incremental heat rate curve, or the incremental cost curve, wherein the kCal/MWhr, or Rs/MWhr are the point by point slopes of the input-output curve. If the input-output curve were a straight line, the incremental cost curve would be a horizontal line. In reality, it rises a little with power, and even may be non-linear, though we usually approximate it as a straight line.

The cost per hour to generate power P_i at the ith unit can be expressed as $C_i(P_i) = a_{0,i} + a_{1,i}P_i + a_{2,i}P_i^2 + \ldots + a_{m,i}P_i^m$. The successive coefficients are smaller and smaller in value. We usually truncate to keep terms upto P_i^2 only. The incremental cost, IC_i is, then $\frac{dC_i(P_i)}{dP_i} = a_{1,i} + 2a_{2,i}P_i$. Since, we are retaining only first three terms in the cost equation, and IC equation is used more often, we change the notation as $C_i(P_i) = c_i + b_i P_i + \frac{a_i}{2} P_i^2$, so that $IC_i = a_i P_i + b_i$.

47.3 EQUAL IC CRITERION

Let there be n units. Then the total cost of operation per hour is $C_T = \sum_{i=1}^{n} C_i(P_i)$, which is to be minimized; and the total generation is $\sum_{i=1}^{n} P_i$, which must equal the plant power demand P_D.

$C_T = \sum_{i=1}^{n} C_i(P_i)$ = is the cost function, or the objective function; and $\sum_{i=1}^{n} P_i - P_D = 0$ is the constraint equation. We construct the augmented cost function as

$$\hat{C}_T = \sum_{i=1}^{n} C_i(P_i) - \lambda\left(\sum_{i=1}^{n} P_i - P_D\right)$$, where λ is called the Lagrange multiplier.

As per the method of Lagrange multipliers, we next find its derivatives wrt P_i and λ, and equate each of the resulting expressions to zero, and solve for P_i and λ. This very simply leads to

$$IC_i = \lambda, \text{ or } a_i P_i + b_i = \lambda; \quad i = 1, 2, \ldots, n.$$

This means that the loading of various units should be in such a manner that they all operate at the same incremental cost point.

λ is the plant cost of supplying additional MW per hour. As P_D increases, λ also increases.

47.4 GENERATION ALLOCATION

Equal incremental costs equations and the equation $\left(\sum_{i=1}^{n} P_i - P_D\right) = 0$, lead to

$$\lambda = \frac{P_D + \sum_{i=1}^{n} \frac{b_i}{a_i}}{\sum_{i=1}^{n} \frac{1}{a_i}} \quad \text{and} \quad P_i = \frac{P_D + \sum_{i=1}^{n} \frac{b_i}{a_i}}{a_i \sum_{i=1}^{n} \frac{1}{a_i}} - \frac{b_i}{a_i}.$$

For a 2-unit plant, we find $P_1 = \frac{a_2 P_D + b_2 - b_1}{a_1 + a_2}$, and $P_2 = \frac{a_1 P_D + b_1 - b_2}{a_1 + a_2}$.

If in a plant, all the units are identical and have rising *IC* curves, the load should be equally divided amongst all the units. If the *IC* lines are identical and horizontal ($a_i = 0$, b's are all equal), there is no problem of economic load dispatch, and the units can be loaded in any manner. If the curves are all horizontal and parallel ($a_i = 0$, b's are different) we should begin loading the lower most curve unit to its P_{Max}, and go up the curves, loading the successive units to their P_{Max} limits.

If we plot the *IC* curves (linear or non-linear) of all the units on the same set of axes, then beginning with the lowest value of *IC*, we can draw several horizontal lines which cut the *IC* curves, and by adding the generations of all the units for a particular *IC* value , we can find the corresponding total generation or the plant power demand P_D. Thus, for each assumed value of *IC* we get individual generations and the total generation. We thus build a unit loading schedule for various plant loads. We also thus have the data of plant load, the corresponding minimum cost of operation and the plant *IC*. This was the reverse direction computing.

47.5 GENERATION LIMITS CONSIDERATIONS

So far, we did not take into account the loading limits of the units. Every unit has to generate at least its P_{Min} power, and as such the plant load is at least $\sum_{i=1}^{n} P_{i,\text{Min}}$. As the plant load is increased, plant λ increases and becomes greater than at least the lowest of $IC_{i,\text{Min}}$.

Thus, all other units operating at their P_{Min} limits have $IC_{\text{Min}} >$ plant λ.

When all the units are crossed their P_{Min} limits, and no P_{Max} limit is crossed, and the load increases further, we load the units as per the criterion $IC_1 = IC_2 = \ldots = IC_n$. The moment we just cross P_{Max} limit for any unit, that unit is henceforth out of equal *IC* criterion, because subsequently *IC* increases and that unit cannot be loaded any further.

Thus, the units which are at their P_{Max} limits, are operating at $IC_{\text{Max}} <$ plant λ.

We can note that the equal *IC* criterion applies only to those units which are within their $P_{i,\text{Min}}$ and $P_{i,\text{Max}}$ limits. The plant maximum load can be only $\sum_{i=1}^{n} P_{i,\max}$.

47.6 TRANSMISSION LOSSES CONSIDERATIONS

The power system plants are spread over a wide geographical area, and the plants are interconnected. In the power system also, one would wish to supply a give system load at the minimum system operating cost. How should the various plants be operated ? Well, here too the same principle of equal *IC*'s applies, if we disregard the system network (the transmission lines) losses. Assuming that the plant *IC* curves are known, we operate the plants in the system, as we operated the units in a plant.

If the transmission losses P_L are to be considered in economic dispatch, the equal *IC* rule has to be revised. The transmission loss P_L is a function of the currents flowing in the lines, their resistances, load power factors etc. Various forms of expressions for P_L exist in literature, two of them being $\sum_{i=1}^{n} \sum_{k=1}^{n} I_i R_{ik} I_k, \sum_{i=1}^{n} P_i \sum_{k=1}^{n} B_{ik} P_k$, .*I*'s and *P*'s are the injected currents, powers at the plant buses, R_{ik} is the resistance part of the element Z_{ik} of system Z_{BUS}, and B_{ik} is the loss coefficient of the system, or simply the *B* coefficient. We shall shortly use the latter form of the loss expression. The *B*'s in this expression can also be expressed in terms of other system variables and parameters, which more or less remain nearly constant, and as such *B*'s are treated as constants, though these are updated every 6 hours to account for significant system operating changes.

The constraint equation now is $\sum_{i=1}^{n} P_i - P_D - P_L = 0$. If we form the augmented cost function and proceed in the same way as we proceeded above, we get

$$IC_i = \lambda \left(1 - \frac{\partial P_L}{\partial P_i}\right); \qquad i = 1, 2, ..., n.$$

$$IC_i = \frac{1}{\left(1 - \frac{\partial P_L}{\partial P_i}\right)} \lambda ; \quad i = 1, 2, ..., n.$$

$\frac{\partial P_L}{\partial P_i}$, also denoted as ITL_i, is called the incremental transmission loss for plant *i*.

$\frac{\partial P_L}{\partial P_i}$ can be + ve, – ve, or zero. Also, it is relatively of small magnitude. Incidentally, *IC* is always positive. $\frac{1}{\left(1 - \frac{\partial P_L}{\partial P_i}\right)}$ is known as the penalty factor, the PF_i of the plant *i*. Its value is around unity, mostly, > 1. Since, $\frac{\partial P_L}{\partial P_i}$ is quite small, $\frac{1}{\left(1 - \frac{\partial P_L}{\partial P_i}\right)} \approx 1 + \frac{\partial P_L}{\partial P_i}$.

$IC_i \times PF_i = \lambda$, $i = 1, 2, ..., n$; along with the constraint equation $\sum_{i=1}^{n} P_i - P_D - P_L = 0$ are called the co-ordination equations.

The expression $\sum_{i=1}^{n} P_i \sum_{k=1}^{n} B_{ik} P_k$ for P_L, leads to the expression for $\frac{\partial P_L}{\partial P_i}$

$$\frac{\partial P_L}{\partial P_i} = 2 \sum_{k=1}^{n} P_k B_{ik}.$$

47.7 ITERATIVE METHOD

The IC_i equations are non-linear equations. We solve them iteratively assuming some initial solution for plant generations. This initial solution can be say, 1.05 times the values of generations found with losses neglected and the corresponding λ. We stop when the constraint equation is satisfied within acceptable accuracy. The iterative expression for generation of plant i is

$$P_i^{(r+1)} = \frac{\lambda^{(r)} \left(1 - 2 \sum_{\substack{k=1 \\ k \neq 1}}^{n} P_k^{(r)} B_{ik} \right) - b_i}{a_i + 2\lambda^{(r)} B_{ii}}$$

$$\Delta P_T^{(r+1)} = -\left(\sum_{k=1}^{n} P_k^{(r+1)} - P_L^{(r+1)} - P_D \right).$$

Positive $\Delta P_T^{(r+1)}$ implies that generation is to be increased.

If $| \Delta P_T^{(r+1)} | > \varepsilon$, we continue the iteration.

$$\lambda^{(r+1)} = \lambda^{(r)} + \Delta\lambda^{(r)}.$$

There is no proper way of finding $\Delta\lambda^{(r)}$. We can begin with some $\lambda^{(0)}$ and find $\Delta P_T^{(1)}$. Next we take another value of λ close to the previous value, say $\lambda^{(1)} = 1.1 \times \lambda^{(0)}$, and find $\Delta P_T^{(2)}$. The next value of λ may be taken as

$$\lambda^{(2)} = \lambda^{(1)} + \Delta\lambda^{(1)}, \text{ where } \Delta\lambda^{(1)} = \frac{\lambda^{(1)} - \lambda^{(0)}}{\Delta P_T^{(2)} - \Delta P_T^{(1)}} (0 - \Delta P_T^{(2)}) = \frac{\Delta \lambda^{(0)}}{\frac{\Delta P_T^{(1)}}{\Delta P_T^{(2)}} - 1}. \text{ Thus,}$$

$$\Delta\lambda^{(r)} = \frac{\Delta \lambda^{(r-1)}}{\frac{\Delta P_T^{(r)}}{\Delta P_T^{(r+1)}} - 1}.$$

47.8 UNIT COMMITMENT

Unit commitment refers to the problem of deciding which set of available units be run during different load slots of the load cycle so that over the cycle the operating cost is minimum. This load cycle can be the daily load cycle or the weekly load cycle. Though the load curve is a continuous curve, but we divide it in suitable number of load intervals; and the load in each interval is assumed to remain constant. As the system load changes, we may be required to shut down or start up one or more units as per the load and economic considerations. Starting a unit, or shutting it down, costs.

In the economic dispatch, the problem was of distributing a given load amongst the running units so that the operating cost for that load is minimum; whereas here, we are to find the complete schedule of the set of units to be run during different load intervals over the load cycle so that the cycle operating cost is minimum. Unit commitment involves economic dispatch. The ratings, cost curves, the start-up, and the shut-down costs of all the units to be considered are supposed to be known.

The straight forward way can be to first find for all the intervals, the various combinations of units which can meet the loads of those intervals keeping in view the P_{Min} and P_{Max} limits of the units. For each load or interval, we carry out economic dispatch and find the combination which costs the least. The total operating cost is the sum of the least costs of all the intervals. The method is simple as we have not considered the unit starting and shutting costs.

If we are to take into account the unit end costs, the number of sets to be tried out for finding the cycle least cost increases manifolds. Say, there are m intervals and the number of combinations available in various intervals are $C_1, C_2, ..., C_m$. Then we have to test each possible set of the combinations for its cost. Thus, there are $C_1 \times C_2 \times ... \times C_m$ sets to be checked. For instance, say there are three intervals; and the combinations in interval 1 are a and b, in interval 2 are c, d and e, and in interval 3 are f and g. Then the arrangements are—*acf, acg, adf, adg, aef, aeg, bcf, bcg, bdf, bdg, bef, beg*. This number is $2 \times 3 \times 2 = 12$. The least cost of any of the sets is sum of the least costs of the constituent combinations for the respective loads and the transition costs from one combination to the next in the set. If there are n units, and their ratings are such that any combination of one, two, ..., n units can meet the load of any interval, then there are a total of $(2^n - 1)$ combinations available in each interval and the total number of sets becomes $(2^n - 1)^m$. For $n = 5$, and $m = 6$, this figure is $31^6 = 887503681$!!

In briging out the above fact, we assumed that all the combinations are feasible in all the load intervals. In actual practice, several of the combinations are infeasible in various intervals and as such the number of arrangements to be tried is not that alarming.

The number of sets to be tried becomes much less if we make use of the ***dynamic programming principle***, according to which every optimal policy or path has only optimal subpolicies or subpaths.

To incorporate this principle, we first define a few terms.

The sequence of the load interval is denoted by k, the combination by x, and the combination number is denoted by i. Thus, x_i is the ith combination of the generating units. $P_i(k)$ is the minimum production cost with x_i in the load interval k. $T_{ij}(k)$ is the transition cost associated with transition from x_i at stage k to x_j at the next stage or load interval $(k + 1)$. $F_i(k)$ is the minimum cumulative cost of final $(n - k + 1)$ intervals in reaching x_i at stage k. Thus, we are working out the minimum operating cost in one go, starting from the last stage.

$$\mathbf{F_i(k) = \underset{\{x_j(k+1)\}}{Min} \{P_i(k) + T_{ij}(k) + F_j\ (k + 1)\}}$$

When we are going from stage $(k + 1)$ to a particular x_i at stage k, we test the transitions and work the costs from all the combinations x_j at stage $(k + 1)$ to x_i at stage k. Finally, we reach stage 1. The least of the costs at stage 1 is the minimum operating cost. We trace the minimum cost path starting at the first stage, passing through the combinations which resulted in the least cost, and reaching the final stage. Therefore, at various load levels the combinations on the Minimum cost path are the ones which should be implemented.

Working out the costs in the forward direction, we shall have the following relation.

$$\mathbf{F_j(k+1) = \underset{\{x_j(k)\}}{Min} \{ F_i\ (\ k\ ) + T_{ij}(k)\ + P_j(k+1)\ \}}$$

$F_i\ (k)$ is the minimum cumulative cost of first k intervals in reaching x_i at stage k.

$T_{ij}(k)$ is the transition cost associated with transition from x_i at stage k to x_j at stage $(k + 1)$.

$P_j(k + 1)$ is the minimum production cost with x_j at stage $(k + 1)$.

In testing $F_j\ (k + 1)$ for the minimum, i, on the RHS takes values corresponding to all feasible combinations at stage k.

We trace the minimum cost path starting at the final stage, passing through the combinations which resulted in the least cost, and reaching the first stage.

The result obtained is independent of the direction of processing.

Chapter 48 *LOAD FREQUENCY CONTROL*

48.1 GENERAL

In power systems, frequency is very closely monitored and kept as close to the nominal value as possible, though a variation of 3% is permitted. Any deviation from the nominal value affects the life of turbine blades of thermal units. The variations in frequency also cause variations in speeds of induction motors and synchronous clocks. Per second time error caused in synchronous clocks is Δf_{pu}.

The deviation in frequency is related to the imbalance in real power generated and the real power load on the system. If generation is short, frequency falls; and vice versa.

We can realize it when we consider an isolated generator having constant excitation and constant mechanical input and supplying a purely resistive load, and examine the effects which occur when we attempt to increase the loading by decreasing the load resistance. These simple relations hold: $E = k\omega$, $z^2 = R^2 + L^2\,\omega^2$, $I = E/z$, Power $P = EI\cos\theta_z$. These lead to $I_{ph}^2 = \frac{P}{R}\left(1 - \frac{L^2P}{k^2R}\right)$, where I_{ph} is the armature current component in phase with E. We find that keeping P fixed, for $k^2R > 2L^2P$, I_{ph} increases as R decreases. As the torque is proportional to the product of field winding current I_f and I_{ph}; we note that for fixed power, as R is reduced torque increases and hence the speed or the frequency falls. Alternatively, from the same equations we can also find the relation for ω^2 itself ; $\omega^2 = \frac{PR^2}{k^2R - L^2P}$. This relation also leads to the same conclusion. Taking $P = 2500$ W, $k = 0.8$, $L = 0.03$ H, $R = 20$ ohms, we find $f = 49$ Hz, $V_R = 223$ V, and with $R = 18$ ohms, $f = 47$ Hz, $V_R = 212$V. Thus, without any increase in input to the generating unit, frequency falls as we attempt to draw more power from it.

If the load is purely inductive, there is no armature current component in phase with E, and hence no torque. Varying L has no effect on the torque or the speed. The machine continues to run at the constant speed. E remains constant. The armature current and the reactive power to the load are inversely proportional to the inductance. There is nothing like input reactive power, in contrast to the input mechanical power.

48.2 SPEED GOVERNING SYSTEM

A generating unit is equipped with speed governing system, which has revolving fly balls and speed changer. Small variations in load are taken care of by the revolving fly balls which can control the steam valve opening by small amounts. For significant changes in load, the steam valve opening and therefore the steam input to the turbine is changed by adjusting the speed changer setting. This, in fact, shifts the governor characteristic. Thus, balance in demand and supply is maintained.

Given below is the schematic of a conventional speed governing system of a thermal unit. As the speed rises, the fly balls fly apart which causes downward displacement of B, C, and D; and an upward displacement of E, thereby causing reduction in the steam valve opening and therefore in the steam input to the turbine. Moving the point A downward through the speed changer mechanism, causes upward displacement of points C and D, and downward displacement of E, thereby causing increase in the steam valve opening and hence in the steam input to the turbine.

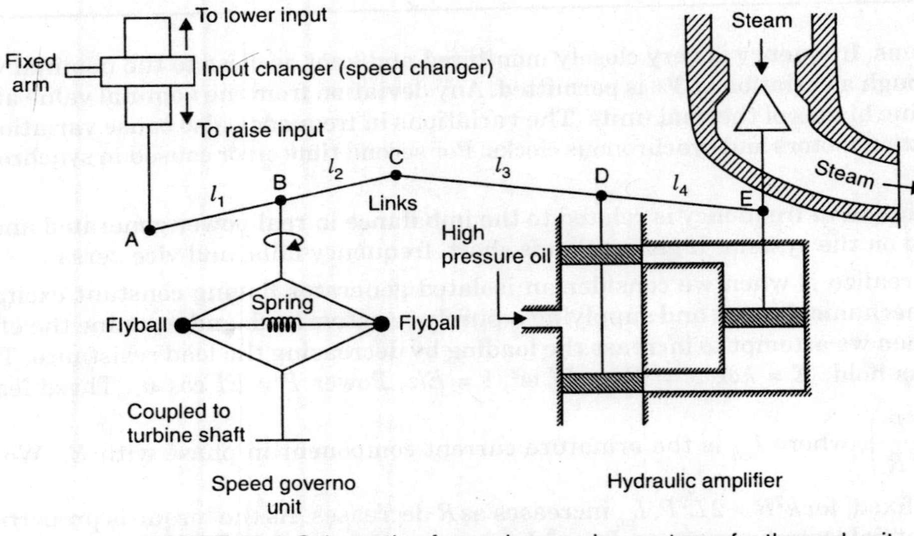

FIGURE 48.1. Schematic of speed governing system of a thermal unit.

48.3 SPEED GOVERNOR BLOCK DIAGRAM DEVELOPMENT

Taking downward displacement of all points positive, we find that the increase in the y-ordinate of the point E has the Laplace transform

$$\Delta Y_E(s) = \frac{k_1 k_3 k_C \,\Delta P_C(s) - k_2 k_3 \,\Delta F(s)}{\left(k_4 + \dfrac{s}{k_5}\right)}$$, which can further be expressed as

$$\left(\Delta P_C(s) - \frac{1}{R}\Delta F(s)\right) \times \left(\frac{K_{sg}}{1 + T_{sg}s}\right)$$; where

$$k_1 = \frac{l_2}{l_1},$$

$$k_2 = \frac{l_1 + l_2}{l_1} k_2', k_2' \text{ is } \frac{\Delta Y_B}{\Delta_f},$$

which is inversely proportional to the fly ball spring stiffness constant,

$$k_3 = \frac{l_4}{l_3 + l_4},$$

$$k_4 = \frac{l_3}{l_3 + l_4},$$

k_5 = a constant proportional to the port size of the high-pressure oil chamber,

$\Delta P_C(s)$ = Laplace transformed commanded increase in the input power,

$k_C = \frac{\Delta Y_A}{\Delta P_C}$,

$\Delta F(s)$ = Laplace transformed increase in frequency,

$R = \frac{k_1 k_C}{k_2} = \frac{l_2}{l_1 + l_2} \frac{k_C}{k_2'}$ = , the speed regulation of the speed governor,

$K_{sg} = \frac{k_1 k_3 k_C}{k_4} = \frac{l_2 l_4}{l_1 l_3} k_C$, the gain of the speed governor, and

$T_{sg} = \frac{1}{k_4 k_5} = \frac{l_3 + l_4}{l_3} \frac{1}{k_5}$, the time constant of the speed governor.

The above equation, relating $\Delta Y_E(s)$ to $\Delta P_C(s)$ and $\Delta F(s)$, enables us to draw the block diagram of the speed governing system.

$\Delta P_T(s)$, the increase in the turbine power output is directly proportional to the increase in the steam valve opening, $\Delta Y_E(s)$, through the turbine transfer function $\frac{K_t}{1 + T_t s}$.

48.4 GENERATOR-LOAD BLOCK DIAGRAM DEVELOPMENT

The increase in the turbine power output, or the increase in the mechanical power input to the shaft is absorbed as increase in generation and as increase in the power accelerating the rotating masses. The accelerating power prior to the changes was zero.

$$\Delta P_T = \Delta P_G + \Delta P_{acc}.$$

The power system load is frequency sensitive, the sensitivity being $\frac{\partial P_D}{\partial f}$. It is denoted by B, and has a small positive value. If the whole of the system load were purely resistive $\frac{\partial P_D}{\partial f}$ would be zero.

$$\Delta P_G = \Delta P_D + \frac{\partial P_D}{\partial f}\Delta f = \Delta P_D + B\Delta f.$$

The accelerating power is $\Delta P_{acc} = \frac{2HS}{f_0}\frac{d}{dt}\Delta f$, where H is the inertia constant of the machine in seconds, S is the machine rating in volt-amperes, and f_0 is the nominal frequency, or the frequency prior to the occurrence of the change in load. Then the power balance equation is $\Delta P_T = \Delta P_D + B\Delta f + \frac{2HS}{f_0}\frac{d}{dt}\Delta f$. This is converted to per unit by dividing all through by the machine rating, S. Then we have $\Delta P_T = \Delta P_D + B\Delta f + \frac{2H}{f_0}\frac{d}{dt}\Delta f$; ΔP_T, ΔP_D, and B now being in per unit. In the s domain, this equation becomes, $\Delta P_T(s) = \Delta P_D(s) + B\Delta F(s) + \frac{2H}{f_0}s\Delta F(s)$.

This leads to $\Delta F(s) = \{\Delta P_T(s) - \Delta P_D(s)\} \times \frac{K_{ps}}{1+T_{ps}s}$; where the power system gain constant, $K_{ps} = \frac{1}{B}$, and the power system time constant, $T_{ps} = \frac{2H}{f_0}K_{ps} = \frac{2H}{f_0 B}$.

48.5 COMPLETE BLOCK DIAGRAM AND TRANSFER FUNCTIONS

The complete block diagram of the system comprising the speed governing unit, the turbine, the generator and the load can be drawn as under:

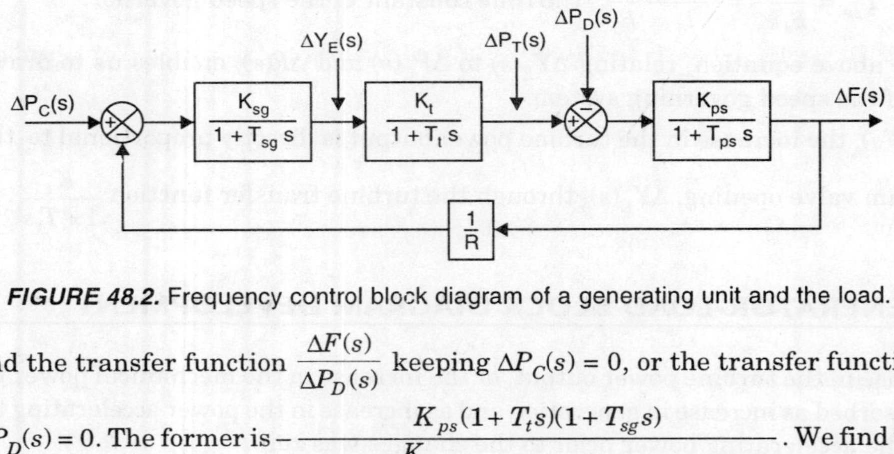

FIGURE 48.2. Frequency control block diagram of a generating unit and the load.

We can find the transfer function $\frac{\Delta F(s)}{\Delta P_D(s)}$ keeping $\Delta P_C(s) = 0$, or the transfer function $\frac{\Delta F(s)}{\Delta P_C(s)}$ keeping $\Delta P_D(s) = 0$. The former is $-\frac{K_{ps}(1+T_t s)(1+T_{sg}s)}{K_{ps}K_{sg}\frac{K_t}{R} + (1+T_{ps}s)(1+T_t s)(1+T_{sg}s)}$. We find the change in frequency on applying a step change of ΔP_D in load. Then $\Delta P_D(s) = \frac{\Delta P_D}{s}$. The transformed change in frequency due to step change in load is

$$\Delta F(s) = -\frac{K_{ps}(1+T_t s)(1+T_{sg}s)}{K_{ps}K_{sg}\frac{K_t}{R} + (1+T_{ps}s)(1+T_t s)(1+T_{sg}s)}\frac{\Delta P_D}{s}.$$

48.6 STEADY STATE ANALYSIS

To find the steady state change in frequency, one can apply final value theorem, *i.e.*, $\lim_{t \to \infty} f(t) = \lim_{s \to 0} sF(s)$ and find that

$$\Delta f \Bigg|_{\substack{t \to \infty \\ \Delta P_C = 0}} = -\frac{K_{ps}\Delta P_D}{1 + K_{ps}K_{sg}K_t/R} = -\frac{1}{1/K_{ps} + K_{sg}K_t/R}\Delta P_D.$$

It is usually possible to set $K_{sg}K_t = 1$, by adjusting the link lengths. Noting that $1/K_{ps} = B$, we have

Δf_{ss}, or for convenience simply $\Delta f = -\dfrac{1}{B + 1/R}\Delta P_D$.

The slope is $\dfrac{\Delta f}{\Delta P_D} = -\dfrac{1}{B + 1/R}$. Further noting that $B << 1/R$, *i.e.*, neglecting the change in load due to change in frequency, we have $\dfrac{\Delta f}{\Delta P_D} = -R$. The operation with $\Delta P_C = 0$, is known as free governor operation.

Also,

$$\Delta P_T \Bigg|_{t \to \infty} = \lim_{s \to 0} -\frac{K_{sg}K_t}{R(1 + T_{sg}s)(1 + T_t s)} s\Delta F(s) = -\frac{1}{R}\left(\lim_{s \to 0} sF(s)\right) = -\frac{1}{R}\Delta f.$$

Under this condition, *i.e.*, the steady state conditions, $\Delta P_T = \Delta P_G$. Thus, under steady state, we have

$$\Delta P_G = -\frac{1}{R}\Delta f = -\frac{1}{R}\left(-\frac{1}{B + 1/R}\Delta P_D\right) = \frac{1}{BR + 1}\Delta P_D,$$ *i.e.*, a little less than ΔP_D.

The increase in system load on account of increase in frequency is $B\Delta f = -\dfrac{B}{B + 1/R}\Delta P_D$.

We see that $\quad \Delta P_D - \dfrac{B}{B + 1/R}\Delta P_D = \dfrac{1}{BR + 1}\Delta P_D = \Delta P_G$.

Neglecting the change in load on account of change in frequency, under *ss* condition we have $\Delta P_T = \Delta P_G = \Delta P_D$. Then $\dfrac{\Delta f}{\Delta P_D} = -R$ can be written as $\dfrac{\Delta f}{\Delta P_G} = -R$. This is the generator governor characteristic.

$\dfrac{\Delta F(s)}{\Delta P_C(s)}$ with $\Delta P_D(s) = 0$ is seen to be $\dfrac{K_{ps}K_{sg}K_t R}{K_{ps}K_{sg}K_t + R(1 + T_{ps}s)(1 + T_t s)(1 + T_{sg}s)}$

$$\Delta f \Bigg|_{\substack{t \to \infty \\ \Delta P_D = 0}} = \frac{K_{ps}K_{sg}K_t R}{R + K_{ps}K_{sg}K_t}\Delta P_C.$$ Again, with $K_{sg}K_t = 1$,

$$= \Delta f \Bigg|_{\substack{t \to \infty \\ \Delta P_D = 0}} = \frac{K_{ps} R}{R + K_{ps}} \Delta P_C = \frac{1}{B + 1/R} \Delta P_C$$

If both the changes occur simultaneously, then the steady state change in frequency is

$$\Delta f = \frac{1}{B + 1/R} (\Delta P_C - \Delta P_D).$$

For Δf to be zero, ΔP_C should equal ΔP_D.

48.7 DYNAMIC STATE ANALYSIS

We see that the system is a third order system. Keeping in view that $T_{sg} < T_t << T_{ps}$, and $K_{sg} K_t = 1$, we get, with $\Delta P_C = 0$, the following simplified expression for the transfer function.

$$\frac{\Delta F(s)}{\Delta P_D(s)} = -\frac{K_{ps}}{(1 + K_{ps}/R) + T_{ps} s}$$

Therefore, $\Delta F(s) = -\dfrac{K_{ps}}{(1 + K_{ps}/R) + T_{ps} s} \dfrac{\Delta P_D}{s} = -\dfrac{K_{ps}/T_{ps}}{s\left(s + \dfrac{R + K_{ps}}{RT_{ps}}\right)} \Delta P_D$. This gives

$$\Delta f(t) = -\frac{RK_{ps}}{R + K_{ps}} \left(1 - e^{\left(\frac{t}{T_{ps}}\left(\frac{R + K_{ps}}{R}\right)\right)}\right) \Delta P_D.$$

The actual response is oscillatory and the oscillations die down in accordance with the time constant.

As $t \to$ infinity, $\Delta f(t) \to \Delta f_{ss} = -\dfrac{RK_{ps}}{R + K_{ps}} \Delta P_D = -\dfrac{1}{B + 1/R} \Delta P_D$.

48.8 INTEGRAL CONTROLLER

As we have seen that by making $\Delta P_C = \Delta P_D$, we reduce Δf to zero.

We achieve this action automatically by introducing integral controller. The input to the integrator block is $\Delta F(s)$, the transfer function of the integrator is $-\dfrac{K_I}{s}$, and the output in time domain is $-K_I \int_0^t \Delta f(t) dt$. This should be $\Delta P_C(t)$. When steady state is reached, *i.e.*, when $\Delta f(t)$ finally falls to zero, there is no further contribution to the integral and this value should equal ΔP_D, *i.e.*, $-K_I \int_0^{t_{ss}} \Delta f(t) dt = \Delta P_D$. With the incorporation of integral controller, the block diagram is under :

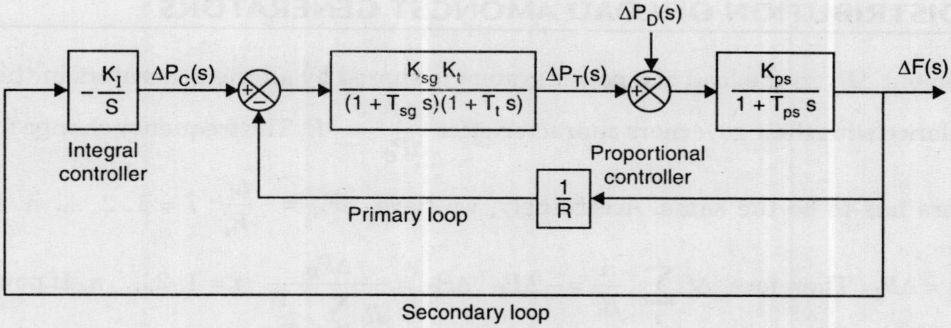

FIGURE 48.3. Frequency control block diagram incorporating integral controller.

The loop with feedback element $1/R$ is the primary loop, and the one with the integrator is the secondary loop.

The transfer function is

$$\frac{\Delta F(s)}{\Delta P_D(s)} = -\frac{RK_{ps}(1+T_{sg}s)(1+t_t s)s}{R(1+T_{sg}s)(1+T_t s)(1+T_{ps}s)s + K_{ps}(K_I R + s)}.$$

The steady state change in frequency on applying a step change of ΔP_D in load is

$$\Delta F\Big|_{ss} = \lim_{s\to 0} s\Delta F(s) = \lim_{s\to 0} -s \times \frac{RK_{ps}(1+T_{sg}s)(1+t_t s)s}{R(1+T_{sg}s)(1+T_t s)(1+T_{ps}s)s + K_{ps}(K_1 R + s)} \times \frac{\Delta P_D}{s} = 0.$$

It is in accordance with the control theory wherein we learn that for a step input, an integral controller brings steady state error to zero.

For bringing out the dynamic features, we apply the approximations which we used earlier too, *i.e.*, we take $T_{sg} = T_t = 0$, and $K_{sg} K_t = 1$. Then for a step change in the load,

$$\Delta F(s) = -\frac{RK_{ps}\Delta P_D}{R(1+T_{ps}s)s + K_{ps}(K_I R + s)} = -\frac{K_{ps}}{T_{ps}} \frac{\Delta P_D}{s^2 + s\dfrac{1+K_{ps}/R}{T_{ps}} + \dfrac{K_I K_{ps}}{T_{ps}}}$$

The characteristic equation of $\Delta F(s)$ is of the form of $s^2 + 2\delta\omega_n s + \omega_n^2 = 0$, $\delta\omega_n$ is also denoted by α; $-\alpha$ being the real part of the complex roots of the equation. By comparing, we see that

$$\omega_n = \sqrt{\frac{K_I K_{ps}}{T_{ps}}}, \text{and } \delta = \frac{1}{2}\frac{1+K_{ps}/R}{T_{ps}}\sqrt{\frac{T_{ps}}{K_1 K_{ps}}}, \alpha = \frac{1}{2}\frac{1+K_{ps}/R}{T_{ps}}$$

For the roots to be real, in other words, for the response to be non-oscillatory,

$\alpha^2 \geq \omega_n^2$. This leads to $K_I \leq \dfrac{(1+K_{ps}/R)^2}{4T_{ps}K_{ps}}$.

48.9 DISTRIBUTION OF LOAD AMONGST GENERATORS

An increment ΔP_D in the load of a power system is shared by all the generators in the system in accordance with their governors characteristics $\dfrac{\Delta f}{\Delta P_G} = -R$. The frequency change for all the generators has to be the same. Accordingly, we have $\Delta F_{Gi} = -\dfrac{\Delta f}{R_i}$; $i = 1, 2, ..., n$. Further, $\sum_i \Delta P_{Gi} = \Delta P_D$. Therefore, $\Delta f \sum_i \dfrac{1}{R_i} = -\Delta P_D$. $\Delta P_{Gi} = \dfrac{\Delta P_D}{R_i \sum_i \dfrac{1}{R_i}}$; $i = 1, 2, ..., n$. If power units are in MW, R should be in Hz/MW.

48.10 CONTROL AREA

A large power system can be regarded as composed of several interconnected smaller power systems. The boundaries of individual power systems are the points on the tielines where the responsibilities of one ends and those of the neighbourer begin. Within the smaller power system, the generators are quite close to each other, *i.e.*, the lines connecting the generators have very small reactances and in the event of any disturbance, all the machines swing in unison. This is called coherency. All the generators in the area can be represented by a single equivalent machine. The smaller power system is also called a control area.

48.11 TIE LINE POWER AND BLOCK DIAGRAM

The tie line power is $P_{\text{tie},\, ij} = \dfrac{V_i V_j}{X_{ij}} \sin \delta_{ij} = P_{\text{Max}} \sin \delta_{ij}$, where $\delta_{ij} = \delta_i - \delta_j$.

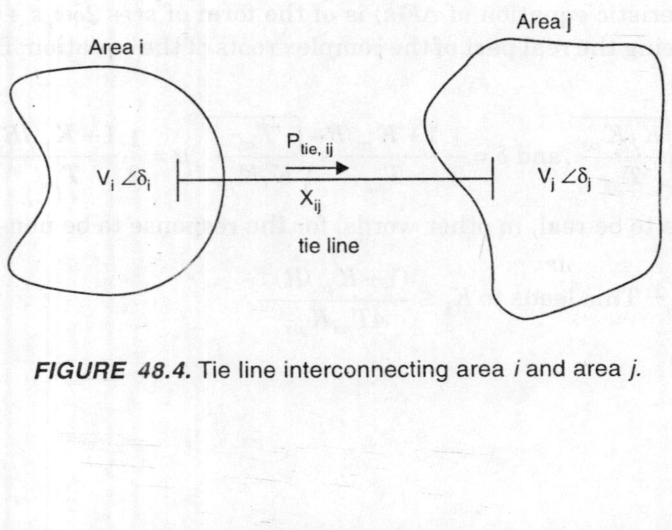

FIGURE 48.4. Tie line interconnecting area *i* and area *j*.

When a disturbance occurs, $\Delta P_{\text{tie},ij} = P_{\text{Max}} \cos \delta_{ij\,0}\, \Delta\delta_{ij} = T_{ij\,0}\Delta\delta_{ij}$, where $\Delta\delta_{ij} = \Delta\delta_i - \Delta\delta_j$.

Now $\quad \Delta\delta(t) = \int \Delta\omega(\bar{t})\, dt = 2\pi \int \Delta f(t)\, dt$. Thus, $\Delta\delta_i = 2\pi \int \Delta f_i\, dt$ and $\Delta\delta_j = 2\pi \int \Delta f_j\, dt$.

Therefore, $\quad \Delta P_{tie,ij} = T_{ij0}\, 2\pi \left(\int \Delta f_i dt - \int \Delta f_i dt \right)$

The total differential power out of area i is

$$\Delta P_{\text{tie},\,i} = \sum_j \Delta P_{\text{tie},\,ij} = 2\pi \sum_j T_{ij0} \left(\int \Delta f_i\, dt - \int \Delta f_j dt \right)$$

Laplace transforming,

$$\Delta P_{\text{tie},i}(s) = \frac{1}{s} \sum_j 2\pi T_{ij0} (\Delta F_i(s) - \Delta F_j(s))$$

The above equation can be represented in the block diagram form.

The power balance equation for area i is

$$\Delta P_{Ti} = \frac{d}{dt} W_{KE,.i} + \Delta P_{Di} + B_i\, \Delta f_i + \Delta P_{\text{tie},\,i}$$

$$= \frac{2H_i S_i}{f_0} \frac{d}{dt} (\Delta f_i) + \Delta P_{Di} + B_i\, \Delta f_i + \Delta P_{\text{tie},i}$$

When expressed in per unit on a power base of area capacity S_i,

$$\Delta P_{Ti} = \frac{2H_i}{f_0} \frac{d}{dt} (\Delta f_i) + \Delta P_{Di} + B_i\, \Delta f_i + \Delta P_{\text{tie},\,i}.$$

On taking the Laplace transform, and arranging, we get

$$(\Delta P_{Ti}(s) - \Delta P_{Di}(s) - \Delta P_{\text{tie},i}(s))\, G_{pi}(s) = \Delta F_i(s),$$

where $\quad G_{pi}(s) = \dfrac{K_{pi}}{1 + T_{pi}s}$, $K_{pi} = \dfrac{1}{B_i}$, and $T_{pi} = \dfrac{2H_i}{B_i f_0}$

This $\Delta F_i(s)$ equation can also be represented in the block diagram form.

The complete block diagram will now have additional input $\Delta P_{tie,i}(s)$.

Earlier, the input to the integrator was only $\Delta F(s)$. Now because of the tie line powers, the input to the integrator is a mix of $\Delta F(s)$ and $\Delta P_{tie,i}(s)$, which is called the area control error, $ACE_i(s)$.

The area control error, $ACE_i = (P_{\text{tie},i\ \text{actual}} - P_{\text{tie},i\ \text{scheduled}}) + B_i\,(f_{\text{actual}} - f_{\text{rated}}) = \Delta P_{\text{tie},\,i} + B_i\, \Delta f_i \neq B_i$ is the frequency bias setting in MW/Hz. The ACE is thus in MW.

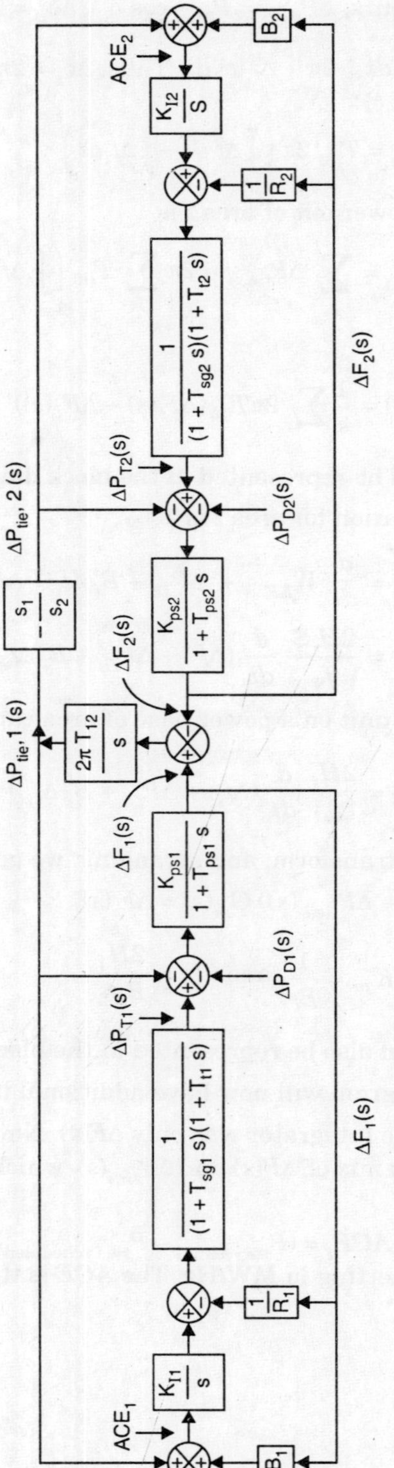

FIGURE 48.5. Frequency control block diagram of two connected areas. Suffixes 1 and 2 have been used for area 1 and area 2 respectively.

48.12 STEADY STATE RESPONSE OF TWO CONNECTED AREAS

Under steady state conditions, the power balance equation for area i is

$$\Delta P_{Gi} = P_{Di} + B_i \,\Delta f_i + \Delta P_{\text{tie},\,ij}$$ and for area j, it is

$$\Delta P_{Gj} = \Delta P_{Dj} + B_j \,\Delta f_j + \Delta P_{\text{tie},\,ji}$$

Noting that $\Delta P_G = -\dfrac{\Delta f}{R}$, and $\Delta P_{tie,ji} = -\Delta P_{\text{tie},\,ij}$, we find

$$\Delta f = -\frac{\Delta P_{D,i} + \Delta P_{D,j}}{\beta_i + \beta_j}, \quad \text{and} \quad \Delta P_{ij} = \frac{\beta_i \Delta P_{D,j} - \beta_j \Delta P_{D,i}}{\beta_i + \beta_j}$$

where $\beta_i = B_i + \dfrac{1}{R_i}$, and $\beta_j = B_j + \dfrac{1}{R_j}$

Consistent units should be used.

48.13 DYNAMIC RESPONSE ASPECTS OF TWO CONNECTED AREAS

$$\Delta P_{Ti} - \Delta P_{Di} = \frac{2H_i S_i}{f_0}\frac{d}{dt}(\Delta Fi) + B_i \,\Delta f_i + \Delta P_{tie,ij}$$

A similar equation can be written for area j. Let $S_j = S_i = S$

Letting the powers to be in pu on a base of S, and working in s domain, we have

$$\Delta P_{Ti}(s) - \Delta P_{Di}(s) = \frac{2H_i}{f_0} s\Delta F_i(s) + B_i \,\Delta F_i(s) + \Delta P_{\text{tie},\,ij}(s)$$

$$\Delta P_{Tj}(s) - \Delta P_{Dj}(s) = \frac{2H_j}{f_0} s\Delta F_i(s) + B_j \,\Delta F_j(s) + \Delta P_{\text{tie},\,ji}(s)$$

Letting $B_i = B_j = 0$,

$$\Delta P_{Ti}(s) - \Delta P_{Di}(s) = \frac{2H}{f_0} s\Delta F_i(s) + \Delta P_{\text{tie},\,ij}(s)$$

$$\Delta P_{Tj}(s) - \Delta P_{Dj}(s) = \frac{2H}{f_0} s\Delta F_j(s) + \Delta P_{\text{tie},\,ji}(s)$$

Subtracting first equation from the second,

$$(\Delta P_{Tj}(s) - \Delta P_{Ti}(s)) + (\Delta P_{Di}(s) - \Delta P_{Dj}(s)) = \frac{2H}{f_0} s(\Delta F_j(s) - \Delta F_i(s)) - 2\Delta P_{tie,\,ij}(s),$$

Letting $H_i = H_j = H, R_i = R_j = R$, *i.e.*, taking that the two areas are identical, and recalling that $\Delta P_T = -\dfrac{\Delta f}{R}$, and $\Delta P_{tie,ij}(s) = \dfrac{2\pi T_{ij0}}{s}(\Delta F_i(s) - \Delta F_j(s))$, we have

$$\left(-\frac{\Delta F_j(s)}{R} + \frac{\Delta F_i(s)}{R}\right) + (\Delta P_{Di}(s) - \Delta P_{Dj}(s)) = -\frac{2H}{f_0} s\Delta P_{tie,\,ij}(s)\frac{s}{2\pi T_{ij,0}} - 2\,\Delta P_{tie,\,ij}(s),$$

or $$\Delta P_{\text{tie},\,ij}(s)\frac{s}{2\pi RT_{ij,\,0}} + (\Delta P_{Di}(s) - \Delta P_{Dj}(s)) = -\,\Delta P_{tie,\,ij}(s)\left[\frac{s^2 H}{\pi f_0 T_{ij,0}} + 2\right],$$

or $$\Delta P_{\text{tie},\,ij}(s)\left[\frac{s}{2\pi RT_{ij,\,0}} + \frac{s^2 H}{\pi f_0 T_{ij,\,0}} + 2\right] = (\Delta P_{Dj}(s) - \Delta P_{Di}(s)),$$

or $$\Delta P_{\text{tie},\,ij}(s)\frac{H}{\pi f_0 T_{ij,\,0}}\left[s^2 + \left(\frac{f_0}{2RH}\right)s + \frac{2\pi f_0 T_{ij,\,0}}{H}\right] = (\Delta P_{Dj}(s) - \Delta P_{Di}(s)),$$

or $$\Delta P_{tie,ij}(s) = (\Delta P_{Dj}(s) - \Delta P_{Di}(s))/\left\{\frac{H}{\pi f_0 T_{ij,0}}\left[s^2 + \left(\frac{f_0}{2RH}\right)s + \frac{2\pi f_0 T_{ij,0}}{H}\right]\right\}$$

$$= \frac{\pi f_0 T_{ij,\,0}}{H}\,\frac{(\Delta P_{Dj}(s) - \Delta P_{Di}(s))}{\left(s^2 + \dfrac{f_0}{2RH}s + \dfrac{2\pi f_0 T_{ij,\,0}}{H}\right)}$$

The characteristic equation for $\Delta P_{\text{tie},\,ij}(s)$ is

$$s^2 + \frac{f_0}{2RH}s + \frac{2\pi f_0 T_{ij,\,0}}{H} = 0.$$

On comparing with $s^2 + 2\alpha s + \omega_n^{\,2} = 0$, we see

$$\alpha = \frac{f_0}{4RH},\ \omega_n^{\,2} = \frac{2\pi f_0 T_{ij,\,0}}{H}.$$

As $-\alpha$ cannot be positive, the roots are in the LH of the complex s-plane and the system is therefore, stable.

The damped frequency of oscillation is $\omega_d = \sqrt{\dfrac{2\pi f_0 T_{ij,0}}{H} - \dfrac{f_0^2}{16R^2H^2}}$

If the speed governor loop is open, *i.e.*, $R = \infty$, then $\alpha = 0$, and the tie line power oscillates continuously, and $\omega_d = \omega_n$.

One can compare with $s^2 + 2\delta\omega_n s + \omega_n^{\,2} = 0$ also, and find that $\delta = \sqrt{\dfrac{f_0}{32\pi R^2 HT_{ij,0}}}$.

$\delta = 0$, implies no damping; $\delta < 1$, implies underdamping ; $\delta = 1$, implies critical damping; and $\delta > 1$, implies overdamping.

INDEX

C

D

E

F

G

H

I

K

L

T

U

V

W

Z